AF361819

Properties and Novel Applications of Recycled Aggregates

Properties and Novel Applications of Recycled Aggregates

Editor

Fernando Lopez Gayarre

MDPI • Basel • Beijing • Wuhan • Barcelona • Belgrade • Manchester • Tokyo • Cluj • Tianjin

Editor
Fernando Lopez Gayarre
University of Oviedo
Spain

Editorial Office
MDPI
St. Alban-Anlage 66
4052 Basel, Switzerland

This is a reprint of articles from the Special Issue published online in the open access journal *Materials* (ISSN 1996-1944) (available at: https://www.mdpi.com/journal/materials/special_issues/Recycled_Aggregates).

For citation purposes, cite each article independently as indicated on the article page online and as indicated below:

LastName, A.A.; LastName, B.B.; LastName, C.C. Article Title. *Journal Name* **Year**, *Article Number*, Page Range.

ISBN 978-3-03943-164-9 (Hbk)
ISBN 978-3-03943-165-6 (PDF)

Contents

About the Editor

Fernando Lopez Gayarre has a Ph.D. in structural engineering. He is professor of the Department of Construction and Manufacturing Engineering of the University of Oviedo, Spain. He has a vast teaching and research experience, being co-author of numerous papers mainly in the fields of recycled aggregates, recycled concrete, engineering failure analysis, and ground engineering. Currently, he is involved in research projects related to recycled aggregate concrete, steel joints, and ground engineering.

Preface to "Properties and Novel Applications of Recycled Aggregates"

The aggregates used in construction are the natural resource consumed the most in the world after air and water. Due to overexploitation, all environmental laws reward the use of recycled materials to guarantee the reduction of consumption of natural aggregates. The use of reclaimed aggregates, reused aggregates, and recycled aggregates increases sustainability in construction activities. Today, they are strategic materials in the manufacturing of green concrete and mortars and as road construction eco-efficient materials. In addition, the use of recycled aggregates from industrial or mining byproducts presents great potential in construction activities as recycled aggregates and/or supplementary cementitious materials. This Special Issue is open to new experiences in construction materials and/or works made with recycled aggregates.

Fernando Lopez Gayarre
Editor

Article

Chloride Induced Reinforcement Corrosion in Mortars Containing Coal Bottom Ash and Coal Fly Ash

Esperanza Menéndez [1], Cristina Argiz [2] and Miguel Ángel Sanjuán [3,*]

[1]   Instituto de Ciencias de la Construcción Eduardo Torroja (CSIC), C/Serrano Galvache, 4, 28033 Madrid, Spain; emm@ietcc.csic.es

[2]   Civil Engineering School, Technical University of Madrid, C/ProfesorAranguren, 3, Ciudad Universitaria, 28040 Madrid, Spain; cg.argiz@upm.es

[3]   Spanish Institute of Cement and its Applications (IECA), C/José Abascal, 53, 28003 Madrid, Spain

*   Correspondence: masanjuan@ieca.es; Tel.: +34-914429166

Received: 30 May 2019; Accepted: 12 June 2019; Published: 15 June 2019

Abstract: Coal bottom ash is normally used as aggregate in mortars and concretes. When it is ground, its characteristics are modified. Therefore, the assessment of its long-term durability must be realized in depth. In this sense, an accelerated chloride ingress test has been performed on reinforced mortars made of Portland cement with different amounts of coal bottom ash (CBA) and/or coal fly ash (CFA). Corrosion potential and corrosion rate were continuously monitored. Cement replacement with bottom and fly ash had beneficial long-term effects regarding chloride penetration resistance. Concerning corrosion performance, by far the most dominant influencing parameter was the ash content. Chloride diffusion coefficient in natural test conditions decreased from 23×10^{-12} m^2/s in cements without coal ashes to 4.5×10^{-12} m^2/s in cements with 35% by weight of coal ashes. Moreover, the time to steel corrosion initiation went from 102 h to about 500 h, respectively. Therefore, this work presents experimental evidence that confirms the positive effect of both types of coal ashes (CBA and CFA) with regard to the concrete steel corrosion.

Keywords: steel reinforced concrete; polarization; coal bottom ash; coal fly ash

---

1. Introduction

Chloride penetration from seawater into concrete in coastal areas and the associated risk for reinforcement corrosion is recognized as the most important deterioration mechanism for offshore and coastal reinforced concrete structures worldwide [1,2]. Concrete pore solution provides protection to the reinforcing steel against corrosion by means of reinforcing steel passivation promoted by its high alkalinity. In chloride-containing environments, the passive layer is destroyed when the chloride ion concentration exceeds to a certain threshold value [3] in the vicinity of steel reinforcement and, therefore, the corrosion began. Then, the reinforced concrete service life can be divided into an initiation period and a propagation period [4]. The initiation one describes chloride ingress into the concrete and is ended by the reinforcement depassivation, which is followed by the propagation period beginning. The chloride amount associated with reinforcement depassivation has been extensively studied [5].

Several chloride threshold values have been proposed because it is affected by several interconnected parameters. Some of them depend on the type of cement used in the mortar and concrete, type of steel and steel/paste interface properties. Among them, the first one affects directly on the pore solution chemistry. Not only the type of cement, but also the concrete quality and concrete cover thickness influence the time needed to the critical chloride content to be reached at the steel reinforcement surface [6].

For electrical power generation by coal combustion to become sustainable, the reuse of coal combustion by-products such as bottom ash and fly ash is necessary. Blended cements made of coal fly ash present a beneficial effect since it has been acknowledged a more compact microstructure in mortars and concretes leading to lower chloride permeability [7]. However, a decrease of the pore solution pH was also found and, therefore, the chloride threshold value could be lower than in cement-based materials without pozzolanic materials. In any case, coal fly ash is frequently used in mortars and concretes exposed to chloride environments because the pozzolanic additions have a significant influence on the chloride ion transportation. This coal fly ash has a pozzolanic reaction and micro-filler effect, which is beneficial in improving the resistance of concrete against the ingress of harmful ions [7].

On the other hand, coal bottom ash (CBA) is regarded as a potential replacement for sand in concrete mixture. CBA can be used for construction purposes as a sand substitute [8,9] and for industrial purposes as alternative filter media [10] applications, amongst other uses. The amount of recycled aggregate is increasing in the world [11]. Given that, manufactured CBA sand with different sizes may be produced in a crushing plant. It is well known that size and particle size distribution will influence the material characteristics [12–14]. Other types of bottom ashes, from a municipal solid waste incinerator [15] or circulating fluidized bed combustion (CFBC) [16], are also reported in the literature. Among the various means of reusing coal bottom ash reported thus far, it is believed that CBA can be also utilized as a Portland cement constituent [17]. Thus, assessment of cement-based materials made of CBA mixes regarding corrosion performance must involve characterization of the penetration resistance against chlorides and the parameters governing the corrosion rate.

The influence of coal fly ash on chloride ingress and resulting reinforcement corrosion in concrete has been reported in many studies over the last decades [18–20]. Nevertheless, reinforcement corrosion studies have not been found with regard to coal bottom ash cement-based materials. The influence of coal fly ash in cement-based materials on the corrosion processes is mainly due to microstructural changes, lime consumption due to the pozzolanic reaction and binding capacity.

In this work, a standardized setup based on chloride ingress by applying an electrical field was used to study the initiation stage of chloride induced reinforcement corrosion in several mortar mixes made of coal bottom ash, coal fly ash and common Portland cement. Corrosion assessment of coal bottom ash in combination with coal fly ash in reinforced mortars was investigated.

2. Materials and Methods

2.1. Materials, Mix Proportions and Specimen Details

The cement used in this research work was a common Portland cement CEM I 42.5 N according to the European standard EN 197-1:2011 [21] produced by HOLCIM in Almería, Spain. Coal fly ash (CFA) Type F according to the American standard ASTM C618-15 [22] and coal bottom ash (CBA) were obtained from a Spanish power station (Carboneras, Spain). CFA and CBA were generated together in the same boiler of a coal-fired power plant, and then, chemical composition is expected to be quite similar. Chemical compositions of the cement, CBA and CFA, determined according to the European standard EN 196-2:2013 [23], are shown in Table 1.

Table 1. Properties of used cements and ashes: Coal bottom ash and coal fly ash (%).

Compositions	SiO_2	Al_2O_3	Fe_2O_3	CaO	MgO	SO_3	TiO_2	P_2O_5	Soluble Residue [1]	Loss on Ignition
Cement	19.04	3.85	3.43	57.16	1.54	3.14	0.17	0.07	2.15	3.93
Bottom ash	48.12	25.55	5.86	7.07	1.28	0.15	1.5	0.96	81.24	1.85
Fly ash	46.84	26.66	4.72	5.55	1.33	0.37	1.5	1.03	76.00	3.63

[1] Insoluble residue determined by the Na_2CO_3 method (European standard EN 196-2:2013).

Siliceous aggregates with a maximum size of 4 mm were used (provided by the IETcc-CSIC, Madrid, Spain). Distilled water was employed in the mixtures. The steel corrugated bars had a 6 mm nominal diameter and had been previously cleaned, to ensure a completely rust-free surface, in a 1:1 water-HCl solution containing 3 g/L of hexametilentetramine (corrosion inhibitor), rinsed in acetone, dried and then weighed. Their ends were covered with a plastic insulating tape leaving an exposed area in the end (Figure 1).

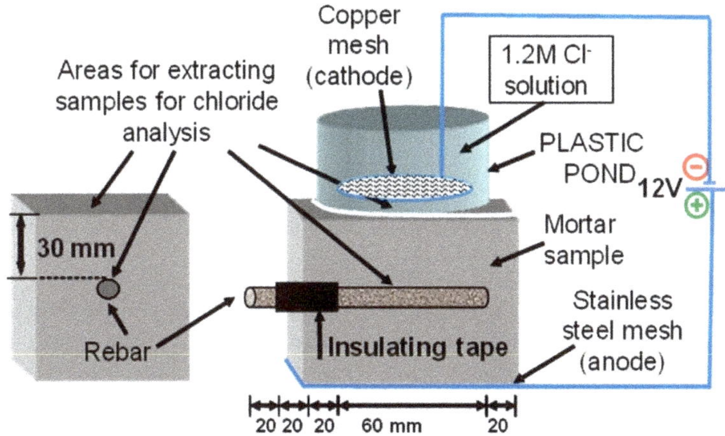

Figure 1. Geometry and embedded rebar and sensors (dimensions in mm).

Coal bottom ash (CBA) and/or coal fly ash (CFA) was used as a partial replacement of cement at 0 wt %, 10 wt %, 25 wt % and 35 wt %. The detailed mix proportions of the mortar specimens are listed in Table 2. 70 mm × 70 mm × 70 mm size mortar cubes as shown in Figure 1 were cast with one steel bar inside. The direction of casting was horizontal to the rebars. The water-to-binder ratio for all tested mortar specimens was 0.50 and the binder-to-sand ratio was 1:3 (by weight). Table 2 shows the mortar dosage. The casting was done in two layers and the mass was consolidated by vibration. Then, the specimens were kept at 100% RH for 24 h and then, they were demolded. Later, the specimens were cured at 25 °C and at 100% of relative humidity for 28 days before placing the plastic pond on the top of the specimen (Figure 1).

Table 2. Mix proportions of coal bottom ash, coal fly ash and cement, CEM I 42.5 N.

Composition [1]	CEM I	10CV	10CVF	10CF	25CV	25CVF	25CF	35CV	35CVF	35CF
Cement	100	90	90	90	75	75	75	65	65	65
Fly ash	0	10	8	0	25	20	0	35	28	0
Bottom ash	0	0	2	10	0	5	25	0	7	35
Sand	300	300	300	300	300	300	300	300	300	300
Water	50	50	50	50	50	50	50	50	50	50

[1] CEM I is the cement without coal ashes; 10CV: 10% of coal fly ash; 10CVF: 10% of coal fly ash and bottom ash; 10CF: 10% of coal bottom ash;; 25CV: 25% of coal fly ash; 25CVF: 25% of coal fly ash and bottom ash; 25CF: 25% of coal bottom ash; 35CV: 35% of coal fly ash; 35CVF: 35% of coal fly ash and bottom ash; 35CF: 35% of coal bottom ash.

2.2. Testing Procedure

The testing procedure was based on the method given in the Spanish standard UNE 83992-2 [24]. A pond with a 0.6 M NaCl and 0.4 M $CuCl_2$ solution, which was prepared by dissolving 35.06 g of NaCl and 68.20 g of $CuCl_2 \cdot 2H_2O$ in distilled water, was located on the top of the mortar specimen (Figure 1). The use of a copper chloride solution was to minimize the pH changes in the exposure solution [25]. A copper electrode (cathode) was submerged in the chloride solution. The anode (stainless steel mesh) was positioned on a water-saturated sponge at the bottom of the specimen. The stainless steel mesh (anode) and the copper electrode submerged in the chloride solution (cathode) were connected to the power source [24]. A 12 V difference in voltage was established between the two electrodes and the electrical current was measured immediately after the connection was made. Both parameters' working voltage and current were recorded throughout the test. The initial and final values were used to calculate the initial resistance and the depassivation resistance according to Equations (1) and (2).

$$Re_{initial} = V/I_{initial} \ [\Omega], \tag{1}$$

$$Re_{depassivation} = V/I_{depassivation} \ [\Omega], \tag{2}$$

where R_e is the electrical resistance, in Ω; V is the voltage applied, in V; $I_{initial}$ is the current circulating in the specimen 5–15 min after connection, in A; $I_{despassivation}$ is the current circulating in the specimen shortly before the end of the test, in A.

Chloride ions penetrated the mortar cover from the top to the bottom face in an accelerated way induced by the electrical field. The corrosion process of the steel bar begins when the chloride ions reached the reinforcing bar. The test ended when the threshold amount of chlorides around the embedded reinforcement was achieved. Corrosion was considered to exist when the voltage, referred to the Ag/AgCl electrode, was less than or equal to −300 mV [24]. The trial was deemed to be over when two consecutive voltage readings more negative than −300 mV, with respect to the silver/silver chloride electrode, were recorded. Nevertheless, it has reported values less than −300 mV in other types of cement-based materials for steel corrosion onset [26].

The time needed for steel depassivation is related to the non-steady state diffusion coefficient, namely apparent diffusion coefficient, D_{ap}. The depassivation time and the electrical charge, measured in coulombs, were recorded.

Electrochemical variables such as the corrosion potential (E_{corr}) and the polarization resistance (R_p) were monitored from the beginning until the end of testing in order to assess the long-term stability of the steel rebar. The corrosion current density (i_{corr}) evolution was determined from R_p measurements as i_{corr} is inversely proportional to R_p, according to Equation (3).

$$i_{corr} = B/R_p \ [\mu A/cm^2], \tag{3}$$

where i_{corr} is the corrosion current density, in $\mu A/cm^2$; R_p is the polarization resistance, in $k\Omega \cdot cm^2$; B is the Tafel constant and usually it has a value of 26 mV.

A three-electrode arrangement was used to carry out the polarization resistance, R_p, measurements: The steel rebar was the working electrode, the stainless steel mesh at the bottom of the specimen was used as the counter-electrode and a silver/silver chloride electrode was used as the reference electrode. The reference electrode was positioned in the solution when the current was shut off to take the measurements. Polarization resistance measurements were performed by applying a linear sweep with a sweep rate scan of 10 mV/min between −20 to +20 mV from the corrosion potential. Compensation of the ohmic drop was done at each measurement to remove the influence of the mortar resistance during the RP measurement [11]. The values were always measured when the voltage between the external electrodes was shut off. The "off-time" period ranged between 15 min and 4 h.

Finally, the specimens were split up to take mortar samples located around the reinforcing steel and at the surface in contact with the chloride solution to verify the extent of the corrosion. Then, the critical chloride concentration and the chloride concentration on the surface were determined by X-ray fluorescence, XRF, with a Bruker S8 TIGER (Bruker Corporation, Billerica, MA, USA), which is a WDXRF (wavelength dispersive X-ray fluorescence) spectrometer for elemental analysis. By varying the angle of incidence from 0° to 147.6°, a single X-ray wavelength was selected. Intensity and voltage were 80 mA and 100 kV, respectively [27]. A silver nitrate solution was also applied to one of the two parts of the concrete sample to determine whether the chlorides reached the steel.

2.3. Calculation of the Non-Steady State (Apparent) Diffusion Coefficient

Diffusion coefficient in natural test conditions was calculated by means of Equation (4).

$$D_{ns} = \frac{e^2}{2 \times t_{lag} \times \varphi} \quad [cm^2/s] \tag{4}$$

where D_{ns} is the diffusion coefficient in natural test conditions, in cm^2/s; t_{lag} is the time to steel corrosion in accelerated test or time lag, in s; e is the cover thickness in the specimen to be tested, in cm; φ is the electrical field acceleration factor, which is calculated according to Equation (5) by using the normalized electrical field $\Delta\phi$, in V, following Equation (6).

$$\varphi = \frac{z \times F}{R \times T} \times \Delta\phi = 40 \times \Delta\varphi \quad \text{for } 22\,°C \tag{5}$$

$$\Delta\phi = \frac{\Delta V}{L}[V] \tag{6}$$

Here, L is the distance between electrodes (specimen thickness), in cm; ΔV is the voltage applied, in V; R is the ideal gas constant, in cal/(mol·K) (1.9872); F is the Faraday constant, in cal/V$_{eq}$ (23,060); T is temperature, in Kelvin; z is the chloride ion valence ($z = 1$).

3. Results and Discussion

3.1. Depassivation Time and Non-Steady State Diffusion Coefficient

Table 3 summarizes the parameters recorded during the test and Figure 2 presents the depassivation time monitored along the time and the calculated non-steady state (apparent) diffusion coefficient. The depassivation time increases with the percentage of coal ash in the mortar regardless the type of ash. With 25% and 35% of coal fly ash this effect is more pronounced than with the same content of bottom ash. Given that, coal fly ash apparently provides a better chloride penetration resistance.

Table 3. Recorded for each mortar mix [1].

Code	t_{lag} (h)	D_{ns} ($\times 10^{-12}$ m^2/s)	I_{corr} at t_{lag} (μA/cm^2)	E_{corr} (mV)	$C_{critical}$ (% wt Dry Sample)	C_s (wt % Dry Sample)	$R_{e,initial}$ (Ω)	$R_{e,final}$ (Ω)	Cover Thickness (cm)
CEM I	102	23.13	1.78	−389	0.19	0.59	1105	1212	3.10
10CV	183	14.61	1.18	−578	0.18	0.97	1432	1660	3.30
10CVF	164	13.47	5.51	−478	0.14	0.53	1423	1863	3.00
10CF	183	12.89	2.65	−384	0.09	0.49	1583	1960	3.10
25CV	455	6.24	1.46	−416	0.03	1.13	3738	3800	3.40
25CVF	420	5.26	1.91	−394	0.09	1.04	3279	4000	3.00
25CF	322	7.81	1.68	−310	0.05	0.77	3053	4633	3.20
35CV	582	4.55	6.94	−326	0.03	0.53	4615	4270	3.60
35CVF	454	4.55	1.43	−342	0.06	1.17	4959	5673	2.90
35CF	567	4.63	1.06	−331	0.03	1.32	3750	7143	3.27

[1] t_{lag} is the time lag or the time to steel corrosion in the accelerated test; D_{ns} is the diffusion coefficient in natural test conditions; I_{corr} at t_{lag} is the corrosion rate measured at the time lag; E_{corr} is the corrosion potential; $C_{critical}$ is the chloride critical concentration for the corrosion onset; Cs is the chloride concentration at the surface of the specimen; $R_{e,initial}$ is the electrical resistance at the beginning of the test and $R_{e,final}$ is the electrical resistance at the end of the test.

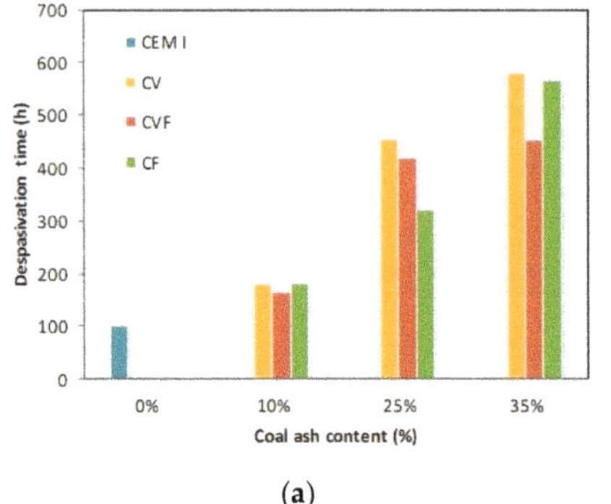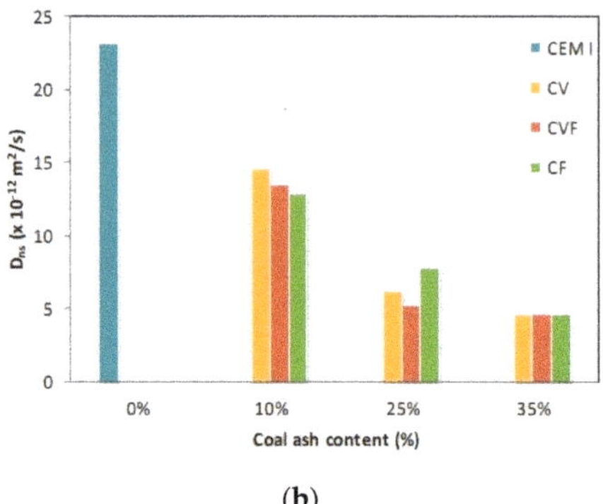

(a)   (b)

Figure 2. Time (**a**) and non-steady state (apparent) diffusion coefficient (**b**) vs. coal ash content.

The most common method widely used to assess the diffusion coefficient of chloride in cement-based materials is the measuring of the chloride profile after a time and fitting it in Fick's second law of diffusion [14,28]. Such a coefficient could either overestimate or underestimate the time to initiation of corrosion due to the great influence of the surface chloride concentration on the result, which changes with time leading to errors in the prediction of the diffusion coefficient of chloride based on Fick's second law.

Pore size redistribution by a pozzolanic reaction and higher chloride binding capacity of ash-cements reduces the non-steady state (apparent) diffusion coefficient in coal bottom ash and coal fly ash mortars by a factor of 0.44 and 0.37 times that of CEM I mortars in 10% ash replacement mortars; 0.66 and 0.73 in 25% ash replacement mortars; 0.80 and 0.80 times in 35% ash replacement mortars, respectively (Figure 2).

3.2. Critical and Surface Chloride Content

The higher the coal ash content in the mortar, the lower the critical chloride content, $C_{critical}$ (Figure 3). Chloride threshold level is affected by several factors [10,29], such as the chloride salt type [30], supplementary cementitious materials in the cement-based materials [11,31], origin of the chloride ions [32] and so on. Therefore, a wide range of threshold values has been reported. In particular, coal fly ash mortar has a lower chloride threshold level than that of the Portland cement mortar [33]. This fact may be attributed to the decrease of pH of the mortar pore solution due to the pozzolanic reaction of coal fly ash [34]. Consequently, the chloride amount needed for the passive film breakdown decreases. On the other hand, coal fly ash can improve the chemical binding ability of the mortars in some particular circumstances. Thomas [35] and Oh et al. [11] reported a decrease in the tolerable chloride content. Conversely, Alonso et al. [32] did not report any influence of coal fly ash content on the chloride threshold. On the other hand, longer depassivation times lead to higher surface concentrations as shown in Figure 3.

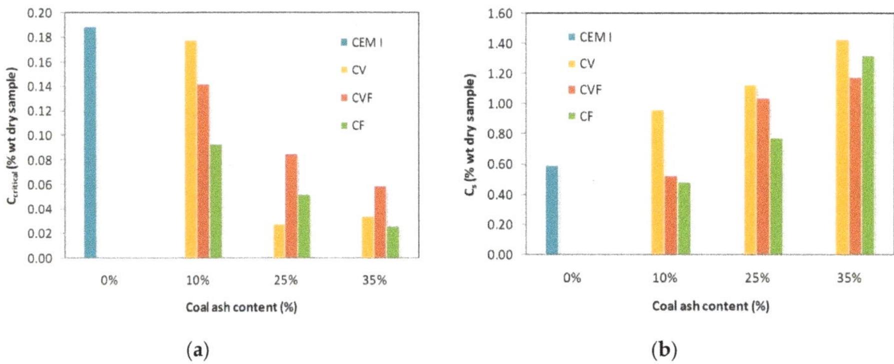

Figure 3. Critical (**a**) and surface (**b**) chloride content in the function of the coal ash content after testing.

3.3. Initial and Final Resistance

The high initial and final electrical resistances, $R_{e,initial}$ and $R_{e,final}$, measured in the bars embedded in blended mortars is attributed to their lower permeability and higher compactness (Figure 4) [12,13]. As expected, both initial and final resistance, $R_{e,initial}$ and $R_{e,final}$, increase with time because the hydration reaction produce a C-S-H gel that fills the pores (Figure 4). Moreover, the pozzolanic reaction produces more calcium silicates at longer times improving the coal ash mortars performance [16].

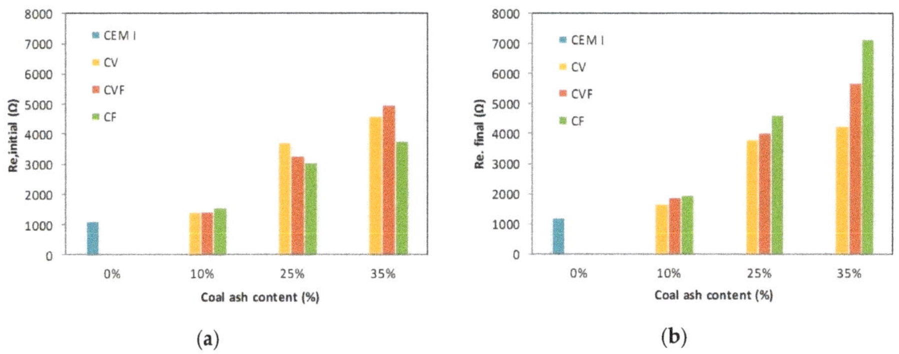

Figure 4. Electrical resistances measured in the bars embedded in blended mortars: (**a**) Initial, $R_{e,initial}$ and (**b**) final, $R_{e,final}$.

The initial resistance, $R_{e,initial}$, results were fitted with the chloride non-steady state (apparent) diffusion coefficient ones (Figure 5). Then, the chloride apparent diffusion coefficient could be approximately estimated by means of Equation (7).

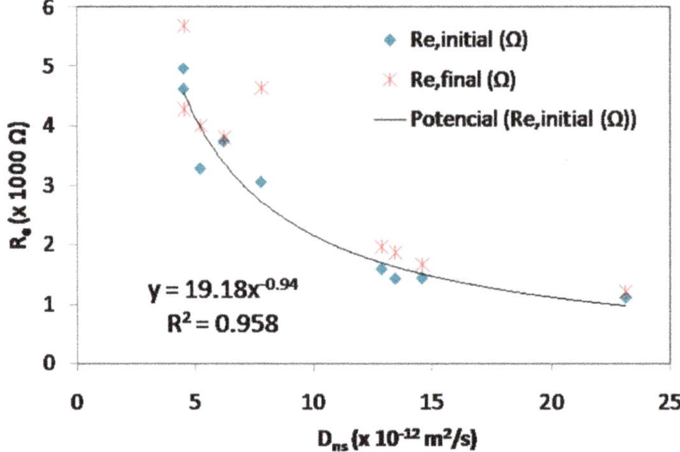

Figure 5. Initial and final resistance vs. non-steady state diffusion coefficient.

$$D_{ns} = (R_{e,initial}\ (\Omega) - 4667)/190. \tag{7}$$

3.4. Potential Monitoring

The steel potential evolution during the subsequent chloride solution exposure of the rebar is presented in Figure 6. In the passive state, the potential was generally stable with a tendency to decrease over time for all mixes. The readings were ranging from −200 to +23 mV Ag/AgCl during the first two weeks of testing corresponding to a state of passivity [24]. Then, large fluctuations between −60 and −230 mV Ag/AgCl were found until one month of testing.

Corrosion onset for the steel rebar embedded in the mortar specimen was apparent from a fall in potential. A reading more negative than −300 mV Ag/AgCl was considered as the corrosion onset threshold. Therefore, mortar specimens with fly ash showed a longer corrosion initiation period [6]. Given that, the more the coal ash amount, the longer the initiation period, regardless of the type of ash used, bottom ash or fly ash.

After 100 days, the potential of the reference mortar specimen without ashes, CEM I, became more negative than −300 mV Ag/AgCl, whereas the potential of the mortar specimens with 10% of ashes, 10CV, 10CVF and 10CF, reached more negative potentials than −300 mV Ag/AgCl after six months. Moreover, amounts of 35% of coal ash in mortars lead to longer initiation periods that ranged between 14 and 16 months.

Traditionally, corrosion potentials have been used as a complement to corrosion rate measurements in studies of steel reinforcement corrosion [36–38]. These measurements are merely qualitative, but are quite useful to detect electrochemical changes on a steel bar when monitored along the time. This technique is also valid to assess corroding zones by comparison with non-corroding ones in the same steel bar.

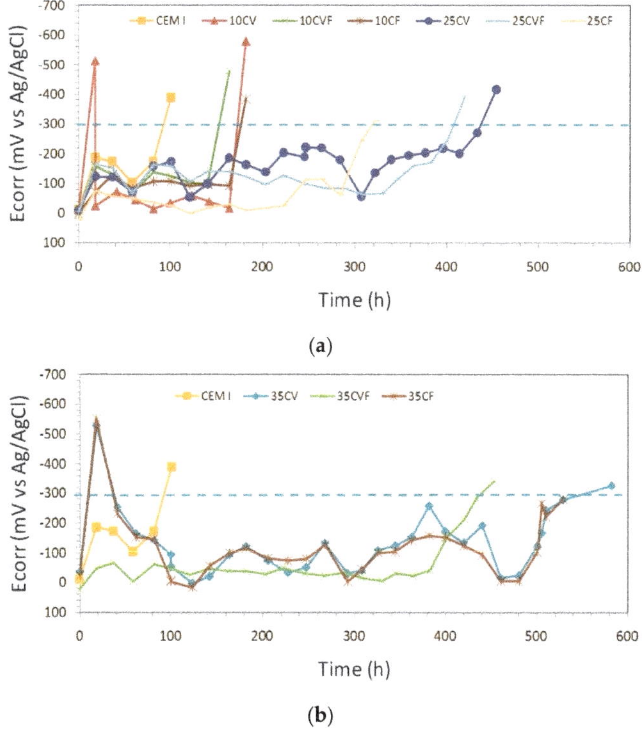

(a)

(b)

Figure 6. E$_{corr}$ vs. time. The different lines per plot represent the samples per mix: (**a**) samples CEM I, 10CV, 10CVF, 10CF, 25CV, 25CVF, 25CF; (**b**) samples CEM I, 35CV, 35CVF, 35CF. The dashed line indicates the E$_{corr}$ value for corrosion initiation (E$_{corr}$ = −300 mV).

3.5. Corrosion Rate Monitoring

Steel potential monitoring could be not enough to assess the effect of the type and amount of coal ashes, since it is affected by a factor, which includes polarization by limited diffusion of oxygen among others [18]. Then, corrosion rate measurements were also undertaken. More stable corrosion rate readings than potential ones were registered throughout the test (Figure 7). According to reference [11], if the corrosion rate of steel in mortar becomes more positive than 0.1–0.3 μA/cm^2, a significant corrosion process occurs. Within this research program, corrosion rates over 1 μA/cm^2 were found at the end of the testing period.

As apparent from Figure 7, after 100 days of chloride exposure, stable corrosion initiated in only CEM I specimen. Later on, after 150 days corrosion initiated in three out of nine blended mortar specimens with 10% of coal ash independent of ash type. More than 300 days were needed in the rest of mortar specimens with 25% or 35% coal ash for corrosion onset. Thus, the samples with coal bottom ash and/or coal fly ash showed longer corrosion initiation periods than the samples without any ash (CEM I). This indicated that a substitution of coal bottom or fly ash increased corrosion resistance of steel in mortar. Moreover, the most important parameter influencing the corrosion onset is the amount of coal ash independent of the type of ash.

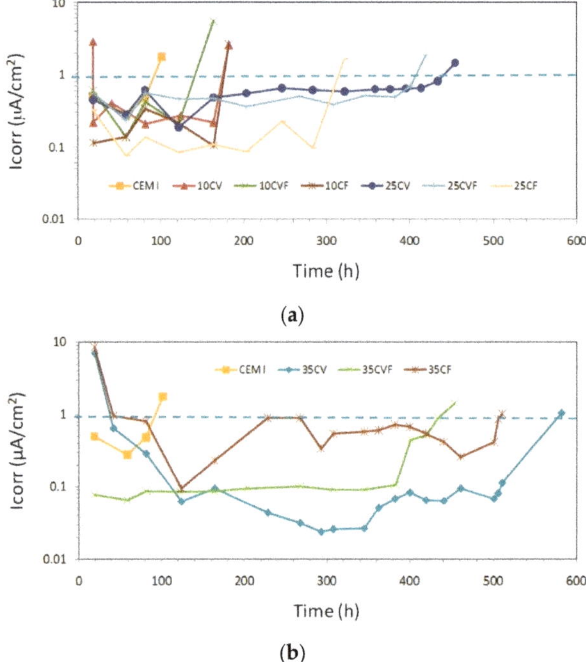

(a)

(b)

Figure 7. I_{corr} vs. time. The different lines per plot represent the samples per mix: (**a**) samples CEM I, 10CV, 10CVF, 10CF, 25CV, 25CVF, 25CF; (**b**) samples CEM I, 35CV, 35CVF, 35CF. The dashed line indicates the I_{corr} value for corrosion initiation (I_{corr} = 1 $\mu A/cm^2$).

Corrosion rate evolution of the steel rebar embedded in mortar specimens was more stable than the corrosion potential. However, both of them showed the same trend. Therefore, Figure 8 shows a clear relationship between the corrosion rate and the corrosion potential, particularly at active corrosion states, regardless the content or type of coal ash. Higher fluctuations were recorded at low corrosion states. On the other hand, a clear correlation between corrosion rate and resistivity does not exist in chloride-induced corrosion [39]. The blue dotted line shows the limit of a high corrosion rate ($1 \mu A/cm^2$).

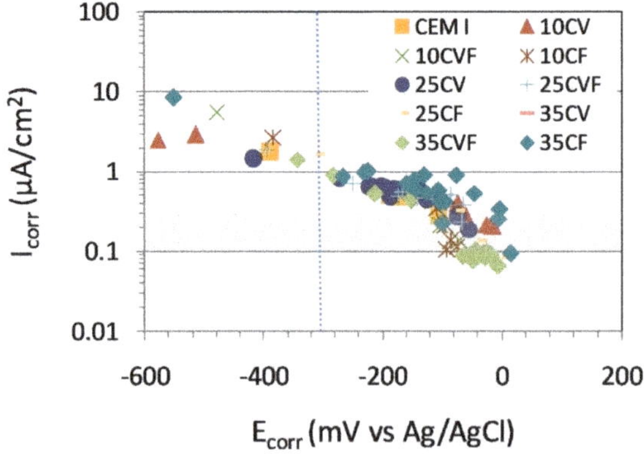

Figure 8. Relationship between I_{corr} and E_{corr} during the transition from a passive to an active state. Data of all the samples are considered before and after depassivation.

Finally, it can be said that the results obtained from electrochemical tests showed that partial replacement of either coal bottom ash or coal fly ash has led to a reduction of corrosion rate and, therefore, an enhancement of corrosion resistance due to the decrease of chloride ions permeability.

3.6. Visual Examination

After splitting the specimens and removing the steel rebar, corrosion was clearly visible in the cases where the electrochemical measurements had indicated depassivation (Figure 9). It was noticed the presence of rust spots on the steel. Red rust was found on both the steel and the mortar at the steel/mortar interface. However, red rust shown in Figure 9 could not show actual corrosion activities. Some works perform corrosion validation by means of measuring the mass loss of steel in corrosion [40] and chloride penetration in the mortar [41,42]. Nevertheless, the experimental procedure followed in this paper has been validated elsewhere [24].

(a)   (b)   (c)

(d)   (e)   (f)

Figure 9. *Cont.*

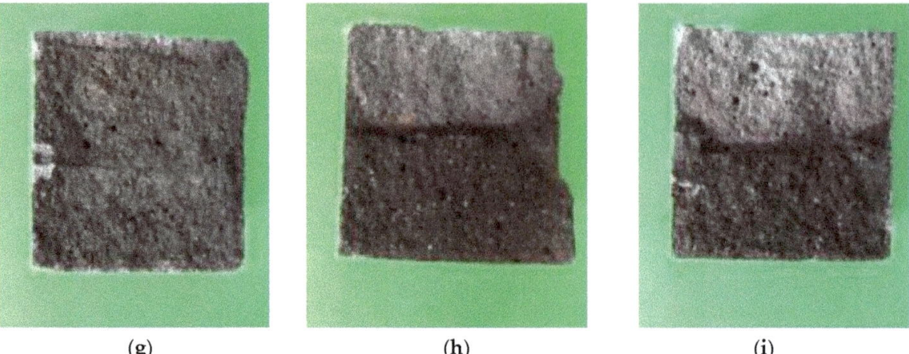

(g) (h) (i)

Figure 9. Sample of mortar extracted from the steel/specimen interface: (**a**) CEM I; (**b**) 25CV; (**c**) 35CV; (**d**) 10CVF; (**e**) 25CVF; (**f**) 35CVF; (**g**) 10CF; (**h**) 25CF and (**i**) 35CF.

4. Conclusions

The main conclusions are summarized as follows:

- Chloride diffusion coefficient in natural test conditions decreased from 23×10^{-12} m^2/s in cements without coal ashes to 4.5×10^{-12} m^2/s in cements with 35% by weight of coal ashes. Moreover, the time to steel corrosion initiation went from 102 h to about 500 h, respectively.
- Coal bottom ash and coal fly ash showed a similar corrosion performance in reinforced mortars. Both of them have a positive effect on the chloride resistance of the reinforced mortars. However, the higher coal ash proportion in the mortar, the lower critical chloride content, $C_{critical}$, was found. This is explained by the lower hydroxyl concentration in blended mortars and, therefore, the lower Cl$^-$/OH$^-$ threshold value than in plain mortars.
- The most important parameter influencing the corrosion onset is the amount of coal ash independent of the type of ash.
- The results reveal that the experimental procedure used being accelerated appears a promising method to arrive at the chloride apparent diffusion coefficient, D_{ap}, in mortars and concretes. It provided reliable information about the quality of the coal bottom ash investigated in this research program with regard to its durability.

Author Contributions: Conceptualization, E.M.; methodology, E.M.; software, M.Á.S.; validation, E.M., M.Á.S. and C.A.; formal analysis, E.M.; investigation, E.M.; resources, E.M.; data curation, M.Á.S.; Writing—Original Draft preparation, M.Á.S.; Writing—Review and Editing, E.M., M.Á.S. and C.A.; project administration, E.M.; funding acquisition, E.M.

Funding: This research was funded bythe CSIC (Intramurales Projects PIE 201460E67 and PIE 201660E054).

Acknowledgments: Authors gratefully acknowledge to Alba María Álvaro García for her valuable work in the experimental phase of this study. The authors also wish to thank to Carmen Andrade for her kind cooperation with this work.

Conflicts of Interest: The authors declare no conflict of interest.

References

1. Tripathi, S.R.; Ogura, H.; Kawagoe, H.; Inoue, H.; Hasegawa, T.; Takeya, K.; Kawase, K. Measurement of chloride ion concentration in concrete structures using terahertz time domain spectroscopy (THz-TDS). *Corros. Sci.* **2012**, *62*, 5–10. [CrossRef]
2. Song, H.-W.; Lee, C.-H.; Ann, K.Y. Factors influencing chloride transport in concrete structures exposed to marine environments. *Cem. Concr. Compos.* **2008**, *30*, 113–121. [CrossRef]

3.   Ann, K.Y.; Song, H.-W. Chloride threshold level for corrosion of steel in concrete. *Corros. Sci.* **2007**, *49*, 4113–4133. [CrossRef]

4.   Tuutti, K. *Corrosion of Steel in Concrete*; Swedish Cement and Concrete Research Institute: Stockholm, Sweeden, 1982.

5.   Muthulingam, S.; Rao, B.N. Non-uniform corrosion states of rebar in concrete under chloride environment. *Corros. Sci.* **2015**, *93*, 267–282. [CrossRef]

6.   Alonso, C.; Castellote, M.; Andrade, C. Chloride threshold dependence of pitting potential of reinforcements. *Electrochim. Acta* **2002**, *47*, 3469–3481. [CrossRef]

7.   Argiz, C.; Menéndez, E.; Moragues, A.; Sanjuán, M.A. Fly ash characteristics of Spanish coal-fired power plants. *Afinidad* **2015**, *72*, 269–277.

8.   Jorat, M.E.; Aziz, M.A.; Marto, A.; Zaini, N.; Jusoh, S.N.; Manning, D.A. Sequestering Atmospheric CO_2 Inorganically: A Solution for Malaysia's CO_2 Emission. *Geosciences* **2018**, *8*, 483. [CrossRef]

9.   Ali, H.L.; Yusuf, B.; Mohammed, T.A.; Shimizu, Y.; Ab Razak, M.S.; Rehan, B.M. Improving the Hydro-Morpho Dynamics of A River Confluence by Using Vanes. *Resources* **2019**, *8*, 9. [CrossRef]

10.   Segismundo, E.Q.; Kim, L.-H.; Jeong, S.-M.; Lee, B.-S. A Laboratory Study on the Filtration and Clogging of the Sand-Bottom Ash Mixture for Stormwater Infiltration Filter Media. *Water* **2017**, *9*, 32. [CrossRef]

11.   Kim, J.-H.; Sung, J.-H.; Jeon, C.-S.; Lee, S.-H.; Kim, H.-S. A Study on the Properties of Recycled Aggregate Concrete and Its Production Facilities. *Appl. Sci.* **2019**, *9*, 1935. [CrossRef]

12.   Chung, S.-Y.; Abd Elrahman, M.; Stephan, D. Effect of Different Gradings of Lightweight Aggregates on the Properties of Concrete. *Appl. Sci.* **2017**, *7*, 585. [CrossRef]

13.   Fan, C.-C.; Huang, R.; Hwang, H.; Chao, S.-J. The Effects of Different Fine Recycled Concrete Aggregates on the Properties of Mortar. *Materials* **2015**, *8*, 2658–2672. [CrossRef]

14.   Valentim, B.; Białecka, B.; Gonçalves, P.A.; Guedes, A.; Guimarães, R.; Cruceru, M.; Całus-Moszko, J.; Popescu, L.G.; Predeanu, G.; Santos, A.C. Undifferentiated Inorganics in Coal Fly Ash and Bottom Ash: Calcispheres, Magnesiacalcispheres, and Magnesiaspheres. *Minerals* **2018**, *8*, 140. [CrossRef]

15.   Kuo, W.-T.; Gao, Z.-C. Engineering Properties of Controlled Low-Strength Materials Containing Bottom Ash of Municipal Solid Waste Incinerator and Water Filter Silt. *Appl. Sci.* **2018**, *8*, 1377. [CrossRef]

16.   Tuan, L.Q.; Thenepalli, T.; Chilakala, R.; Vu, H.H.T.; Ahn, J.W.; Kim, J. Leaching Characteristics of Low Concentration Rare Earth Elements in Korean (Samcheok) CFBC Bottom Ash Samples. *Sustainability* **2019**, *11*, 2562. [CrossRef]

17.   Argiz, C.; Menéndez, E.; Sanjuán, M.A. Effect of mixes made of coal bottom ash and fly ash on the mechanical strength and porosity of Portland cement. *Mater. Construcc.* **2013**, *309*, 49–64.

18.   Choia, Y.-S.; Kima, J.-G.; Lee, K.-M. Corrosion behavior of steel bar embedded in fly ash concrete. *Corros. Sci.* **2006**, *48*, 1733–1745. [CrossRef]

19.   Andrade, C.; Buják, R. Effects of some mineral additions to Portland cement on reinforcement corrosion. *Cem. Concr. Res.* **2013**, *53*, 59–67. [CrossRef]

20.   Faustino, P.; Chastre, C.; Nunes, Â.; Brás, A. Lifetime modelling of chloride-induced corrosion in concrete structures with Portland and blended cements. *Struct. Infrastruct. Eng.* **2016**, *12*, 1013–1023. [CrossRef]

21.   European Committee for Standardization. EN 197-1:2011. *Cement—Part. 1: Composition, Specifications and Conformity Criteria for Common Cements*; European Committee for Standardization (CEN) Standards: Brussels, Belgium, 2011.

22.   ASTM C618-15. *Standard Specification for Coal Fly Ash and Raw or Calcined Natural Pozzolan for Use as a Mineral Admixture in Concrete*; ASTM Standards: West Conshohocken, PA, USA, 2001.

23.   European Committee for Standardization. EN 196-2:2013. *Method of Testing Cement. Chemical Analysis of Cement*; European Committee for Standardization (CEN) Standards: Brussels, Belgium, 2013.

24.   UNE—Asociación Española de Normalización. UNE 83992-2:2012 EX. *Durability of Concrete. Test Methods. Chloride Penetration Tests on Concrete. Part 2: Integral Accelerated Method*; UNE: Madrid, Spain, 2012.

25.   Paul, S.C.; van Zijl, G.P. Crack Formation and Chloride Induced Corrosion in Reinforced Strain Hardening Cement-Based Composite (R/SHCC). *J. Adv. Concr. Technol.* **2014**, *12*, 340–351. [CrossRef]

26.   Paul, S.C.; van Zijl, G.P.; Babafemi, A.J.; Tan, M.J. Chloride ingress in cracked and uncracked SHCC under cyclic wetting-drying exposure. *Constr. Build. Mater.* **2016**, *114*, 232–240. [CrossRef]

27.   King, F. *Corrosion of Copper in Alkaline Chloride Environments*; Technical Report TR-02-25; Swedish Nuclear Fuel and Waste Management Co.: Stockholm, Sweden, 2002; p. 71.

28. Liu, J.; Qiu, Q.; Chen, X.; Wang, X.; Xing, F.; Han, N.; He, Y. Degradation of fly ash concrete under the coupled effect of carbonation and chloride aerosol ingress. *Corros. Sci.* **2016**, *112*, 364–372. [CrossRef]

29. Hussain, S.E.; Al-Musallam, A.; Al-Gahtani, A.S. Factors affecting threshold chloride for reinforcement corrosion in concrete. *Cem. Concr. Res.* **1995**, *25*, 1543–1555. [CrossRef]

30. Jiang, L.; Liu, R.; Mo, L.; Xu, J.; Yang, H. Influence of chloride salt type on critical chloride content of reinforcement corrosion in concrete. *Mag. Concr. Res.* **2013**, *65*, 319–331. [CrossRef]

31. Oh, B.H.; Jang, S.Y.; Shin, Y.S. Experimental investigation of the threshold chloride concentration for corrosion initiation in reinforced concrete structures. *Mag. Concr. Res.* **2003**, *55*, 117–124. [CrossRef]

32. Xu, J.; Jiang, L.; Wang, W.; Jiang, Y. Influence of $CaCl_2$ and NaCl from different sources on chloride threshold value for the corrosion of steel reinforcement in concrete. *Constr. Build. Mater.* **2011**, *25*, 663–669. [CrossRef]

33. Liu, R.; Jiang, L.; Huang, G.; Zhu, Y.; Liu, X.; Chu, H.; Xiong, C. The effect of carbonate and sulfate ions on chloride threshold level of reinforcement corrosion in mortar with/without fly ash. *Constr. Build. Mater.* **2016**, *113*, 90–95. [CrossRef]

34. Härdtl, R.; Schiessl, P.; Wiens, U. Limits of pozzolanic additions with respect to alkalinity and corrosion protection of reinforcement. In *Durability of High Performance Concrete*; Sommer, H., Ed.; RILEM International Workshop: Vienna, Austria, 1994; pp. 189–193.

35. Thomas, M. Chloride thresholds in marine concrete. *Cem. Concr. Res.* **1996**, *26*, 513–519. [CrossRef]

36. Hausmann, D.A. Electrochemical behaviour of steel in concrete. *ACI J. Proc.* **1964**, *61*, 171–188.

37. Gouda, V.K. Corrosion and corrosion inhibition of reinforcing steel I. Immersed in alkaline solutions. *Brit. Corros.* **1970**, *5*, 198–203. [CrossRef]

38. Andrade, C.; Alonso, C. Corrosion rate monitoring in the laboratory and on-site. *Constr. Build. Mater.* **1996**, *10*, 315–328. [CrossRef]

39. Hornbostel, K.; Angst, U.M.; Elsener, B.; Larsen, C.K.; Geiker, M.R. Influence of mortar resistivity on the rate-limiting step of chloride-induced macro-cell corrosion of reinforcing steel. *Corros. Sci.* **2016**, *110*, 46–56. [CrossRef]

40. Paul, S.C.; van Zijl, G.P. Corrosion Deterioration of Steel in Cracked SHCC. *Int. J. Concr. Struct. Mater.* **2017**, *11*, 557–572. [CrossRef]

41. Kobayashi, K.; Iizuka, T.; Kurachi, H.; Rokugo, K. Corrosion protection performance of High Performance Fiber Reinforced Cement Composites as a repair material. *Cem. Concr. Compos.* **2010**, *32*, 411–420. [CrossRef]

42. Alonso, C.; Andrade, C.; Castellote, M.; Castro, P. Chloride threshold values to depassivate reinforcing bars embedded in a standardized OPC mortar. *Cem. Concr. Res.* **2000**, *30*, 1047–1055. [CrossRef]

Article

Recycled Mineral Raw Materials from Quarry Waste Using Hydrocyclones

Menéndez-Aguado L.D. [1], Marina Sánchez M. [2], Rodríguez M.A. [1], Coello Velázquez A.L. [3] and Menéndez-Aguado J.M. [1,*]

[1]   Escuela Politécnica de Mieres, Universidad de Oviedo, 33600 Oviedo, Spain
[2]   Idonial Centro Tecnológico, Parque Tecnológico de Asturias, 33428 Llanera, Spain
[3]   CETAM, Universidad de Moa Dr. Antonio Núñez Jiménez, Bahía de Moa 83300, Cuba
*   Correspondence: maguado@uniovi.es

Received: 19 May 2019; Accepted: 21 June 2019; Published: 26 June 2019

Abstract: Mining activities in general, and quarrying processes in particular, generate huge amounts of tailings with a considerable presence of fine particles and with a variable composition of minerals, which could limit the direct application of those wastes. Under the paradigm of a circular economy, more effort has to be made to find adequate applications for those secondary raw materials. In this study, a process was proposed and tests were performed to valorise fine particle product as a raw material for the building and construction industry. Samples were taken from wastes in several aggregate production plants, being characterized and processed to remove the clayey components to obtain the cleanest quartz fraction. Then, different characterization and validation tests were carried out to analyse the application of these products as raw materials in the building and construction industry (cement and ceramics). Results showed that with no complex technologies, the tailings can be considered as a mineral raw material in different applications.

Keywords: waste; aggregates; fines processing

---

1. Introduction

The decrease in the grade of ore in the mineral deposits, with lower liberation sizes, and the growing demand for mineral products results in increasing tonnage of processed ore in the mineral processing plants at a lower grinding size [1,2]. In most of the unit operations, the efficiency of recovering such fine particles is very low [3]. The quarrying process creates a great amount of tailings with considerable presence of particles under 100 μm and of a variable mineral composition, along with other components, such as iron oxides or clayey materials, which could limit the application of those wastes. Under the paradigm of a circular economy, several researchers have emphasised the need for rethinking mining wastes and tailings as secondary raw materials [2,4,5].

Following this concept, some authors reported the use of mine tailings on mortars [6–11], ceramics [12], and other applications [13]. In this study, a process was proposed that can valorise a product as a raw material to the building and construction industry [14,15]. Samples were taken from wastes in several quartz aggregate production plants in Asturias (Spain); after mineral sample characterization, they were processed to remove the clayey component and to obtain the cleanest quartz fraction. The designed process includes unitary operations requiring low capital and operating costs, as is the case of hydrocyclone classification. Subsequently, the application of the products in the cement and ceramics industry was investigated.

2. Materials and Methods

2.1. Characterization Tests

To perform the mineralogical assays, a Bruker XRD model D8 Advance (Billerica, MA, USA) was used. XRF assays were made in a Bruker XRF S-4 Pioneer Advance Billerica, MA, USA), with sample preparation in a Claisse perler, model M-4 (Malvern, Worcestershire, UK). Trace element detection was carried out using inductively coupled plasma atomic emission spectroscopy (ICP-OES), in a Varian Vista-PRO.

To perform specific gravity measures, a Micromeritics He was used. Particle size distribution tests were performed in a laser diffraction device, a Beckman Coulter LS 13 320 (Brea, CA, USA).

2.2. Separation Tests

Hydrocyclones are classifying devices that can perform an efficient particle size separation to remove clayey particles. Furthermore, hydrocyclones are commonly used in mineral processing plants when there is a need of particle classification under 200 microns. In this study, the hydrocyclone was run in closed circuit using a Mozley C124 test rig that incorporates a bypass line to control the flow-rate/pressure to the hydrocyclone. The vortex finder and spigot sizes were 14.3 and 6.4 mm, respectively.

The hydrocyclone feed sample was pulped to about 20% solids, measured using a Marcy gauge. The hydrocyclone operating pressure was adjusted to 7 kPa to obtain a rightly shaped underflow. Then, samples of feed, underflow, and overflow slurry were collected.

Batch tests were conducted to examine the required number of cleaning stages (treatments) to get the underflow product (coarser particles) as clean as possible of clay contents, thereby increasing the quartz content. The cleaning stages means that the underflow product of a treatment is afterwards used as feed for the following treatment or cleaning stage.

2.3. Application Tests

The use of the recycled product as a secondary mineral raw material in mortars, white cement, and ceramics was studied.

To study the application of the final product in mortars, a reference was established as follows: a series of cast cubes (5 × 5 × 5 cm) were prepared with 75% normalized sand from the Eduardo Torroja Institute (Madrid, Spain) and 25% Portland cement CEM II/B-V 32.5R (Cementos Tudela Veguin, Asturias, Spain), the cement/water ratio being 1:0.5. Once we had prepared the mortar, it rested in the mold for 1 day and afterwards it was introduced in a wet chamber for 28 days. Finally, various groups of three cast cubes were tested in a 250 ton press with rapid breakage. Compression strength tests were carried according to the norm UNE-EN 196-1:2018 (www.une.org).

Once the reference was established, the process was repeated, replacing the sand to different degrees with the recycled product. Table 1 lists the weights used in each case.

Table 1. Mortars prepared replacing sand by cleaned product.

Sand Substitution (%)	Normalized Sand (g)	Cement (g)	Water (g)	Cleaned Product (g)
0	700	233	116.5	0
5	665	233	116.5	35
10	630	233	116.5	70
15	595	233	116.5	105
20	560	233	116.5	140

To test the ability of application in white cement production, in which a major importance parameter is the brightness index, a colorimeter HAC model DR/890 was used to obtain the whiteness index, according to the norm EN ISO 787-16:1995 (www.une.org).

Finally, the use of the recycled product in ceramics was validated through characterization of plastic properties. In this case, the application of the clean product as a raw material in ceramics was studied, with the economy mainly conditioned by transportation costs. Table 2 shows the specification of quartz sand to be used in some ceramic materials.

Table 2. Quartz sand specifications.

Material	Fe_2O_3 (%)	TiO_2 (%)	Al_2O_3 (%)	K_2O (%)	CaO (%)	D_{90} (µm)
Enamel	<0.06	<0.06	<0.7	<0.2	<0.2	150
Ingobbio	<0.06	<0.06	<0.7	<0.2	<0.2	63–125
Ceramic pastes (porous/gres/porcelain)	<0.15	<0.1	3–7	2–5	<0.3	100 µm–6 mm *

* Depending on the preparation process.

It can be envisaged that, in this case, the main problems will come from the Al_2O_3 and TiO_2 contents. The alumina content can be reduced because it is associated with clayey components, but this cannot be said in the case of titanium oxide. This oxide can be undesirable because it enhances the color of other oxides in the kiln (Fe, Cr, Mn, Co, Ni, Cu); therefore, an additional cleaning processes (froth flotation, leaching) will probably be necessary.

To carry out the study in ceramic mixtures, a basis clay material was selected with the composition shown in Table 3. Mixtures at different compositions were prepared to study the liquid limit, plastic limit, and plasticity index, following the norm ISO 17892-12:2018 (https://www.iso.org/standard/72017.html) Additionally, the behaviour under different thermal treatments at 300 °C, 600 °C, and 900 °C was studied.

Table 3. Chemical composition in the basis clay material.

Al_2O_3	SiO_2	Fe_2O_3	TiO_2	CaO	MgO	Na_2O	K_2O	P2O5	LOI
19.05	64.71	3.89	0.84	0.15	0.61	0.18	4.03	0.13	6.42

Dimensional variation with temperature was also studied on cylinder probes with different compositions, up to 30%. Samples in the form of cast cylinders (37.7 mm diam. and 22.1 mm long) were prepared, as summarized in Table 4.

Table 4. Samples prepared to the thermal study.

Cast Cylinder	Clay (g)	Recycled Product (g)	Water (g)
1	80	0	30
2	100	0	30
3	95	0	30
4	90	0	32
5	70	0	33
6	95	5	32
7	90	10	31
8	80	20	30
9	70	30	30

Special care was taken in the preparation procedure to avoid air bubble formation within the mass. The probes were dried at 70 °C for 18 h, then weighed and measured. Then, the cylinders were dried again at 300 °C for 18 h, and afterwards, again weighed and measured. This operation was repeated again for the treatments at 600 °C and 900 °C.

3. Results and Discussion

3.1. Material Characterization

The mineralogical analysis results are shown in Figure 1, which demonstrates the presence of quartz and the presence of clay constituents such as illite and nacrite. Both complex silicates were the main problem for the direct reuse of this waste and needed to be removed in an earlier treatment.

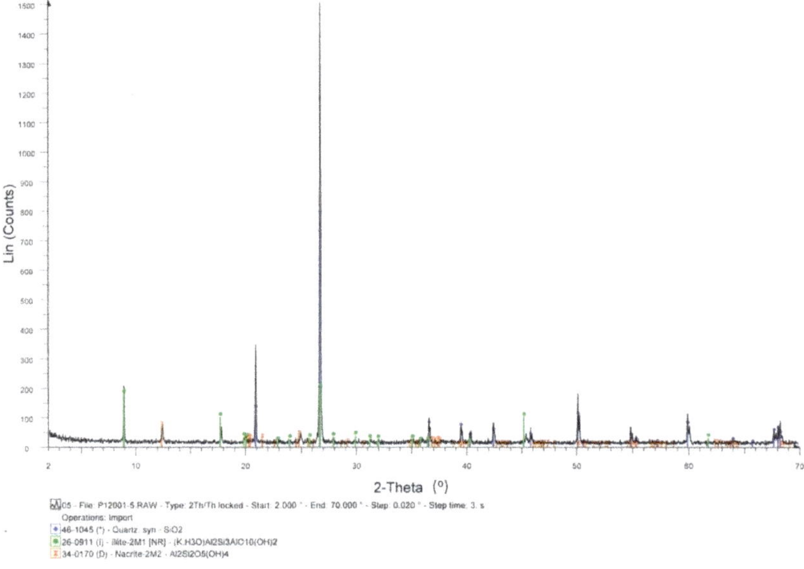

Figure 1. Mineralogical analysis result.

In Figure 2, the size distribution graph obtained in the test is shown, revealing that the top size in the feed was 50 microns.

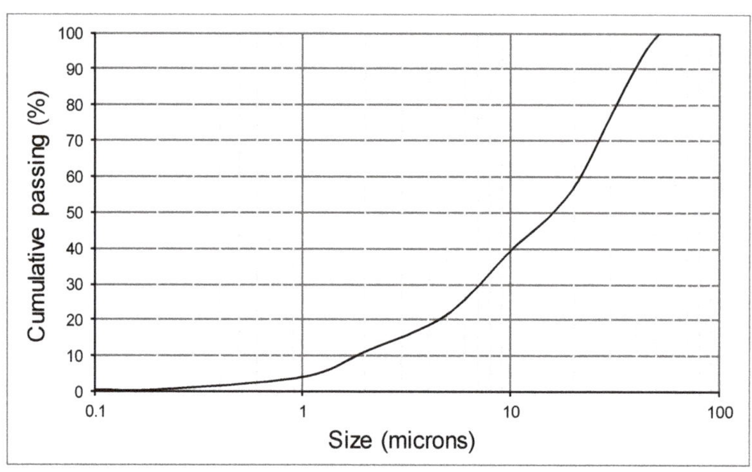

Figure 2. Particle size distribution.

Due to the fineness of the sample, XRF analysis could be performed without further comminution, and results for major components are shown in Table 5.

Table 5. Major components in the sample (%).

Al_2O_3	SiO_2	Fe_2O_3	TiO_2	CaO	MgO	Na_2O	K_2O	P_2O_5	LOI
6.49	89.42	0.73	0.67	<0.1	<0.1	0.03	1.10	<0.1	1.54

To carry out the trace elements detection, ICP-OES was performed; the sample was digested with aqua regia and microwave action. Results are shown in Table 6.

Table 6. ICP-OES trace elements results (mg/kg).

As	Ba	Sr	Sb	Co	Cr	Cu	Cd	Hg	Pb	Zn	Zr	Ni	Mn	Sn
18	142	15	<10	14	31	22	<10	<10	<10	<10	85	<10	17	<10

To clear up any possible doubt about the environmental aspects of this secondary raw material, leaching tests were carried out to evaluate the toxicity risk; sub-samples of 10 g were taken and introduced in 50 mL water during 12 h. The liquid obtained was tested with the spectrometer ICP-OES, and the results (Table 7) showed that toxicity levels were very low.

Table 7. Leach test results (mg/kg).

As	Ba	Sr	Sb	Co	Cr	Cu	Cd	Hg	Pb	Zn	Zr	Ni	Mn	Sn	Al	Fe	K
<5	<5	<5	<5	<5	<5	<5	<5	<5	<5	<5	<5	<5	<5	<5	17	6	27

The aforementioned presence of clay components (illite and nacrite), which are the main problems for the direct reuse of this waste, suggested the requirement for prior treatment.

3.2. Separation Tests

As can be seen in Table 8, the quartz content increased from 89% to 98% with four hydrocyclone treatments under the abovementioned operating conditions.

Table 8. Underflow fraction results after 3 treatments.

Product	Al_2O_3	SiO_2	Fe_2O_3	TiO_2	CaO	MgO	K_2O	P_2O_5
Feed	6.49	89.42	0.73	0.67	<0.1	<0.1	1.10	<0.1
Underflow, treatment 1	2.19	96.51	0.11	0.48	<0.1	<0.1	0.31	<0.1
Underflow, treatment 2	1.16	96.92	0.07	0.45	<0.1	<0.1	0.15	<0.1
Underflow, treatment 3	0.92	98.05	0.05	0.42	<0.1	<0.1	0.1	<0.1
Underflow, treatment 4	0.89	98.12	0.04	0.41	<0.1	<0.1	<0.1	<0.1

Mean grain size obtained in the underflow and overflow products are shown in Figure 3. The increasing particle size with the treatments in the underflow products can be explained as follows: in each treatment, a certain amount of fine product was removed in the overflow, so the re-treatment of the underflow product meant a slightly coarser feed, which led (with the same conditions) to a slightly coarser cut size. This effect was attenuated with the number of treatments because in each step, the finer fraction was removed from the underflow product.

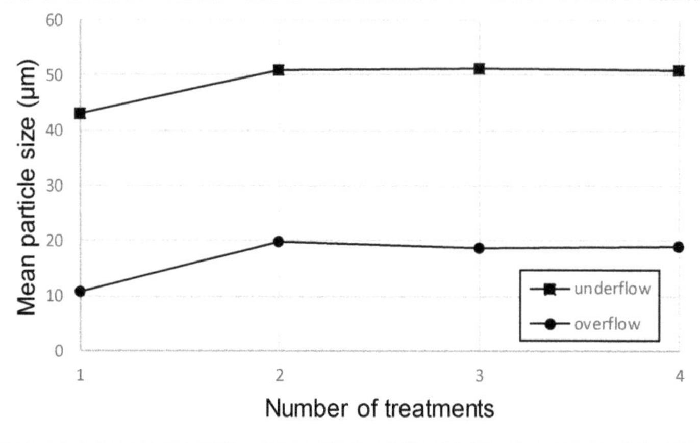

Figure 3. Variation of mean particle size with treatments.

Results after four cleaning stages show that quartz was concentrated in the underflow product (coarser particles), while clays were concentrated in the finer fraction, reducing the Al_2O_3 and Fe_2O_3 levels in the underflow product with each treatment. Once the treatment procedure was defined, enough recycled product had been prepared under the same separation conditions to carry out the mortar tests. Due to the limited capacity of the hydrocyclon test bench (10 kg), the necessary amount of material was obtained in several batches. In all cases, feed material was adequately homogenized and sampled before batch separation. The final product obtained was also homogenized and sampled to ensure sample representativeness.

3.3. Application Tests

3.3.1. Recycled Product in Mortars

Table 9 and Figure 4 show the results obtained in the compression strength tests developed in each case. The reference value was considered to be 35 MPa, without sand substitution. It must be noticed that in all cases, the difference in the strength values were lower than 5%.

Table 9. Compression tests results.

	Substitution of Recycled Sand (%)				
	0	5	10	15	20
Compression tests (Mpa)	35.4	35.1	34.9	34.7	34.6
	35.2	35.2	35.1	34.8	34.5
	35.3	35.1	35.1	34.9	34.7
Average (Mpa)	35.3	35.1	35.0	34.8	34.6
Variation (%)	–	0.5%	0.8%	1.4%	2.0%

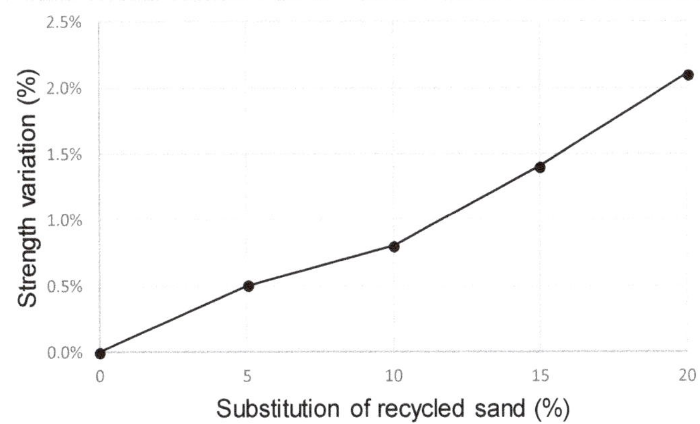

Figure 4. Variation in compression strength due to product addition.

3.3.2. Recycled Product in White Cement Production

Initially, the underflow product from the first hydrocyclone treatment was used; its chemical composition is shown in Table 8. Due to the high SiO_2 content (96.51%), the objective was to study the effect of replacing the quartz contribution of raw materials with this recycled product. As a result, it was found that this product was suitable for white clinker production in terms of compressive strength: strength values registered only a slight decrease and the setting times were only slightly longer. However, the whiteness level reached was not enough. The L* index of the clean product reached only 88, whereas the necessary L* additive index in the cement production process is at least 94. After obtaining this result, further studies where conduction to establish whether more hydrocyclone treatments could improve the L* index. In Figure 5, the variation of L* index is shown with the number of treatments, reaching the value 94 in the third treatment, with no sensible improvement in the fourth one.

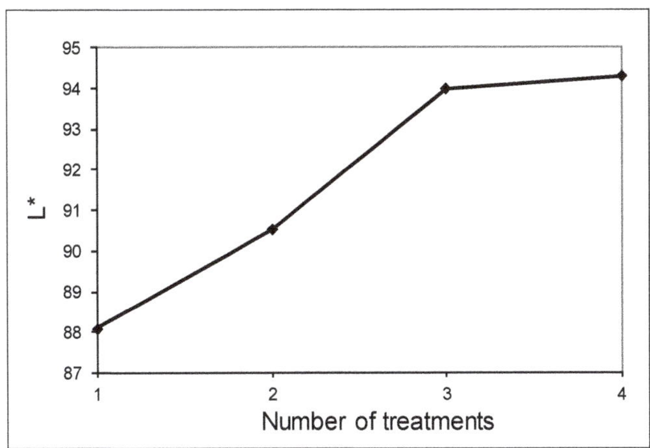

Figure 5. Variation of whiteness index L* with number of cleaning stages.

3.3.3. Recycled Product in Ceramics

Figure 6 shows the results obtained in the plasticity index, from which it can be concluded that the mixtures could contain up to 30% recycled product without great variation in the plasticity index.

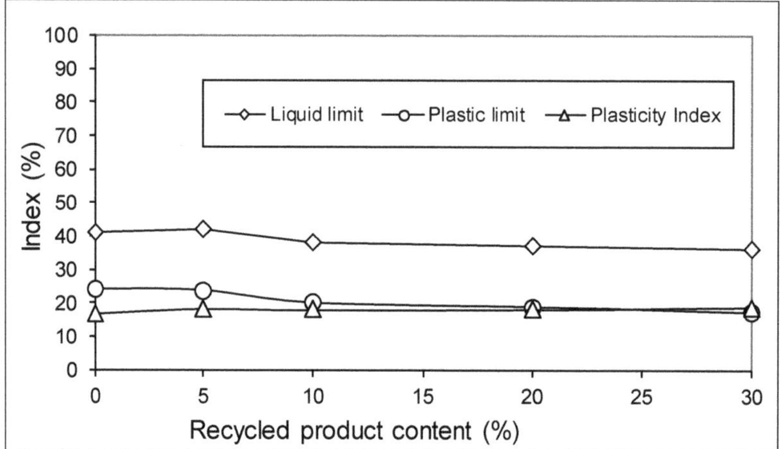

Figure 6. Variation of plasticity parameters.

A similar mass variation was obtained in each composition, with the higher relative variation obtained at 300 °C, probably due to the evaporation of water crystallization in the clayey minerals. The mass loss at 900° was probably due to carbide decomposition and organic matter being removed from the clays. All of these changes are shown in Figures 7 and 8.

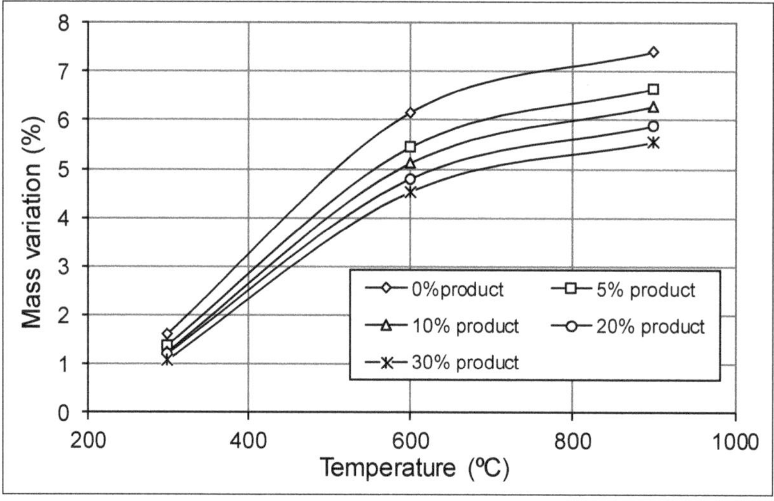

Figure 7. Mass variation with temperature at different recycled product content.

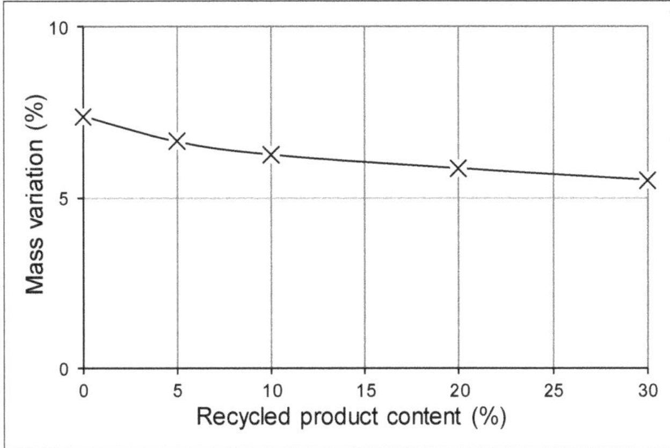

Figure 8. Mass variation at 900 °C.

Regarding dimensional variation, results observed in each temperature range are depicted in Figure 9. It must be clarified that in the temperature ranges 0–300 °C and 600–900 °C, shrinkage was observed, while in the interval 300–600 °C, an expansion was recorded.

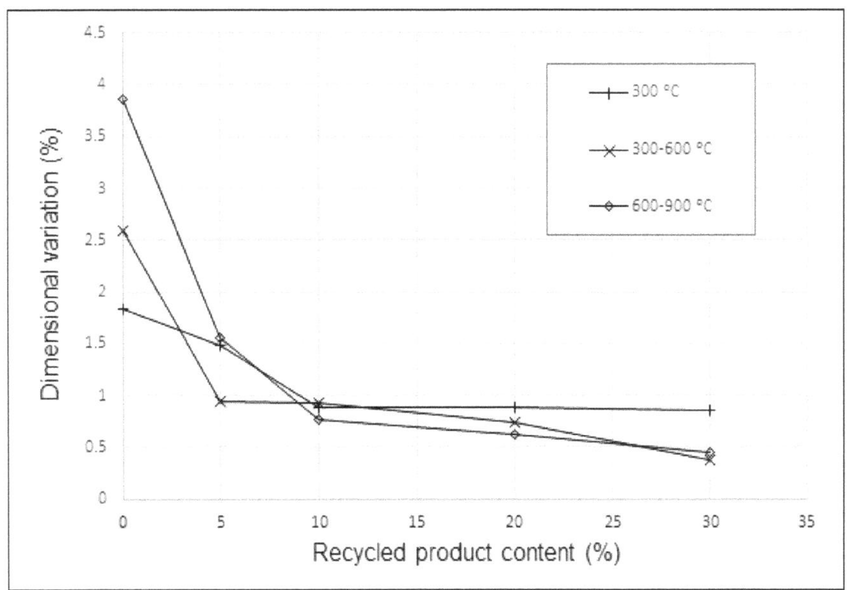

Figure 9. Dimensional variation at 300 °C, 300–600 °C, and 600–900 °C.

As observed in the curves represented in Figure 9, within the 0–10% range of sand substitution by recycled sand, more differences in the properties were measured; this effect of physical properties variations within this range was reported by Almeida et al. [10]. It must be noted here that the addition must be performed carefully to ensure the properties in the mixture; a 10% sand addition reduced the dimensional variation significantly without the appearance of cracks after thermal treatment, but this was not the case when the addition was raised until 30%.

4. Conclusions

- Treatment of quarry tailings with hydrocyclones is an effective methodology to turn a waste into a by-product. The underflow product can be considered a cleaned quartz fraction, with a wide range of applications under circular economy principles. In the case of some specific applications, the recycled product could not meet the necessary standard of purity due to the concentration of titanium oxides in the underflow product. In that case, further processing steps could be performed, as is the case with froth flotation.

- Use in mortars and concrete: The recycled product can replace up to 20% of the normalized sand with a compression strength reduction below 3%. Furthermore, in all cases, the recycled product substitution had no noticeable impact on the mechanical properties or the setting time in the cement.

- Use in grey clinker production: In terms of mechanical properties, the recycled product with only one cleaning stage could be used directly in grey clinker production.

- Use in white cement production: With 3–4 cleaning stages, the recycled product reached the necessary whiteness level to be a raw material for white cement production.

- Use in ceramics: The addition of up to 10% of recycled product did not have a relevant influence in the plasticity properties of the ceramic paste relative to the basis clay selected; in this case, an effect of decreasing the dimensional variation during the process within the kiln was reported, without the appearance of cracks when the addition level was set to around 10%.

Author Contributions: Conceptualization, M.-A.J.M., C.V.A.L.; Formal analysis, M.-A.L.D. and M.S.M.; Investigation, M.S.M. and R.M.A.; Methodology, M.-A.L.D. and M.-A.J.M.; Resources, R.M.A.; Supervision, R.M.A. and C.V.A.L.; Validation, M.S.M.; Writing – original draft, M.-A.L.D.; Writing—review & editing, M.-A.J.M.

Funding: This research received no external funding.

Conflicts of Interest: The authors declare no conflict of interest.

References

1. Chryss, A.; Fourie, A.B.; Monch, A.; Nairn, D.; Seddon, K.D. Towards an integrated approach to tailings management. *J. South. Afr. Inst. Min. Met.* **2012**, *112*, 965–969.
2. Tayebi-Khorami, M.; Edraki, M.; Corder, G.; Golev, A. Re-Thinking Mining Waste through an Integrative Approach Led by Circular Economy Aspirations. *Minerals* **2019**, *9*, 286. [CrossRef]
3. Das, A.; Sarkar, B. Advanced Gravity Concentration of Fine Particles: A Review. *Miner. Process. Extr. Metall. Rev.* **2018**, *39*, 359–394. [CrossRef]
4. Lebre, É.; Corder, G. Integrating industrial ecology thinking into the management of mining waste. *Resources* **2015**, *4*, 765–786. [CrossRef]
5. Adiansyah, J.S.; Rosano, M.; Vink, S.; Keir, G. A framework for a sustainable approach to mine tailings management: Disposal strategies. *J. Clean Prod.* **2015**, *108*, 1050–1062. [CrossRef]
6. Argane, R.; Benzaazoua, M.; Hakkou, R.; Bouamrane, A. Reuse of base-metal tailings as aggregates for rendering mortars: Assessment of immobilization performances and environmental behavior. *Constr. Build. Mater.* **2015**, *96*, 296–306. [CrossRef]
7. Argane, R.; Benzaazoua, M.; Hakkou, R.; Bouamrane, A. A comparative study on the practical use of low sulfide base-metal tailings as aggregates for rendering and masonry mortars. *J. Clean Prod.* **2016**, *112*, 914–925. [CrossRef]
8. Bentz, D.P. Replacement of "coarse" cement particles by inert filler in low w/c ratio concretes II. Experimental validation. *Cem. Concr. Res.* **2005**, *35*, 185–188. [CrossRef]
9. Lawrence, P.; Cyr, M.; Ringot, E. Mineral admixtures in mortars effect of type, amount and fineness of fine constituents on compressive strength. *Cem. Concr. Res.* **2005**, *35*, 1092–1105. [CrossRef]
10. Almeida, N.; Branco, F.; de Brito, J.; Santos, J.R. High-performance concrete with recycled stone slurry. *Cem. Concr. Res.* **2007**, *37*, 210–220. [CrossRef]

11. Gayana, B.C.; Chandar, K.R. Sustainable use of mine waste and tailings with suitable admixture as aggregates in concrete pavements—A review. *Adv. Concr. Constr.* **2018**, *6*, 221–243.
12. Liu, T.Y.; Tang, Y.; Han, L.; Song, J.; Luo, Z.W.; Lu, A.X. Recycling of harmful waste lead-zinc mine tailings and fly ash for preparation of inorganic porous ceramics. *Ceram. Int.* **2017**, *43*, 4910–4918. [CrossRef]
13. Driussi, C.; Jansz, J. Technological options for waste minimisation in the mining industry. *J. Clean Prod.* **2006**, *14*, 682–688. [CrossRef]
14. Ahmed, E.A.; El-Kourd, A.A. Properties of concrete incorporating natural and crushed stone very fine sand. *ACI Mater. J.* **1989**, *86*, 417–424.
15. Topçu, I.B.; Ugurlu, A. Effect of the use of mineral filler on the properties of concrete. *Cem. Concr. Res.* **2003**, *33*, 1071–1075. [CrossRef]

 *materials*

Article

SEM Image Analysis in Permeable Recycled Concretes with Silica Fume. A Quantitative Comparison of Porosity and the ITZ

Manuel J. Chinchillas-Chinchillas [1], Carlos A. Rosas-Casarez [1], Susana P. Arredondo-Rea [1], José M. Gómez-Soberón [2,*] and Ramón Corral-Higuera [1]

[1] Faculty of Engineering Mochis, Autonomous University of Sinaloa, Fuente de Poseidón y Ángel Flores s/n, Col. Jiquilpan, Module B2, Los Mochis, Sinaloa 81210, Mexico
[2] Barcelona School of Building Construction, Polytechnic University of Catalonia, Av. Doctor Marañón 44-50, 08028 Barcelona, Spain
*   Correspondence: josemanuel.gomez@upc.edu; Tel.: +34-934016242

Received: 4 June 2019; Accepted: 3 July 2019; Published: 8 July 2019

Abstract: Recycled aggregates (RA) from construction and demolition can be used in permeable concretes (PC), improving the environment. PCs have a significant porous network, their cement paste and the interaction between the paste and the RA establishing their strength. Therefore, it is important to evaluate the porosity in the interfacial transition zones. The porosity of the cement paste, the aggregate and the interfacial transitional zones (ITZ) of a PC with recycled coarse aggregates (RCA) and silica fume (SF) is measured by means of image analysis–scanning electron microscope (IA)-(SEM) and by mapping the chemical elements with an SEM-EDS (energy dispersive spectrometer) detector microanalysis linked to the SEM and, as a contrast, the mercury intrusion porosimetry technique (MIP). In the IA process, a "mask" was created for the aggregate and another for the paste, which determined the porosity percentage (for the anhydrous material and the products of hydration). The results showed that using SF caused a reduction (32%) in the cement paste porosity in comparison with the PC with RA. The use of RA in the PC led to a significant increase (190%) in the porosity at different thicknesses of ITZ compared with the reference PC. Finally, the MIP study shows that the use of SF caused a decrease in the micropores, mesopores and macropores.

Keywords: porosity; recycled aggregates; permeable concrete; interfacial transition zone; image analysis; porosimetry mercury intrusion

---

1. Introduction

PC is a special concrete with a high gap content and high permeable capacity when compared with conventional concrete [1,2]. This PC offers several environmental benefits, among which are rainwater runoff control, replenishment of underground water supplies, an improvement in water quality and a reduction in water and subsoil contamination [3]. Therefore its use is habitual in parks, reduced traffic areas, tennis courts, etc. [4]. This PC contains interconnected macropores of sizes that oscillate from 2 to 8 mm and with gap content of between 18% and 35%; therefore, its mechanical strength is closely linked to the strength of the cementing matrix itself and the interaction of the matrix with the aggregates that form it. Consequently, it is necessary to determine and study the porosity of these areas of interest [5].

The use of natural aggregates is currently widespread in the manufacture of PC [6], but few studies have used RA [7]. As important quantities of construction and demolition waste are contaminating many ecosystems worldwide, it is imperative to recycle them and produce new building materials, recycled concrete [8]. The production of RA from construction and demolition waste has been studied widely in the last two decades and the studies agree that these aggregates are more heterogeneous, less dense and more porous than the natural aggregates [9,10]. Those studies which have analyzed the microstructure of the RA show that there are two ITZ a) between the old natural aggregates and the old matrix of the cement and b) between the RA and the new cement matrix [11]. The ITZ is considered by many researchers to be the weak link in the concrete [12,13], in which the normal concrete fails, significantly compromising both mechanical properties and permeability; this is also characterized by being an area of high porosity [14]. This differentiated zone can form due to the "wall" effect caused by the looser packing of the cement granules against the surface of the aggregate, or because of the accumulation of free water around the surface of the aggregate [15]. Some models of the ITZ have established that it is a space with 'uniform' porosity, located 5–30 μm from the aggregate surface; in other models, it has been found up to 50 μm from the aggregate [16]. Porosity in the ITZ is one of the basic factors influencing the strength and durability of concrete; the more porosity in this zone, the less mechanical strength and the greater the vulnerability to environmental degradation [17]. Additionally, the porosity in the cementing matrix can be determined as the sum of the volume of the capillary pores, the ice pores, the macropores formed by the presence of air and macropores caused by compacting [18]. This porosity can also be considered as the spaces not filled by the solid components of the hydrated cement paste, which depend to a great extent on the ratio of water/cement (w/c) and the degree of hydration reached by the cement [18].

Nowadays several alternatives that allow a reduction in the porosity percentage in concrete have been established, such as the inclusion of additives, the setting of low w/c ratios, the inclusion of fibers or the application of pozzolans [19,20]. Regarding the last alternative, SF is the most used pozzolan for improving the properties of concrete and reducing the porosity [21] when the ITZ is densified; this can react with the cement hydration products (calcium hydroxide—CH–) to form calcium silicate hydrate (CSH), which thereby improves the mechanical strength. Similarly, this SF also acts as a filler which reduces porosity, as the granule size of this material is between 0.1 and 0.2 μm [22].

Having seen the importance of porosity in the binding matrix of concrete and in its ITZ, several studies have used various methods to try to establish and calculate the percentage of porosity. Among them are computerized tomography X-ray scans [23], mercury intrusion porosimetry technique (MIP) [24,25], mathematical models of finite elements [26] and X-ray microtomography [9]. However, the IA technique using software is noteworthy, as the treatment of images obtained from IA-SEM is a non-destructive, precise and low-cost method [27]. This method breaks the image down to a scale of gray tones to obtain a binary image of the area of interest; it is also applicable when the aggregate or cement paste is made up of various phases, making it possible to identify the hydration products such as CSH, CH or ettringite [28]. The IA makes it possible to make a mask—a selection filter—of the aggregate or the cement paste, allowing these elements to be distinguished, then establishing the porosity with exactitude [29]. The importance of this technique is reflected in those studies where the porosity of the ITZ is calculated at different thicknesses of the aggregate [28], quantifying the percentage of pores and micro-cracks in the concrete [30]. This leads to the identification and quantifying of the hydration products and the anhydrous materials of the concrete [31,32]. However, the majority of these studies have focused on conventional concretes and there is no evidence of studies of image analysis in PCs. Therefore, this study aimed to evaluate and correlate the microporosity of PC made with RA and SF using MIP and the IA of SEM; this was done through studying the porosity of the aggregates, the porosity of the cement paste, the percentage of hydration products, the percentage of anhydrous materials and the percentage of porosity of the ITZ at different zones of the aggregate.

2. Materials and Methods

2.1. Materials

CEMEX® brand composite Portland cement (30R) [33] was used to prepare the PCs. Local companies supplied the distilled water, the natural coarse aggregate (NCA) and the SF. The RCA was obtained by means of grinding waste paving concrete, with a strength of 23 MPa, in a jaw crusher in the sustainable materials laboratory of the Universidad Autónoma de Sinaloa (UAS), Mexico.

The properties of the NCA and RCA are shown in Table 1; these were determined from the ASTM C127 and ASTM C29 [34,35] standards. It should be mentioned that the maximum size of the aggregates was 12.7 mm, obtained after sieving. Of the studied properties, the high absorption of the RCA—more than 3.5 times that of the NCA—is noteworthy. In order to determine the chemical compounds present in the cement and the SF, an X-ray fluorescence analysis was carried out with a set of ARL 8680 (ThermoFisher Scientific, Waltham, MA, USA.), using the wave dispersion method according to the ASTM C114 [36] standard (see Table 2). Both the ASTM C150 cement [33] and the SF (ASTM C1240) [37] showed typical percentages of compounds. Finally, the physical properties of the Portland cement and the SF are to be found in Table 3, establishing that both are within the specifications [33,37].

Table 1. Properties of the aggregates.

Properties	NCA	RCA
Aggregate size (mm)	12.70	12.70
Concrete strength of origin (MPa)	-	23.00
Absorption (%)	1.67	6.28
Humidity (%)	1.67	1.80
Density (kg/cm^3)	2.65	2.35
Porosity (%)	11.00	16.31
Rebound index (MPa)	33.20	29.90

Table 2. Chemical composition of Portland cement and silica fume.

Oxide (%)	SiO$_2$	Al$_2$O$_3$	Fe$_2$O$_3$	CaO	SO$_3$	K$_2$O	Na$_2$O	MgO
Cement	19.94	4.40	2.97	63.5	3.08	0.42	0.12	-
Silica Fume	95.22	0.08	2.37	0.26	0.11	0.56	0.30	0.24

Table 3. Properties of Portland cement and silica fume.

Material	Density (kg/m^3)	Specific Surface (m^2/kg)	Average Size (µm)
Cement	3150	1400	15–25
Silica Fume	2270	19,600	0.1–0.2

2.2. Design of the Permeable Concrete Mixture

Table 4 shows the components and quantities of the different materials that make up the four dosages of the PC mixtures to be used in this study. The criteria of the RCA content percentages have been established with reference to previous similar research [38,39], as well as in the case of SF (10%) [40,41]. The values shown are those needed to make a beam of 15 × 15 × 50 cm, from which the study samples will later be taken (see Figure 1).

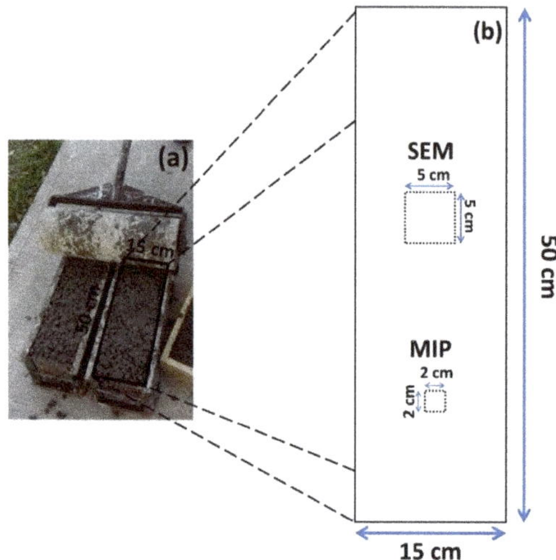

Figure 1. General scheme of the test samples. (**a**) real photo of the specimen beam and (**b**) areas of study for analysis of scanning electron microscope (SEM) and mercury intrusion porosimetry (MIP).

The nomenclature used is classified as follows: the first three letters (NCA or RCA) refer to the origin of the coarse aggregate; the following value shows the percentage of recycled aggregate used (100% or 50%); finally the last letter indicates the cement used (C = 100% Portland and SF = 90% Portland + 10% SF). A value of 0.35 was used for the w/c ratio, as is common in this type of study [42,43]. As this was a PC, no fine aggregates (FA) were added to the mixture.

Table 4. Previous concrete mixtures for a specimen of $15 \times 15 \times 50$ cm (0.0113 m^3).

Materials	NCA 100 C	NCA 100 SF	RCA 50 C	RCA 50 SF
Water (L)	1.9781	1.9781	1.9467	1.9467
NCA (kg)	17.5568	17.5571	8.7785	8.7785
RCA (kg)	-	-	8.7785	8.7785
Cement (kg)	5.5621	5.0059	5.5621	5.0059
Silica fume (kg)	-	0.5562	-	0.5562

The mixing process of the PC followed the methods laid down by previously reported studies [44], as well as the recommendations for using RCA, such as the need for pre-saturation [4]. When the mixture was ready, the concrete was poured into metal $15 \times 15 \times 50$ cm beam molds, in accordance with the ASTMC31 norms [45]. The beams were kept in laboratory conditions for 24 h before being removed from the molds and immersed in a curing tank (ASTM C192) [46], until being tested or processed.

2.3. Test Methods

2.3.1. Mechanical and Physical Properties of the PC

After 28 days of curing, the PC beams were tested for their flexural strength, in accordance with ASTM C 348 [47], and then the compressive strength was evaluated (ASTM C-349 [48]). In addition, the water permeability coefficient was assessed (ASTM D-2434-68 [49]) with a variable charge permeameter.

2.3.2. Mercury Intrusion Porosimetry (MIP)

The porosimeter used for the MIP tests was a Micromeritics Autopore IV 9500 (Micromeritics Corporate Headquarters, Norcross, Atlanta, GA, USA.), with a precision of 33.000 psi (228 MPa) and pore diameter detection capacity ranging from 0.006 to 175 µm. As there are no specific norms for concrete, the norms of ASTM D4404 (for soil and rock) [50] were followed in the test, along with the evidence from previously completed studies [51]. The test age of the specimens was 90 days; they were extracted from the PC beams (approximate volume 2 cm^3) using a Buehler Isomet 4000 precision saw9500 (Buehler, Lake Bluff, IL, USA.), with a blade cutting thickness of 20 µm. This prevents the possible formation of microfissures during the extraction process. The test tubes were dried in the permeameter vacuum chamber for 60 min at 24 °C before the inclusion of the mercury (Hg). The Hg inclusion process was started at low pressure so that the test team could identify the biggest pores; then the pressure was increased so that the Hg penetrated all the simple pore cavities. Each test took approximately 7 h

2.3.3. Scanning Electron Microscope (IA-SEM) and Microanalysis of Chemical Elements with an Energy Dispersive Spectrometer (SEM-EDS)

Small representative samples of the concrete beams were taken (with an exposed face of approximately 5 cm long by 5 cm wide) in order to obtain the IA-SEM images of each of the PC variables in the study. These samples were then placed within circular molds and covered with transparent epoxy resin (EpoFix Resin). They were then put in an oven for 40 min at 40 °C (catalyzer), by which time the resin had hardened and sealed the sample inside. The samples in the resin were then subjected to progressive smoothing and polishing on their exposed study face until they shone. The "areas of interest" were then chosen visually, to be studied later in the IA-SEM. As the RCA used in the PC introduced a different component to the study matrix [52,53], they were located as (a) ITZ between the new aggregate–old aggregate, (b) ITZ between new paste–old paste and (c) ITZ between new aggregate–new paste (see Figure 2); they were present in the RCA 50 SF sample. Once the "areas of interest" had been located, the samples were subjected to drying in an oven at a low temperature until they had attained a consistent weight; they were then metallized with graphite powder (obtaining a conductive surface) [54] and placed in a dryer until just before being studied in the IA-SEM.

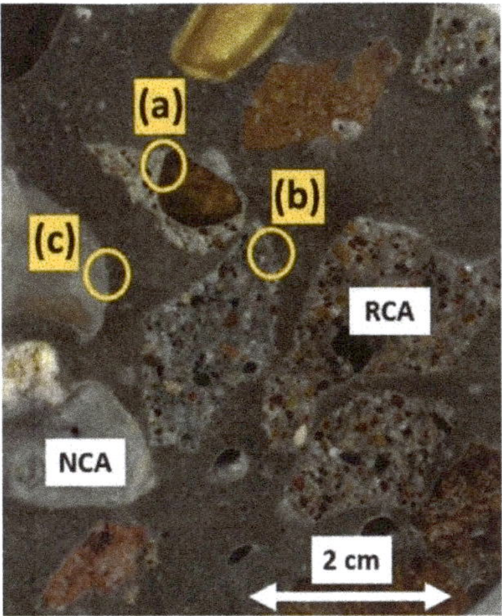

Figure 2. RCA 50 SF embedded in epoxy resin with a polished surface. (a–c) "areas of interest" for study in IA-SEM.

The IA-SEM used a JEOL JSM-6510 microscope (Jeol Ltd., Tokyo, Japan) for each image obtained from the study. The images were taken with (1) high-resolution detector of secondary electrons SEI (secondary electron image) for high-resolution images; (2) backscattered electron detector BEI (backscattered electron image) with lower image resolution but greater contrast for examining the surface topography. The zooms used for taking the images were: 70× for the general composition of the matrix, 200× for the areas of interest and 500× for mapping. The mapping involved determining the chemical composition of the samples by detecting basic elements through microanalysis, which was done with an energy dispersive spectrometer (EDS) (Inca 200, Oxford Instruments, Abingdon-on-Thames, UK) connected to an SEM unit. The EDS study required 14 complete scans for each image in BEI mode to determine the chemical elements present in the samples chosen: Si, Al, Ca, Mg, K, C, Fe and Na (typical in the chemistry of concrete [55,56]). Figure 3 shows the locations of the chemical elements present in one of the micrographs.

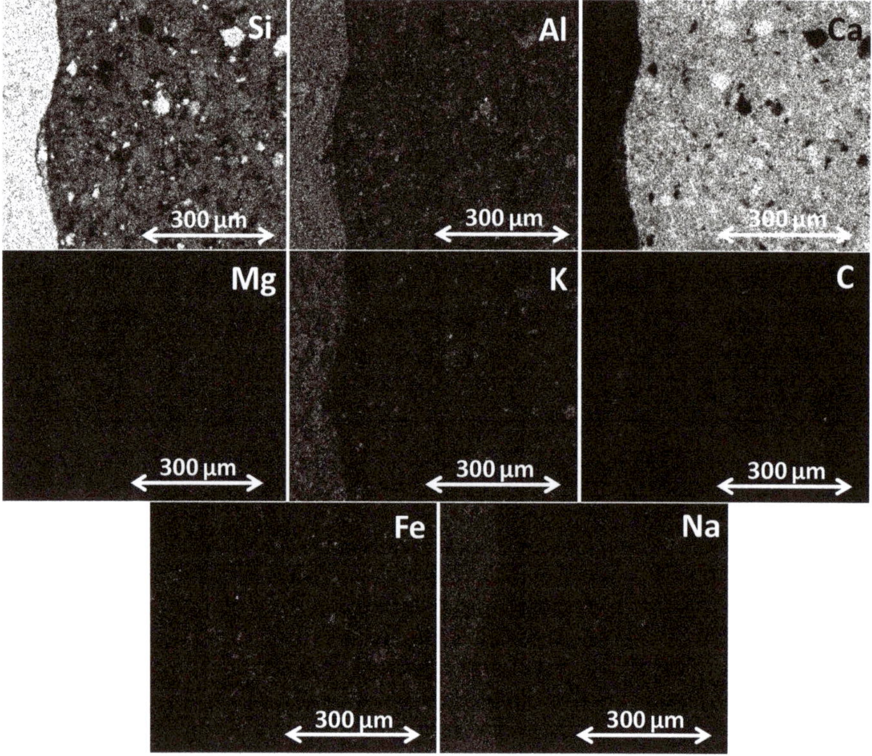

Figure 3. Elemental mapping of the RCA 50 SF sample between the natural coarse aggregate (NCA) and the new paste.

2.3.4. Image Analysis Using NI Vision Assistant

The image analysis process was carried out using National Instrument's NI Vision Assistant 2018 software (LabWindows/CVI Version 7.1, Austin Texas, TX, USA.), as this has proven itself to be a useful image processing tool in industry and among the scientific community [57]. This software's full set of processing functions and digital image capture improves project efficiency by reducing the programming effort and producing better results in less time [58]. Among this software's functions are the study of particles, pattern recognition, detection of edges and barely visible colors, calculus of segmentation and frequency histograms [59]. Image analysis involves examining an image in detail, separating the components in order to know their characteristics and qualities and obtain results. As this software allows separate images of the PC components to be captured, it was possible to analyze the porosity and calculate the amount of anhydrous material and hydration products contained in the PC. Using operations integrated into the software (addition, subtraction, erosion, aggregation, etc.) it was possible to create the SCRIP (batch processing file or command file), which was used in treating the images, creating the masks for the aggregate and the paste.

Making the Masks

The procedure of carrying out an image analysis is the result of applying complex numerical algorithms that process the incidences in the pixels of an image. This was performed by the software, which controlled each numerical operation in an exact, balanced and precise manner for the specific area of interest of each image.

The applied image analysis procedure began with the calibration of the image's dimensions and the relationship between this and its pixels. Then, as the first step in the analysis, the aggregate mask was made. As this was composed mainly of Si, K, Na and Al (Figure 3), the "operating instructions" of NI Vision Assistant (addition, subtraction, multiplication and division, etc.) (Figure 4a) were used to form an outline image. Subsequently, morphological filtering was applied. This is a numerical process based on the definition of the connectivity of pixels. Each pixel may be connected horizontally, vertically or diagonally with its neighbors (these morphological filters may be based on a matrix of pixels with a pre-established structure, e.g., hexagonal, square, etc.). Therefore, in a morphological filter, it is possible to define several versions in order to surround the pixel to be studied [29]. This means that the process has sufficient 'sensitivity' to allow the applied procedure to adjust better to reality, thereby preventing any significant alteration to the original image.

In accordance with the result of the operating instructions, the morphological filters applied to the images were as follows: (a) filtering (noise eradication), (b) opening of objects (eroding and dilating the image—always with the same structural element and permitting the image to keep its original size), (c) elimination of small objects (removing those found inside larger ones) and (d) filling and closing objects (joining and harmonizing elements). The image of the aggregate (aggregate mask, Figure 4b) was established from the results of this group of filters.

Using the same analysis procedure (operating instructions and morphological filtering), but on this occasion with the images obtained from SEM-EDS which define the basic compounds of Ca, C, Fe and Mg (incidences present in the paste), it was possible to obtain the masks that outline the paste, as shown in Figure 5. Finally, for the samples with RCA three masks were made from the aggregate and the paste in order to evaluate the porosity in different areas of the ITZ (a) new aggregate–new paste, (b) old aggregate–new paste and c) old aggregate–old paste.

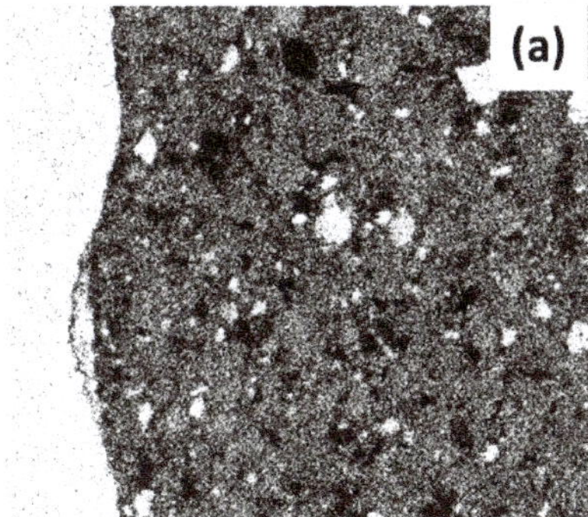
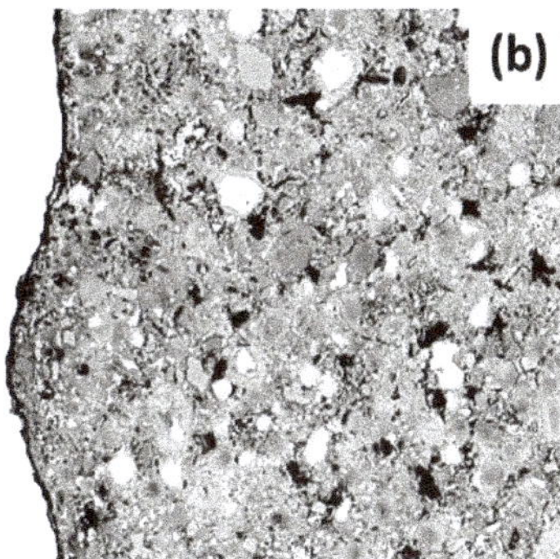

Figure 4. Preparation of the aggregate mask for the RCA 50 C sample (**a**) image after the operating instructions and (**b**) after the morphological filters.

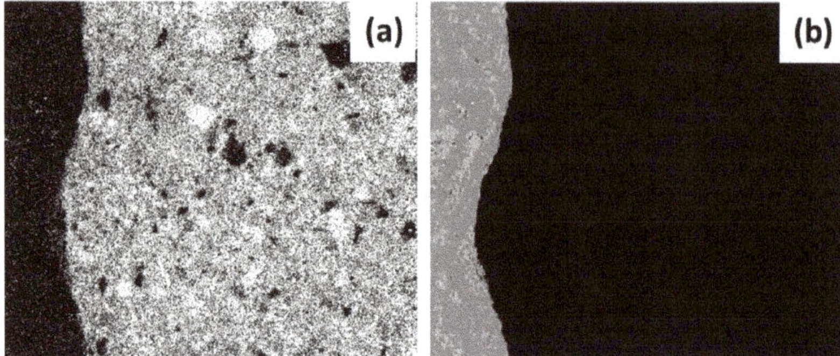

Figure 5. Elaboration of the mask of the paste for the sample RCA 50 C (**a**) image after the operating instructions and (**b**) after the morphological filters.

The criterion of segregation or "threshold" was applied to the initial images of the study in order to know the values of porosity; this allowed sectioning of a specific range of tones in the gray scale—which can have values of between 0 and 255—while ensuring that this range of values was dependent on the materials. A histogram was used to determine the range of application (a pixel-frequency graph vs the value on the grayscale), encompassing the pixels included in the range of values from the initial (equal to zero) to the decrease in the last of the histogram frequency curves. The range of pixels from 0 to 75 (threshold) was chosen for the NCA samples; from 0 to 50 was chosen for the elements with RCA and SF. Therefore, all the image areas containing pixels within these ranges were considered pores or gaps. Figure 6 shows the final binarized image (segmentation of pixels in only two tones: black and white) of this process.

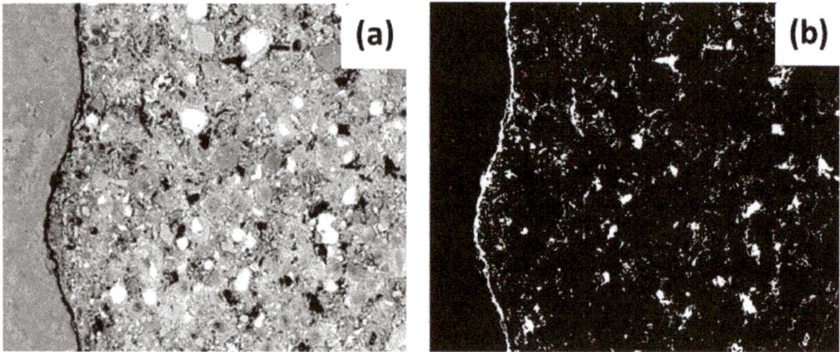

Figure 6. The total porosity of the image (**a**) initial image and (**b**) segmentation and binarization of image after application of threshold segregation.

The multiplication operation with the coarse aggregate mask and the paste mask was applied to the binarized image, thereby obtaining the images that allowed the porosity identified in the aggregate (Figure 7a) and in the paste (Figure 7b) to be observed and quantified. Additionally, by applying a threshold (range 225–255) the elements that did not react chemically (anhydrous material) could be identified; in the (SEM-EDS) mapping of elements, they are shown as white pixels with a single basic chemical element (Figure 7c). Finally, once the porosity and the anhydrous material had been identified it was possible to obtain the percentage of hydration products; this is the result of removing the previous sum of porosity and anhydrous material from the original image.

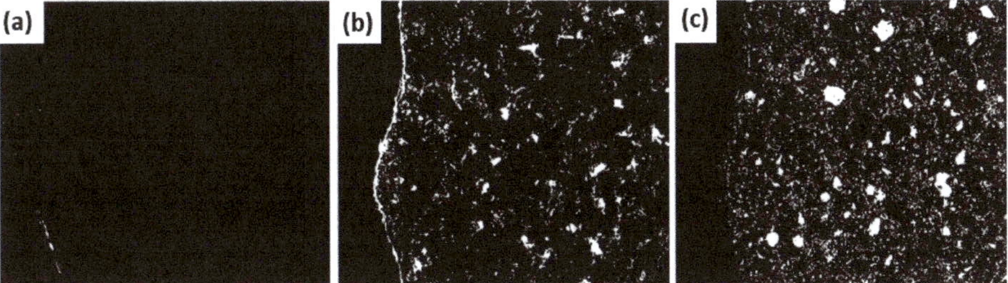

Figure 7. (**a**) Porosity of the aggregate, (**b**) porosity of the paste and (**c**) anhydrous material.

After these procedures and the image calibration, the sector in the scope of the ITZ was built (theoretical ITZ thickness located between the aggregate and paste throughout all areas). To do so, the sector previously established as the filter mask was used to establish the (theoretical) ITZ thicknesses of 10, 20, 30, 40 and 50 μm [14,27]. The image process used to achieve this was "dilation of objects", applied to the original image of the aggregate mask and establishing the various thicknesses within the reach of the ITZ as "desired" (measured in pixels). If the pixel measurements were known it was possible to establish the thickness with the desired distance. Finally, this was overlaid onto the original binarized image. The results of the previous process that determined the sectors of ITZ can be seen in Figure 8. These were ultimately used as a mask to establish the percentage (histogram of the image) of porosity in the ITZ.

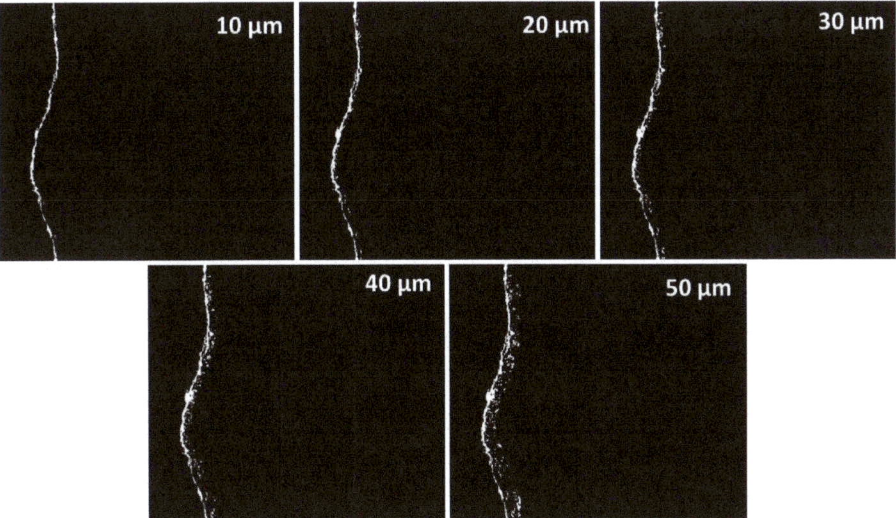

Figure 8. Images resulting from the sectors of the ITZ with the porosities determined at different theoretical thicknesses.

3. Results and Discussion

3.1. Mechanical and Physical Properties of the PC

The results of the tests are shown in Table 5, where it can be seen that the reference concrete (NCA 100 C) had the best mechanical properties in terms of compression and flexure (21.9 and 4.9 MPa respectively). It was also observed that the inclusion of RCA in PC causes a reduction in its mechanical properties of 31% and 30% (compression and flexure respectively) with regard to NCA 100 C. Finally, it was established that the PC with SF did not generate increases in the mechanical properties. This may be because the SF does not react with the CH generated during the cement hydration, as due to the high porosity of this type of cement they were dissolved in the curing water [60].

The permeability coefficients established that there was a connection between the mechanical properties studied. This can be explained by the frequent association of permeability and porosity with the mechanical properties of concrete [61]. The PC (NCA 100 C) showed the lowest permeability coefficient in comparison with the other variables. This indicates that it has lower porosity and is, therefore, more capable of attaining better mechanical properties.

Table 5. Properties of PC.

Study Variables	Compression Strength (MPa)	Flexural Strength (MPa)	Water Permeability Coefficient (m/s)
NCA 100 C	21.9	4.9	8.9×10^{-03}
NCA 100 SF	20.8	4.4	1.1×10^{-02}
RCA 50 C	15.5	3.6	1.0×10^{-02}
RCA 50 SF	15.1	3.4	1.1×10^{-02}

3.2. Mechanical and Physical Properties of the PC

Figure 9a shows the results obtained from the MIP tests in the enrichment phase (positive intrusion). The curves or isotherms of this graph show that their form is similar in all cases, their path being as follows: starting with slight increases from a common origin (the theoretical diameter of maximum pores $\varphi_{p\,max}$ = 500,000 nm, macropores [62,63]), they continue to an abrupt inflexion point, located in the range of critical pore diameter $60 \leq \varphi_{pc} \leq 140$ nm. Finally, they continue to an accelerated increase in the intrusion volume, which causes a steep rise until the points of maximum intrusion value are reached, minimum pore diameter $\varphi_{p\,min}$ in the mesopore zone.

The PC sample—NCA 100 C—establishes the most representative pore samples as $7 \leq \varphi_p \leq 110$ nm, with its inflexion point located in Vp = 0.01 mL/g. This means that a lower amount of Hg is needed to fill the large diameter ($100 \leq \varphi_p \leq 400,000$ nm) pores. In contrast, the NCA 100 SF sample shows fewer pores, which are significant in the size range of $6 \leq \varphi_p \leq 90$ nm; the reason being the insertion of SF, as this is made up of small particles and may be used as a 'filler' for pores in the microstructure of concretes [64]. The RCA 50 C version establishes the highest number of pores of all the samples analyzed—set at $30 \leq \varphi_p \leq 200$ nm—due to the inclusion of the RA; as this material is classified as low density and high porosity [1] the Hg can pass easily through its porous structure. Finally, it can be seen that adding SF to a sample of PC with RCA (RCA 50 SF) causes a reduction in the number of pores of $6 \leq \varphi_p \leq 80$ nm; the cause is similar to that of the NCA 100 SF, although on this occasion with less importance (porous RCA).

To sum up, the order of the increase in porosity of PC with NCA was as follows: first NCA 100 SF and then NCA 100 C; regarding the PC with RA, first RCA 50 SF and finally RCA 50 C. The second group is affected by the RA with high original porosity. In both groups, the inclusion of SF (filling effect) is connected to the reduction of porosity.

The extrusion curves in Figure 9b help to interpret the morphology of the "trapped porosity", identified when the intrusion–extrusion gaps were compared (a hysteresis phenomenon or the ability of the material to keep to the initial intrusion curve line as the Hg pressure returns to its initial state in the test). The difference between both curves established the quantity of Hg retained in the sample, a consequence of the number of bottleneck pores in the system [65,66]. The extrusion curves shown in the test samples have similar lines and the same relative positioning on the intrusion curves. Table 6 shows the values of trapped porosity (P_a) obtained after determining the difference in volumes between both curves. It can be seen that the samples with RCA have twice as many bottleneck pores as the NCA samples. This is understandable if we consider that the RCA has open pores which are partially blocked by the new CP cement paste and that the use of C or SF is not significant in either sample group.

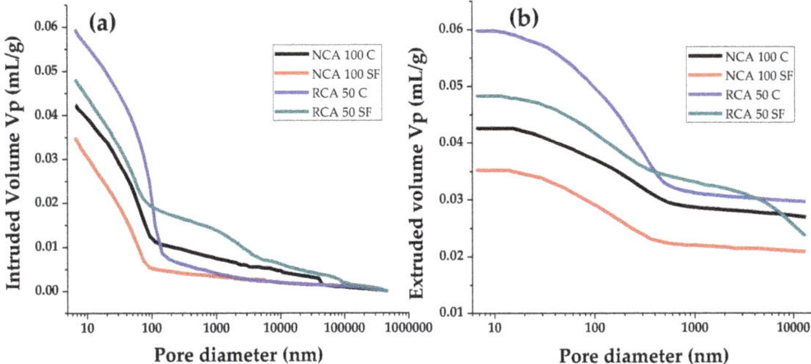

Figure 9. Mercury intrusion porosimetry, (**a**) intrusion and (**b**) extrusion.

Tables 6 and 7 show the values of the porous network indicators determined according to the previously established procedures [51,67,68]: maximum pore radius (r_{max}), minimum radius (r_{min}), critical radius (r_c), average radius (r_m), total pore area (A_{tp}), intruded and extruded volume (V_i y V_e) and P_a. Vi = Volume intruded, Ve = Volume extruded.

Table 6. Porous network indicators in Hg extruded stage.

Sample	r_{max} (nm)	r_{min} (nm)	r_c (nm)	r_m (nm)	V_e (mL/g)	P_a (mL/g)
NCA 100 C	6708.05	3.4	273.50	90.918	0.0269	0.0154
NCA 100 SF	6730.10	3.4	121.25	64.522	0.0208	0.0144
RCA 50 C	6702.55	3.4	273.50	90.650	0.0296	0.0299
RCA 50 SF	6637.60	3.4	273.50	150.591	0.0236	0.0246

Table 7. Porous network indicators in Hg intruded stage.

Sample	r_{max} (nm)	r_{min} (nm)	r_c (nm)	r_m (nm)	V_i (mL/g)	A_{tp} (mL/g)
NCA 100 C	244,514.75	3.400	110.700	28.333	0.0424	4.15×10^{02}
NCA 100 SF	246,402.35	3.400	57.600	18.522	0.0351	4.26×10^{02}
RCA 50 C	242,670.09	3.378	136.691	39.307	0.0595	3.77×10^{02}
RCA 50 SF	245,620.64	3.400	137.000	29.588	0.0482	7.41×10^{02}

Regarding the r_{max} in the intrusion stage, the average radius established in the study samples (244,802 nm) highlights the ability of the PC to cause large pores—up to 40 times bigger than those noted in the usual recycled concretes [51,69]. However, in the extrusion stage, the average (6695 nm) is similar to the usual recycled concretes in their intrusion stage. In terms of the varieties studied, the order of radii variation between them is not considered important (less than 1% in both stages of the test). In both stages of the test, and all the varieties studied, r_{min} is considered a constant (equal to 3.4 nm). This raises two points for consideration: the MIP technique may be able to establish the network of pores in the micropore zone, and this minimum value probably has more in common with the limits or parameters of experimental techniques than with the varieties of this research.

The average values attained for the varieties studied in the case of r_c were 110.5 and 235.4 nm (intrusion and extrusion respectively). However, unlike the previous indicators (in the intrusion stage) the NCA 100 SF sample established an r_c some 50% less than that of the NCA 100 C, which shows that the use of SF is able to shift the r_c toward lower r_c values (closing the porous network). With regard to the RCA 50 C and RCA 50 SF versions (intrusion stage), the r_c variation between them is not important (a percentage of 50% SF is not important, the change of effect in r_c requires a high percentage of SF to be noticeable). Finally, the effect of the type of aggregate establishes that the use of RCA regarding NCA (intrusion stage) causes the former to establish a greater r_c than the latter (about 60% more), a result of the high porosity of the RCA itself. Concerning the extrusion stage, the previous trends have the same application, although the values established are greater than in the intrusion stage.

Regarding the indicators of r_m, $V_{i,e}$, and A_{tp}, the values shown in both stages of the study have, in percentage terms, similar interpretations to those established for r_c; it can be seen that these are the most appropriate and sensitive for establishing the correlation between the study varieties and the porous network produced in those PCs.

3.3. Porosity, Anhydrous Material and Hydration Products in Cement Paste by mMeans of IA-SEM

After performing an IA-SEM on the study samples, it was possible to obtain the percentages of porosity, anhydrous material and hydration products (PH) (see Figure 10). Three specifically different zones of interest were also evaluated for the RCA 50 C and RCA 50 SF samples: (a) ITZ between the new aggregate and the new paste, (b) ITZ between the old aggregate and the old paste and (c) ITZ between the old paste and the new paste.

As for porosity, and by comparing the samples that analyze the ITZ, it can be seen that this is higher in the samples containing 100% CP (possible paralysis of the hydration process); in the case of SF it is prolonged with age reducing the porosity, or is favored by the filling effect of pores which it causes. Among the zones of interest, b) (ITZ between old aggregate and old paste) establishes the greatest porosity (the variety linked to the less dense and more porous RCA concrete). If the use of RCA in the PC is eliminated (NCA 100 C and NCA 100 SF), it can be seen that there are minor variations of porosity between them, with the former being the best (in this case the filling effect of pores by SF is not favored, as the use of NCA does not include them). With regard to the anhydrous material and the PH, the use of SF establishes low percentages of anhydrous material and high ones of PH in the samples that contain RCA (the SF reacts with the CH present in the cement hydration to form more SCH). The sample RCA 50 SF is in the zone of interest b), the best of all (possibly due to the prolonged hydration over time of the old concrete from which the RCA originated). For the samples that use NCA, the use of C in the PC increases the formation of PH (the use of SF does not favor the creation of more PH, due to the possible effects of a blockage in hydration caused by a lack of pores or spaces wherein crystals may form).

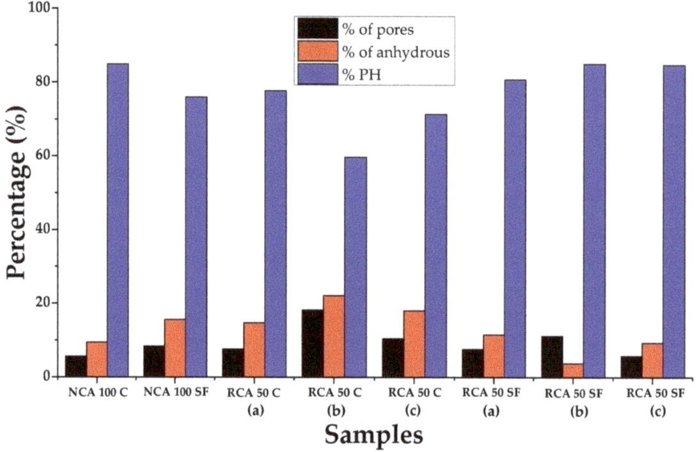

Figure 10. Porosity, anhydrous material and pH of the cement paste in the PC.

3.4. Evaluation of Porosity at Different ITZ Thicknesses by Means of IA of SEM

Figure 11 shows the number of pores (unitary accumulated porosity, μm) determined in the study samples. It also considers the different theoretically possible thicknesses of the aggregate ITZ (ITZ of 10, 20, 30, 40 and 50 μm), which allows the transition from the porous network of the PC to be verified. It can be seen that there is a direct connection in most of the samples between the increase in the percentage of porosity and the increase in the theoretical ITZ thickness. The theoretical thickness is 10 μm, which establishes the best percentage of porosity compared with the other thicknesses evaluated; this confirms that the percentage of porosity of the ITZ decreases as its theoretical thickness increases. Another way to understand the curves is that the ITZ has finished when the curves of the samples tend toward a horizontal asymptote, with a constant—or non-increasing—percentage of porosity equal to that of the paste zone. Significantly, the three evaluated samples which had the lowest porosity in the ITZ are those with 50% of RCA and SF; this confirms that the use of SF leads to the filling effect in the ITZ porosity, making it denser and stronger. The RCA 50 C (a) sample shows the highest percentage of unitary porosity when the theoretical thickness is set at 10 μm (0.0216%/μm); at 50 μm reaches 0.032%/μm (the SEM image analyzed showed that this sample contained considerable porosity). This may have been due to the more porous nature of the RCA. Similarly, the effects of using SF were also observed in the NCA 100 SF, which had lower porosity than the NCA 100 C sample.

Therefore, it can be stated that the use of RCA in making a PC will cause increases in the ITZ porosity. However, the addition of SF may compensate for this, and achieve a denser cement matrix.

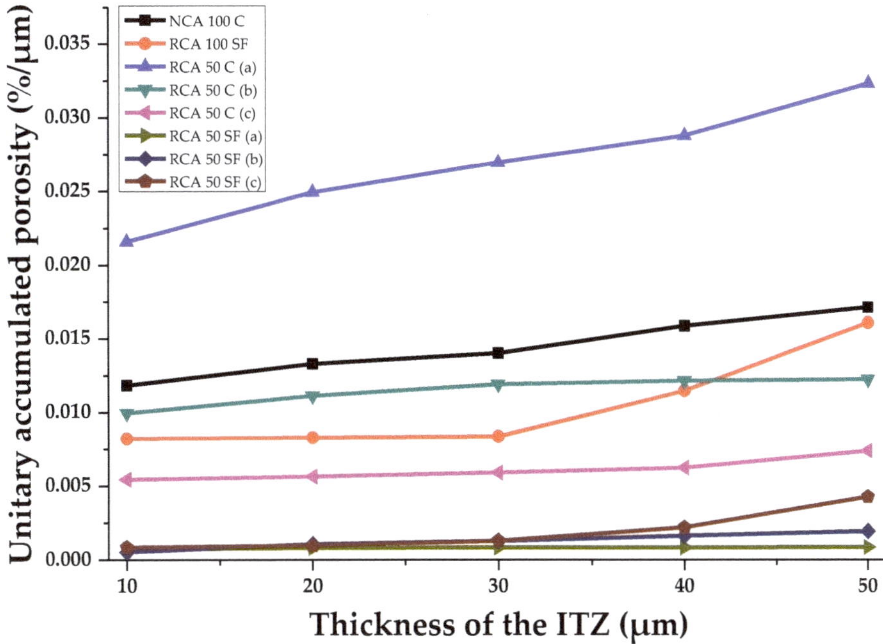

Figure 11. The porosity of the samples analyzed in the ITZ with several theoretical thicknesses.

Figure 12 shows the percentage of unitary accumulated anhydrous material at different theoretical ITZ thicknesses in the PC samples studied. The trend shows a direct, rising, connection between the increase of this parameter and the increase in the theoretical ITZ thickness. It should be noted that the samples with SF have lower percentages of anhydrous material. The SF can react with the cement hydration products and form new ones, which show low percentages of non-reacting material. The RCA 50 C sample (b), has the most anhydrous material: at 10 μm an important concentration of

0.0062%/μm can be seen, reaching 0.0106%/μm when the theoretical thickness is 50 μm. Additionally, comparing NCA 100 C and NCA 100 SF, the former is shown to contain a far greater percentage of anhydrous material at all theoretical ITZ thicknesses than the latter. Therefore, to sum up, it can be said that the use of SF contributes to the formation of more hydration products in the cement paste; this material can fill pores, thereby leading to lower percentages of anhydrous material.

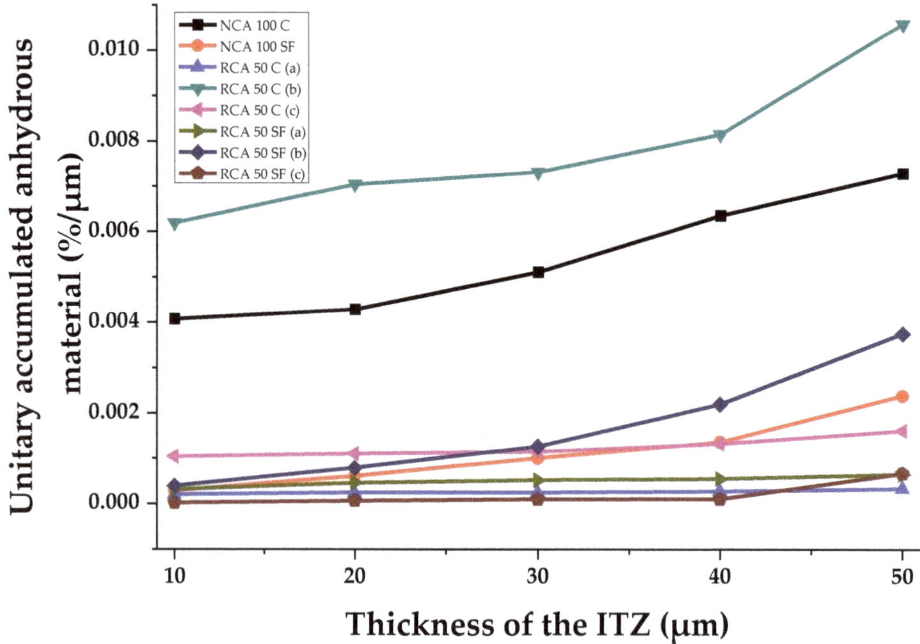

Figure 12. Anhydrous material of the samples analyzed in the IZT with various theoretical thicknesses.

The PH (the result of subtracting the porosity and the anhydrous material from the paste) that can be detected in the ITZ studied usually has a low value, as porosity dominates in an ITZ. Figure 13 shows the percentage of unitary pH at different theoretical ITZ thicknesses (%/μm). The RCA 50 C samples (a and b) have the highest percentage of PH, which corroborates previous paragraphs. This sample showed low percentages of porosity and anhydrous material. The NCA 100 C has a high amount of PH and shows a significant increase at theoretical ITZ thicknesses of between 40 and 50 μm.

The samples with SF (NCA 100 SF, RCA 50 SF (a and b)) have the lowest amounts of pH and correlate the results of the mechanical properties (they are the least resistant). Their mechanical properties were low compared with the samples made from Portland.

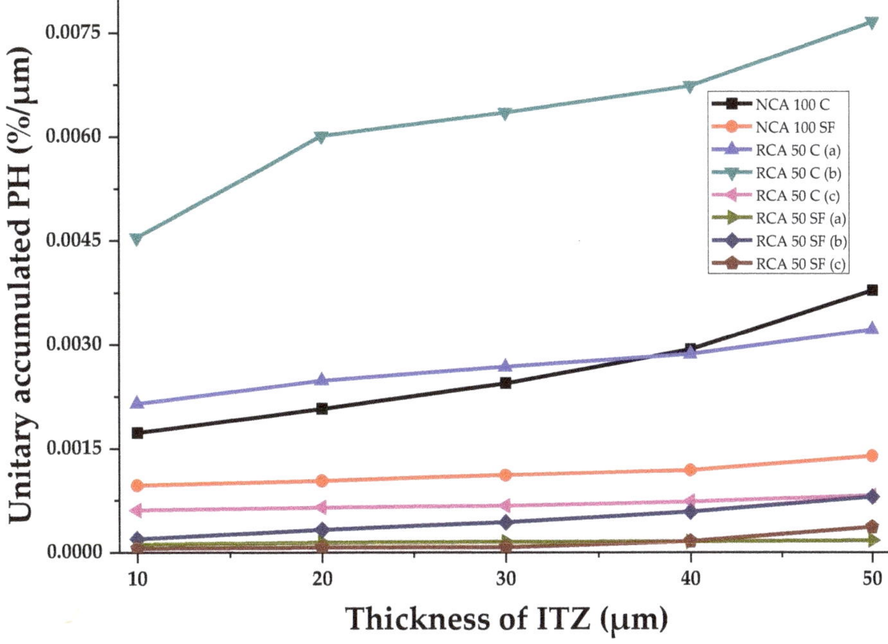

Figure 13. Hydration products of the samples analyzed in the IZT with various theoretical thicknesses.

4. Conclusions

The physical properties evaluated show that using RA in the PC leads to a reduction in mechanical properties when compared to the reference PC (32%, 30% and 6% for compression, flexure and permeability, respectively), as these old aggregates are nearing the end of their life cycle and have less mechanical strength than natural aggregates. It is also shown that the use of SF in these concretes does not lead to an improvement in compressive strength, flexural strength or permeability, as they do not react with the CH crystals (which dissolve due to the high porosity of this concrete).

By means of image analysis with the NI Vision Assistant software, it was possible to prepare a script to make a mask of the aggregate and the cement paste. This was then used to analyze the percentage of porosity, the anhydrous material and the hydration products of the ITZ of several PC samples. This analysis confirmed that the PC with RA increased the porosity by 32% at 50 μm ITZ thickness between the new aggregate and the new paste. This is due to the RA being less dense, more porous and higher water-absorbent. It also shows that the use of SF caused an 89% decrease in porosity, mainly in the reference concrete (at 50 μm ITZ thickness in the new aggregate-new paste area), acting as a filling material for microporosity. The above information has shown that porosity increases when there is a thicker ITZ and that the most porous area of the PC is to be found between the new aggregate and the new paste. This method is an option for analyzing the microporosity of many materials. The evidence of the image analysis was corroborated by mercury intrusion porosimetry. The use of RCA in making permeable concrete leads to higher porosity in the cement paste and the ITZ; but the use of SF in the concrete also helps reduce the porosity, as it fills the gaps in the microstructure.

5. Future Research

To find a correlation of XRD with SEM to locate specific compounds and their relationship with mechanical behavior.

To obtain a numerical quantification of the amount of old mortar adhered to the recycled aggregates and its connection with other mechanical and durability properties.

Regarding the technique of image analysis, there are parameters that will be necessary to calibrate in the definition of these masks; among the most complex and therefore the most demanding future study due to its great diversity of application alternatives are morphological filtering and noise filtering.

The use of scripts for image analysis can be used to measure porosities and failure sites (cracking) for a wide range of materials (ceramic, metallic and polymeric).

Images can be obtained in different layers and at different depths of the sample and develop scripts so that the material can be considered statistically homogeneous (the results are representative of the sample volume).

Author Contributions: Conceptualization, J.M.G.-S. and S.P.A.-R.; methodology, J.M.G.-S., R.C.-H. and S.P.A.-R.; validation, S.P.A.-R. and R.C.-H.; formal analysis, M.J.C.-C. and C.A.R.-C.; investigation, M.J.C.-C. and R.C.-H.; resources, J.M.G.-S. and R.C.-H.; data curation, C.A.R.-C. and M.J.C.-C.; writing—original draft preparation, M.J.C.-C. and J.M.G.-S.; writing—review and editing, J.M.G.-S. and M.J.C.-C.; supervision, S.P.A.-R. and C.A.R.-C.

Funding: This research received no external funding

Acknowledgments: The authors would like to thank CONACYT for its master and doctoral scholarship program, the Barcelona School of Building Construction-UPC, the Department of Architecture Technology-EPSEB-UPC and the Faculty of Engineering Mochis-UAS.

Conflicts of Interest: The authors declare no conflict of interest.

References

1. Zaetang, Y.; Sata, V.; Wongsa, A.; Chindaprasirt, P. Properties of pervious concrete containing recycled concrete block aggregate and recycled concrete aggregate. *Constr. Build. Mater.* **2016**, *111*, 15–21. [CrossRef]
2. Ho, H.L.; Huang, R.; Hwang, L.C.; Lin, W.T.; Hsu, H.M. Waste-based pervious concrete for climate-resilient pavements. *Materials* **2018**, *11*, 900. [CrossRef] [PubMed]
3. Yap, S.P.; Chen, P.Z.C.; Goh, Y.; Ibrahim, H.A.; Mo, K.H.; Yuen, C.W. Characterization of pervious concrete with blended natural aggregate and recycled concrete aggregates. *J. Clean. Prod.* **2018**, *181*, 155–165. [CrossRef]
4. Güneyisi, E.; Gesoğlu, M.; Kareem, Q.; İpek, S. Effect of different substitution of natural aggregate by recycled aggregate on performance characteristics of pervious concrete. *Mater. Struct.* **2016**, *49*, 521–536. [CrossRef]
5. Ibrahim, A.; Mahmoud, E.; Yamin, M.; Patibandla, V.C. Experimental study on portland cement pervious concrete mechanical and hydrological properties. *Constr. Build. Mater.* **2014**, *50*, 524–529. [CrossRef]
6. Tsang, C.; Shehata, M.H.; Lotfy, A. ptimizing a test method to evaluate resistance of pervious concrete to cycles of freezing and thawing in the presence of different deicing salts. *Materials* **2016**, *9*, 878. [CrossRef] [PubMed]
7. Shahidan, S.; Koh, H.B.; Alansi, A.M.S.; Loon, L.Y. Strength development and water permeability of engineered biomass aggregate pervious concrete. In Proceedings of the 3rd International Conference on Manufacturing and Industrial Technologies, Istanbul, Turkey, 25–27 May 2016; pp. 2–7.
8. Braga, A.M.; Silvestre, J.D.; de Brito, J. Compared environmental and economic impact from cradle to gate of concrete with natural and recycled coarse aggregates. *J. Clean. Prod.* **2017**, *162*, 529–543. [CrossRef]
9. Leite, M.B.; Monteiro, P.J.M. Microstructural analysis of recycled concrete using x-ray microtomography. *Cem. Concr. Res.* **2016**, *81*, 38–48. [CrossRef]
10. Wu, X.; Zhou, J.; Kang, T.; Wang, F.; Ding, X.; Wang, S. Laboratory investigation on the shrinkage cracking of waste fiber-reinforced recycled aggregate concrete. *Materials* **2019**, *12*, 1196. [CrossRef]
11. Li, W.; Xiao, J.; Sun, Z.; Kawashima, S.; Shah, S.P. Interfacial transition zones in recycled aggregate concrete with different mixing approaches. *Constr. Build. Mater.* **2012**, *35*, 1045–1055. [CrossRef]
12. Scrivener, K.L.; Crumbie, A.K.; Laugesen, P. The interfacial transition zone (itz) between cement paste and aggregate in concrete. *Interface Sci.* **2004**, *12*, 411–421. [CrossRef]
13. Kong, Y.; Wang, P.; Liu, S.; Zhao, G.; Peng, Y. SEM analysis of the interfacial transition zone between cement-glass powder paste and aggregate of mortar under microwave curing. *Materials* **2016**, *9*, 733. [CrossRef] [PubMed]

14. Zhu, X.; Gao, Y.; Dai, Z.; Corr, D.J.; Shah, S.P. Effect of interfacial transition zone on the young's modulus of carbon nanofiber reinforced cement concrete. *Cem. Concr. Res.* **2018**, *107*, 49–63. [CrossRef]

15. Vargas, P.; Restrepo-Baena, O.; Tobón, J.I. Microstructural analysis of interfacial transition zone (itz) and its impact on the compressive strength of lightweight concretes. *Constr. Build. Mater.* **2017**, *137*, 381–389. [CrossRef]

16. Zhu, X.; Yuan, Y.; Li, L.; Du, Y.; Li, F. Identification of interfacial transition zone in asphalt concrete based on nano-scale metrology techniques. *Mater. Des.* **2017**, *129*, 91–102. [CrossRef]

17. Zhang, L.; Zhang, Y.; Liu, C.; Liu, L.; Tang, K. Study on microstructure and bond strength of interfacial transition zone between cement paste and high-performance lightweight aggregates prepared from ferrochromium slag. *Constr. Build. Mater.* **2017**, *142*, 31–41. [CrossRef]

18. Nambiar, E.K.K.; Ramamurthy, K. Air-void characterisation of foam concrete. *Cem. Concr. Res.* **2007**, *37*, 221–230. [CrossRef]

19. Xu, J.; Wang, B.; Zuo, J. Modification Effects of nanosilica on the interfacial transition zone in concrete: A multiscale approach. *Cem. Concr. Compos.* **2017**, *81*, 1–10. [CrossRef]

20. Chinchillas-Chinchillas, M.J.; Orozco-Carmona, V.M.; Gaxiola, A.; Alvarado-Beltrán, C.G.; Pellegrini-cervantes, M.J.; baldenebro-lópez, F.J.; castro-beltrán, A. evaluation of the mechanical properties, durability and drying shrinkage of the mortar reinforced with polyacrylonitrile microfibers. *Constr. Build. Mater.* **2019**, *210*, 32–39. [CrossRef]

21. Mohamed, O.A. A review of durability and strength characteristics of alkali-activated slag concrete. *Materials* **2019**, *12*, 1198. [CrossRef]

22. Chahal, N.; Siddique, R. Permeation Properties of Concrete Made with Fly Ash and Silica Fume: Influence of Ureolytic Bacteria. *Constr. Build. Mater.* **2013**, *49*, 161–174. [CrossRef]

23. du Plessis, A.; Olawuyi, B.J.; Boshoff, W.P.; le Roux, S.G. Simple and fast porosity analysis of concrete using x-ray computed tomography. *Mater. Struct.* **2016**, *49*, 553–562. [CrossRef]

24. Da Silva, P.R.; De Brito, J. Experimental study of the porosity and microstructure of self-compacting concrete (scc) with binary and ternary mixes of fly ash and limestone filler. *Constr. Build. Mater.* **2015**, *86*, 101–112. [CrossRef]

25. Choi, H.; Koh, T.; Choi, H.; Hama, Y. Performance evaluation of precast concrete using microwave heating form. *Materials* **2019**, *12*, 1113. [CrossRef] [PubMed]

26. Beneš, M.; Štefan, R. Hygro-Thermo-Mechanical analysis of spalling in concrete walls at high temperatures as a moving boundary problem. *Int. J. Heat Mass Transf.* **2015**, *85*, 110–134. [CrossRef]

27. Diamond, S. Considerations in image analysis as applied to investigations of the itz in concrete. *Cem. Concr. Compos.* **2001**, *23*, 171–178. [CrossRef]

28. Diamond, S.; Huang, J. The ITZ in concrete—A different view based on image analysis and sem observations. *Cem. Concr. Compos.* **2001**, *23*, 179–188. [CrossRef]

29. Head, M.K.; Buenfeld, N.R. Measurement of aggregate interfacial porosity in complex, multi-phase aggregate concrete: binary mask production using backscattered electron, and energy dispersive X-Ray images. *Cem. Concr. Res.* **2006**, *36*, 337–345. [CrossRef]

30. Soroushian, P.; Elzafraney, M.; Nossoni, A. Specimen preparation and image processing and analysis techniques for automated quantification of concrete microcracks and voids. *Cem. Concr. Res.* **2003**, *33*, 1949–1962. [CrossRef]

31. Mouret, M.; Bascoul, A.; Escadeillas, G. Study of the degree of hydration of concrete by means of image analysis and chemically bound water. *Adv. Cem. Based Mater.* **1997**, *6*, 109–115. [CrossRef]

32. Yang, R.; Buenfeld, N.R. Binary segmentation of aggregate in sem image analysis of concrete. *Cem. Concr. Res.* **2001**, *31*, 437–441. [CrossRef]

33. *ASTM C150. Historical Standard: Standardized Specification for Portland Cement*; ASTM Int.: West Conshohocken, PA, USA, 1999; 4, 1–5; Available online: https://www.astm.org/ (accessed on 3 May 2016).

34. *ASTM C127. Standard Test Method for Specific Gravity and Absorption of Coarse Aggregate*; ASTM Int.: West Conshohocken, PA, USA, 1993; 88, 1–5; Available online: https://www.astm.org/ (accessed on 3 May 2016).

35. *ASTM C29. Standard Test Method for Bulk Density (" Unit Weight ") and Voids in Aggregate*; ASTM Int.: West Conshohocken, PA, USA, 2003; 97, 3–6; Available online: https://www.astm.org/ (accessed on 3 May 2016).

36. *ASTM C114. Standard Test Methods for Chemical Analysis of Hydraulic Cement*; ASTM Int.: West Conshohocken, PA, USA, 2008; pp. 1–9. Available online: https://www.astm.org/ (accessed on 3 May 2016).

37. *ASTM C1240. Standard Specification for Silica Fume Used in Cementitious Mixtures*; ASTM Int.: West Conshohocken, PA, USA, 2009; 15, 1–6; Available online: https://www.astm.org/ (accessed on 3 May 2016).

38. Güneyisi, E.; Gesoglu, M.; Algın, Z.; Yazıcı, H. Rheological and fresh properties of self-compacting concretes containing coarse and fine recycled concrete aggregates. *Constr. Build. Mater.* **2016**, *113*, 622–630. [CrossRef]

39. Pedro, D.; Brito, J.D.; Evangelista, L. Structural concrete with simultaneous incorporation of fine and coarse recycled concrete aggregates: mechanical, durability and long-term properties. *Constr. Build. Mater.* **2017**, *154*, 294–309. [CrossRef]

40. Khodabakhshian, A.; Ghalehnovi, M.; De Brito, J.; Shamsabadi, E.A. Durability performance of structural concrete containing silica fume and marble industry waste powder. *J. Clean. Prod.* **2018**, *170*, 42–60. [CrossRef]

41. Liu, H.; Luo, G.; Wang, L.; Wang, W.; Li, W.; Gong, Y. Laboratory evaluation of eco-friendly pervious concrete pavement material containing silica fume. *Appl. Sci.* **2019**, *9*, 73. [CrossRef]

42. Yeih, W.; Fu, T.C.; Chang, J.J.; Huang, R. Properties of pervious concrete made with air-cooling electric arc furnace slag as aggregates. *Constr. Build. Mater.* **2015**, *93*, 737–745. [CrossRef]

43. Chang, J.J.; Yeih, W.; Chung, T.J.; Huang, R. Properties of pervious concrete made with electric arc furnace slag and alkali-activated slag cement. *Constr. Build. Mater.* **2016**, *109*, 34–40. [CrossRef]

44. Ngohpok, C.; Sata, V.; Satiennam, T.; Klungboonkrong, P.; Chindaprasirt, P. Mechanical properties, thermal conductivity, and sound absorption of pervious concrete containing recycled concrete and bottom ash aggregates. *KSCE J. Civ. Eng.* **2018**, 1–8. [CrossRef]

45. *ASTM c31. Standard Practice for Making and Curing Concrete Test Specimens in the Field*; ASTM Int.: West Conshohocken, PA, USA, 2010; 4, 1–5; Available online: https://www.astm.org/ (accessed on 3 May 2016).

46. *ASTM C192. Standard Practice for Making and Curing Concrete Test Specimens in the Laboratory*; ASTM Int.: West Conshohocken, PA, USA, 2007; Available online: https://www.astm.org/ (accessed on 3 May 2016).

47. *ASTM C348. Standard Test Method for Flexural Strength of Hydraulic-Cement Mortars*; ASTM Int.: West Conshohocken, PA, USA, 1998; Available online: https://www.astm.org/ (accessed on 3 May 2016).

48. *ASTM C349. Standard Test Method for Compressive Strength of Soil-Cement Using Portions of Beams Broken in Flexure (Modified Cube Method)*; ASTM Int.: West Conshohocken, PA, USA, 1997; Available online: https://www.astm.org/ (accessed on 3 May 2016).

49. *ASTM D2434. Standard Test Method for Permeability of Granular Soils (Constant Head)*; ASTM Int.: West Conshohocken, PA, USA, 2006; 10, 1–8; Available online: https://www.astm.org/ (accessed on 3 May 2016).

50. *ASTM D4404. Standard Test Method for Determination of Pore Volume and Pore Volume Distribution of Soil and Rock by Mercury Intrusion Porosimetry*; ASTM Int.: West Conshohocken, PA, USA, 2004; 12 (Reapproved), 1–6; Available online: https://www.astm.org/ (accessed on 3 May 2016).

51. Gómez-Soberón, J.M.V. Porosity of recycled concrete with substitution of recycled concrete aggregate: An experimental study. *Cem. Concr. Res.* **2002**, *32*, 1301–1311. [CrossRef]

52. Laneyrie, C.; Beaucour, A.L.; Green, M.F.; Hebert, R.L.; Ledesert, B.; Noumowe, A. Influence of recycled coarse aggregates on normal and high performance concrete subjected to elevated temperatures. *Constr. Build. Mater.* **2016**, *111*, 368–378. [CrossRef]

53. Hao, L.; Liu, Y.; Wang, W.; Zhang, J.; Zhang, Y. Effect of salty freeze-thaw cycles on durability of thermal insulation concrete with recycled aggregates. *Constr. Build. Mater.* **2018**, *189*, 478–486. [CrossRef]

54. Melgarejo, J.C.; Proenza, J.A.; Galí, S.; Llovet, X. Técnicas de caracterización mineral y su aplicación en exploración minera. *Bol. Soc. Geol. Mex.* **2010**, *62*, 1–23. [CrossRef]

55. Evangelista, L.; Guedes, M.; Brito, J.D.; Ferro, A.C.; Pereira, M.F. Physical, chemical and mineralogical properties of fine recycled aggregates made from concrete waste. *Constr. Build. Mater.* **2015**, *86*, 178–188. [CrossRef]

56. Waly, E.A.; Bourham, M.A. Annals of nuclear energy comparative study of different concrete composition as gamma-ray shielding materials. *Ann. Nucl. Eng.* **2015**, *85*, 306–310. [CrossRef]

57. Chandra, K.V.; Rashmi, K. Detection of Corneal Diseases Using NI VISION Assistant in LabVIEW. In Proceedings of the 2015 International Conference on Control, Instrumentation, Communication and Computational Technologies (ICCICCT), Kumaracoil, India, 18–19 December 2015; pp. 791–795.

58. Posada-Gómez, R.; Salvador-Gonzales, O.; Martinez-Sibaja, A.; Portillo-Rodriguez, O. Practical Aplications and Solutions Using LabView™ Sofware. In *Digital Image Processing Using LabView*; Folea, S., Ed.; Intech: Rijeka, Croatia, 2011; pp. 311–330.

59. Kwon, K.-S.; Ready, S. *Practical Guide to Machine Vision Software: An Introduction with LabVIEW.*; Wiley: Weinheim, Germany, 2014.

60. Carde, C.; François, R. Effect of the leaching of calcium hydroxide from cement paste on mechanical and physical properties. *Cem. Concr. Res.* **1997**, *27*, 539–550. [CrossRef]

61. Lian, C.; Zhuge, Y.; Beecham, S. The relationship between porosity and strength for porous Concrete. *Constr. Build. Mater.* **2011**, *25*, 4294–4298. [CrossRef]

62. Alderete, N.; Villagrán, Y.; Mignon, A.; Snoeck, D.; Belie, N.D. Pore structure description of mortars containing ground granulated blast-furnace slag by mercury intrusion porosimetry and dynamic vapour sorption. *Constr. Build. Mater.* **2017**, *145*, 157–165. [CrossRef]

63. Larsen, M.B.; Van Horn, J.D.; Wu, F.; Hillmyer, M.A. Intrinsically hierarchical nanoporous polymers via polymerization-induced microphase separation. *Macromolecules.* **2017**, *50*, 4363–4371. [CrossRef]

64. Rostami, M.; Behfarnia, K. The effect of silica fume on durability of alkali activated slag concrete. *Constr. Build. Mater.* **2017**, *134*, 262–268. [CrossRef]

65. Zhang, Z.L.; Cui, Z.D. Effects of freezing-thawing and cyclic loading on pore size distribution of silty clay by mercury intrusion porosimetry. *Cold Reg. Sci. Technol.* **2018**, *145*, 185–196. [CrossRef]

66. Hong, S.; de Bruyn, K.; Bescher, E.; Ramseyer, C.; Kang, T.H.K. Porosimetric features of calcium sulfoaluminate and portland cement pastes: testing protocols and data analysis. *J. Struct. Integr. Maint.* **2018**, *3*, 52–66. [CrossRef]

67. Mendivil-Escalante, J.M.; Gómez-Soberón, J.M.; Almaral-Sánchez, J.L.; Cabrera-Covarrubias, F.G. Metamorphosis in the porosity of recycled concretes through the use of a Recycled Polyethylene Terephthalate (PET) additive correlations between the porous network and concrete properties. *Materials* **2017**, *10*, 176. [CrossRef] [PubMed]

68. Kumar, R.; Bhattacharjee, B. Study on Some Factors Affecting the Results in the Use of MIP Method in Concrete Research. *Cem. Concr. Res.* **2003**, *33*, 417–424. [CrossRef]

69. Awoyera, P.O.; Akinmusuru, J.O.; Dawson, A.R.; Ndambuki, J.M.; Thom, N.H. Microstructural characteristics, porosity and strength development in ceramic-laterized concrete. *Cem. Concr. Compos.* **2018**, *86*, 224–237. [CrossRef]

Article

Lime Treatment of Coal Bottom Ash for Use in Road Pavements: Application to El Jadida Zone in Morocco

Souad El Moudni El Alami [1], **Raja Moussaoui** [1], **Mohamed Monkade** [2], **Khaled Lahlou** [3], **Navid Hasheminejad** [4], **Alexandros Margaritis** [4], **Wim Van den bergh** [4] and **Cedric Vuye** [4,*]

[1] Applied Geoscience Laboratory (LGA), Department of Geology, Mohamed First University, 60000 Oujda, Morocco
[2] Department of Physics, Chouaib Doukkali University, 24000 El Jadida, Morocco
[3] Department of Roads and Bridges, Hassania School of Public Works, 20000 Casablanca, Morocco
[4] Energy and Materials in Infrastructure and Buildings (EMIB) Research Group, Faculty of Applied Engineering, University of Antwerp, 2020 Antwerp, Belgium
* Correspondence: cedric.vuye@uantwerpen.be; Tel. +32-495-44-0707

Received: 11 July 2019; Accepted: 18 August 2019; Published: 22 August 2019

Featured Application: In this paper waste bottom ash from a thermal power station was treated with lime and sand. This improves certain material characteristics leading to a potential use in a subbase in road construction.

Abstract: Industrial waste causes environmental, economic, and social problems. In Morocco, the Jorf Lasfar Thermal Power Station produces two types of coal ash with enormous quantities: fly ash (FA) and Bottom ash (BA). FA is recovered in cement while BA is stored in landfills. To reduce the effects of BA disposal in landfills, several experimental studies have tested the possibility of their recovery in the road construction, especially as a subbase. In the first phase of this study, the BA underwent a physicochemical and geotechnical characterization. The results obtained show that the BA should be treated to improve its mechanical properties. The most commonly used materials are lime and cement. In the selected low-cost treatment, which is the subject of the second phase of the study, lime is used to improve the low pozzolanicity of BA while calcarenite sand is used to increase the compactness. Several mixtures containing BA, lime, and calcarenite sand were prepared. Each of these mixtures was compacted in modified Proctor molds and then subjected to a series of tests to study the following characteristics: compressive strength, dry and wet California Bearing Ratio (CBR), dry density and swelling. The composition of each mixture was based on an experimental design approach. The results show that the values of the compressive strength, the dry density, and the CBR index have increased after treatment, potentially leading to a valorization of the treated BA for use in a subbase.

Keywords: coal bottom ash; lime treatment; modified Proctor; CBR; subbase; road construction

---

1. Introduction

The fast expansion of infrastructure in developing countries, such as Morocco, increases the need for raw materials. In addition, the continued production of waste, in particular solid waste, requires the implementation of a waste storage system. This storage causes several economic, urban, and environmental problems. Therefore, there have been multiple studies over the possibility of using these by-products in the field of construction, including road construction [1,2]. In past years, different types of waste such as coal ash, steel slag and municipal solid waste incinerator ash [1,3,4] and other more recent waste types such as silicon waste [5] have been investigated in order to test the possibility of their valorization in civil engineering applications.

Coal ash, which is a residue of coal combustion in thermal power plants, has been the subject of several research projects, distinguishing between fly ash and bottom ash [1,2]. Fly Ash (FA) is a fine residue which is recovered at the level of the electrofilters and then sent to storage hoppers. They have been used for years as a substitute for cement in the field of construction [6]. On the other hand, the Bottom Ash (BA) has a sandy appearance [2]. It should be noted that the BA is recovered in a wet state (after cooling under water) at the bottom of the boiler.

The coal ash characterization work carried out by several researchers has shown the possibility of their use in the field of infrastructure and construction [1,2]. For instance, the BA has been used as a substitute for aggregates for the production of the bituminous mix [1]. However, the use of BA in asphalt mixtures decreases some mechanical properties such as tensile strength. To solve this issue, lime was used in the mixtures, which was able to significantly improve these mechanical properties [1]. BA has also been used in concrete as a partial substitute for sand [7]. Although, this causes a decrease in compressive strength. For example, for a concrete containing 10% BA as a sand substitute, the decrease in compressive strength exceeds 10% compared to conventional concrete. This decrease is attributable to the higher porosity of the BA, which causes a higher demand for water [2]. To solve this problem, a super plasticizer was added, which allows the reduction of the water demand and leading to an increase of the compressive strength [8]. BA was mixed with ordinary Portland cement and even rice husk ash in [9] to develop self-leveling hybrid mortars. In [10] a mixture of soil, BA, FA, crumb rubber, cement and water was investigated as flowable backfill material for an underground pipeline. Finally, geopolymers including BA have been investigated as well [11–13]. The actual performance of the geopolymer-based materials is strongly dependent on the specific local materials (mix of FA and BA), processing conditions and used blend.

Jorf Lasfar Energy Company (JLEC) is an energy company located in El Jadida, Morocco. JLEC consumes 5.4 million tons/year of coal to deliver more than 40% of Morocco's electricity demands. This generates a large amount of ash (both FA and BA). FA accounts for 80% of the solid residues produced, an annual average exceeding 470,000 tons. As for the BA, they constitute 20% of the solid residues and represent more than 40,000 tons/year [6,14]. The majority of FA is used in Moroccan cement, while BA is still stored in landfills [6,14]. In addition to the economic burden, in terms of transport costs and storage of these residues, there is an environmental risk in terms of leaching of the elements, polluting the water underground. Therefore, an alternative solution to the landfill is investigated. As shown in [15–17] solidification/stabilization (S/S) can be considered as an effective action to prevent and minimize the release of contaminants into the environment.

This research focuses on an experimental study of JLEC Bottom Ash to test the possibility of their reutilization in the foundation layer or subbase of the road. First, the BA has undergone a complete characterization to determine the following properties: particle size, cleanliness, density, pozzolanic activity, and both chemical and mineralogical composition. Following the BA characterization results, it was suggested to treat BA with lime (L) and sand (S). The percentage of lime varies from 1 to 5% of the total weight of the mixture while that of sand is between 5 and 25%. The water (W) content of the mixture is between 21 and 22.4%. For each mixture, cylindrical specimens were compacted in a modified Proctor mold. In order to have reliable and statistically representative results, all the tests are performed with three repetitions, leading to a total number of test specimens of 192. The material properties (responses) that were tested are: compressive strength (CS), dry and wet California Bearing Ratio (CBR_d, CBR_w), dry density (γ), and swelling (G). For the design of the experiments the Design-Expert software was used. This makes it possible to define each characteristic output (response Y_i) according to the influencing input factors (X_i).

2. Materials and Methods

2.1. The Origin of the Materials

The BA is obtained from the JLEC plant EL Jadida, Morocco. They have been water-cooled and, therefore, they are recovered at the outlet of the boiler in a very humid (submerged) state, comparable to the moisture state of sea sand at the time of its extraction. Before being transported to the laboratory by truck (quantity necessary for all our tests), they were left outside the plant for three weeks in order to reduce their high water content sufficiently. Next, smaller quantities for each test batch were dried at the road geotechnical laboratory at the Hassania school of public works (Casablanca, Morocco).

To evaluate the pozzolanic reactivity of BA, mortar molds containing cement, sand, and water were prepared. Next, BA was used as a partial substitute for standardized sand. The compressive strength (CS) of the specimens containing the BA was compared with those of the conventional sand-based mortar. The cement used is the CPJ45 produced by the Holcim Lafarge Group in Morocco. Its chemical composition is given in Section 3.1.

As will be presented in the results, the BA has a porous structure, low compactness, and low pozzolanic reactivity. These weak characteristics do not allow for their use in a road foundation layer. To improve their performance, the BA was treated with lime and calcarenite sand. We note that the choice of these two treatment materials is made, taking into consideration the results of similar research [1]. They have proven the effectiveness of these two materials in improving the properties of BA. Lime is used to improve the pozzolanic activity of BA, while calcarenite sand increases its compactness. The lime used is produced at the Tetouan plant of the Lafarge group in Casablanca (Morocco). Its chemical composition is marked by a free lime (CaO) content greater than 80% whereas that of MgO does not exceed 8%. For the sand, we used a granular class 0/3.

2.2. Methodology

This work was carried out in three phases, summarized as follows: the first phase of this study consists of a complete characterization of the BA. The characterization tests cover density, chemical and mineralogical composition, cleanliness, granulometric analysis, Proctor test, and wet and dry CBR indices. The second part describes the experimental program including results obtained for the treatment of the BA using lime and sand. In this part, we also explain the experimental design, obtained by the Design-Expert software [18]. Lastly, the third phase of this study consists of the theoretical design of a pavement structure, with and without BA treated with lime and sand. The design is done according to the classical method of the 1998 new pavement type structure catalogue [19]. This method takes into consideration: traffic, climate and geotechnical environment of the El Jadida area in Morocco, which represents the location envisaged for the future pavement. The used methodology is shown in Figure 1.

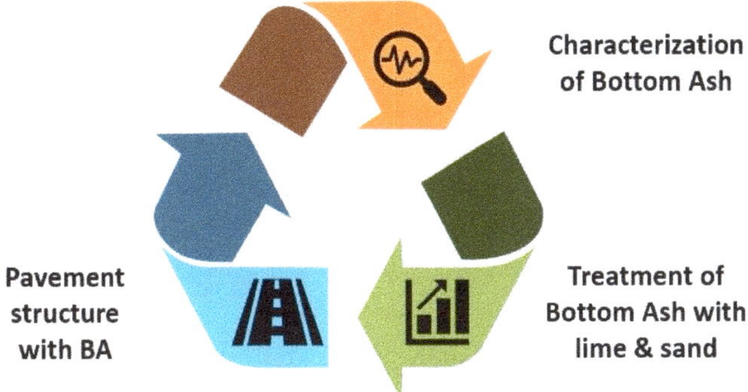

Figure 1. Methodology of this study comprising of three phases: characterization, treatment with lime and sand, and implementation in the theoretical design of a pavement structure.

2.3. Characterization Methods

To test the possibility of using the BA in a foundation layer, several physicochemical and mechanical tests were carried out. The BA particles have a very porous shape. Their density is determined by the technique of the graduated cylinder [20], which consists in introducing a quantity of bottom ash in a known volume of water. The density is thus obtained by determining the volume occupied by this mass. The chemical composition of the BA Section 3.1 is obtained using X-ray fluorescence spectroscopy, while the mineralogical composition Section 3.1 is established using X-ray diffraction.

These two tests were carried out at the chemistry department at the Faculty of Sciences of the University of Hassan II in Casablanca, Morocco.

Next, the cleanliness of the BA is evaluated by two methods: sand equivalent (SE) method and methylene blue method according to the standards NF P18-598 [21] and NF P94-068 [22]. The first test gives us an idea of the rate of the clay part in our sample, the second determines the activity of this clay part and the degree of its sensitivity to water [23]. For the particle size analysis, the BA of class 0/20 mm was sieved according to the NF P18-560 standard [24], using 22 sieves from 0.04 up to 20 mm. The results obtained are compared to those specified by the ASTM standard [25] for BA used as a sand substitute [23]. In order to investigate the possibility of using BA in a road structure, the optimal compaction conditions, puncture resistance and swelling have been evaluated. In order to determine the optimal compaction conditions, the BA is sieved with a sieve of 20 mm and then subjected to a compaction test in a modified Proctor mold. The puncture resistance is measured twice: once compacted at natural water content as the dry CBR test (CBR_d) and then after immersion for four days, called the wet CBR or CBR after immersion (CBR_w). CBR_d characterizes the ability of BA to withstand traffic flow. The swelling test is used to evaluate the behavior of BA under unfavorable humidity conditions. The swelling is measured during four days of immersion in order to test the volume stability of the BA [26]. The experimental protocols describing Proctor, CBR (dry and wet) and swelling test are detailed in the NF P94-078 standard [27].

Finally, the pozzolanic activity was judged. A material is pozzolanic if it can, in the presence of moisture, chemically react with calcium hydroxide to form compounds with binding properties. To evaluate the pozzolanic activity of BA, three series of mortar prisms of dimension $4 \times 4 \times 16$ mm were prepared in the building materials laboratory of the Hassania School of Public Works (Casablanca, Morocco), see Figure 2a. The production procedure for mortars is described in the Moroccan standard NM 10.1.004 [20] for hydraulic binders. The first series is the standard mortar, containing 450 g of cement, 1350 g of standardized sand and 225 g of water, which gives three mortar prisms. In the second and third series, part of the normalized sand mass is replaced by BA of size 0/2 mm with percentages of respectively 25% and 50%. The prepared mixtures are introduced into the mortar molds, which will be stored in a wet cupboard at a temperature of 20 °C. After 24 h, they are demolded, then stored in a water bath at a constant temperature of 20 °C. The samples will be crushed after 7, 15, 28, 60, and 90 days to evaluate their compressive strength (CS).

Two different types of specimens were manufactured:

- Figure 2a: mortar prisms ($4 \times 4 \times 16$) containing BA to study their pozzolanic activity (BA is used as a substitute for normalized sand);
- Figure 2b: compacted test tubes of the BA specimens for the other characterization tests.

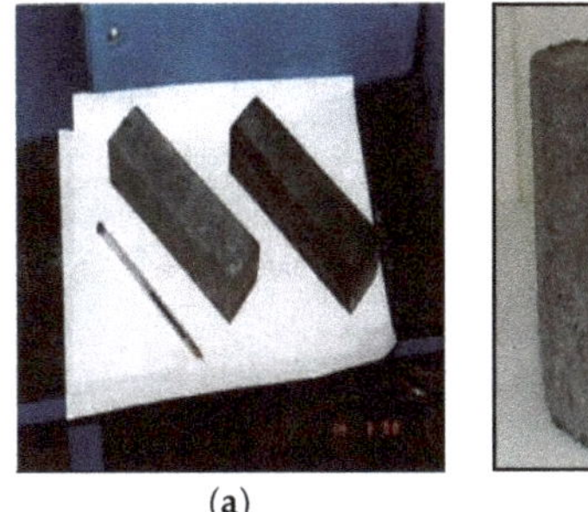 

(a)   (b)

Figure 2. (a) Mortar prism (4 × 4 × 16 cm) (b) Compacted test tubes of bottom ash ($R = 7$ cm, $h = 15$ cm).

2.4. Treatment of Bottom Ash with Lime and Sand

2.4.1. Overview of the Experimental Approach Adopted for the Treatment of BA

In order to improve the properties of the BA and with the intention of using them as foundation materials, they were subjected to a lime and sand treatment. The experimental approach adopted to establish mixing dosages and also to model the results is the design of experiments method [6]. The experimental design consists of selecting and ordering tests to identify the effects of parameters (factor X_i) on the desired properties (response Y_i) of a material. These are statistical methods using simple mathematical concepts. The implementation of this method involves the following main steps:

1. Identification of the properties to be studied (responses);
2. Identification of the parameters (factors) that influence these responses;
3. Definition of the ranges of variation of each factor, with two levels of extreme variation (±1);
4. Carrying out the experiments planned by the model;
5. Analysis of the results (answers) and mathematical modeling.

2.4.2. Experimental Plan Adopted for the Treatment of BA

The final material studied is a mixture of BA treated with lime and sand. The factors considered to influence the properties of this mixture are:

- The percentage of BA;
- The water content noted as W;
- The percentage of lime noted as L;
- The percentage of calcarenite sand noted as S.

Other factors can be mentioned, such as the nature of the lime and sand, the compaction energy, the temperature during implementation, the mixing, and the storage conditions of the specimens. In this study, all of these were kept constant. In addition, the total solid mass of the mixture satisfies the following condition, given by Equation (1):

$$\sum (\% \, BA + \% \, S + \% \, L) = 100\% \tag{1}$$

The responses studied to evaluate the possibility of using the mixture as a foundation layer are:

- Compressive strength at 60 days noted as CS_{60};
- Dry density noted as γ_d;
- Swelling noted as G;
- Dry CBR index noted as CBR_d;
- The CBR index after immersion noted as CBR_w.

There are several types of experimental design [18]. In this study, we opted for a centered composite factorial design composed of the following elements:

- A two-level factorial plan (−1, +1);
- An experimental point located in the center of the field of study;
- Experimental points located on the axes of each of the 3 factors. These points belong to the interval (−α, +α).

The experimental design adopted will make it possible to evaluate the effects of the three factors at five different levels: −α, −1, 0, 1, and +α. The value of α is calculated as follows: $\alpha = NP^{1/4}$, with NP the number of points of the factorial plane. In our study this leads to the following value, Equation (2):

$$\alpha = 8 \times \frac{1}{4} = (23) \times \frac{1}{4} = 1.683 \tag{2}$$

The three main factors Lime (*L*), Water (*W*), and Sand (*S*) are associated with the reduced centered variables: *A, B* and *C*. The levels +α and −α are assigned to the extreme values of *A, B* and *C*. The levels −1, 0, and +1 are obtained by linear interpolation (see Table 1). The relationships between the reduced centered variables and the real variables are given by Equation (3)–(5):

$$A = \frac{L - 2,5}{1.485} \tag{3}$$

$$B = \frac{W - 22}{0.367} \tag{4}$$

$$C = \frac{S - 12,5}{7.427} \tag{5}$$

Table 1. Factor Levels of the Centered Composite Plane.

Factor Levels	−α = −1.683	−1	0	+1	+α = +1.683
Lime dosage (%)	0	1.015	2.5	3.985	5
Water dosage (%)	21.4	21.762	22	22.238	22.6
Sand dosage (%)	0	5.073	12.5	19.927	25

To obtain statistically representative results, each test is performed with three replicates. The number of specimens manufactured is distributed as follows:

- $24 = 2^3 \times 3$ tests in the factorial design, in which the factors were adjusted to the −1 and +1 levels.
- $18 = (2 \times 3) \times 3$ axial tests in which the factors were adjusted to the ±α levels, to estimate the quadratic effect of the different parameters on the responses.
- $6 = 2 \times 3$ center tests for model verification and determination of the experimental error.

In order to obtain results for CS_{60}, γ_d, G, CBR_d and CBR_w this study required $(24 + 18 + 6) \times 4 = 192$ cylindrical specimens, see Figure 2b. All the tests are represented by the matrix included in Table 2.

Table 2. Matrix of the tests of the composite centered factorial plan adopted for the treatment of bottom ash (BA).

Tests	Lime Dosage	Water Dosage	Sand Dosage
	A	B	C
1	-1	-1	-1
2	1	-1	-1
3	-1	1	-1
4	1	1	-1
5	-1	-1	1
6	1	-1	1
7	-1	1	1
8	1	1	1
9	$-\alpha$	0	0
10	$+\alpha$	0	0
11	0	$-\alpha$	0
12	0	$+\alpha$	0
13	0	0	$-\alpha$
14	0	0	$+\alpha$
15	0	0	0
16	0	0	0

3. Results

3.1. Chemical and Mineralogical Characteristics of Bottom Ash

The elemental composition of the studied BA and those found in the literature are given in Table 3. The results show that the BA mainly consists of silicon dioxide (SiO_2), aluminum oxide Al_2O_3 and ferric oxide (Fe_2O_3) with small amounts of calcium oxide, magnesium oxide, and potassium hydroxide. The sum of the percentages of the elements SiO_2, Fe_2O_3, and Al_2O_3 exceeds 80% of the total mass, while the percentage of CaO remains low. Therefore, BA can be classified as a silico-aluminous material (class F) according to the ASTM 225 standard [25]. This table also shows that the BA used in this study has a chemical composition that is very close to the BA found in literature [2,28,29]. In Table 1 of Reference [2] an overview is included showing the chemical analyses from different studies.

Table 3. Elemental composition (%) of bottom ash and used cement.

Chemical Element	CaO	SiO$_2$	Fe$_2$O$_3$	Al$_2$O$_3$	K$_2$O	Na$_2$O	P$_2$O$_5$	SO$_3$	MgO	Free CaO
(JLEC) BA	1.9	52.1	8.9	23.3	1.9	0.4	0.1	<1	0.9	0.3
BA [28]	4.2	50.5	10.9	27.6	0.8	0.6	0.2	0.1	1.2	-
BA [29]	7.0	46.1	5.8	23.7	1.2	0.7	-	-	1.2	-
Cement CPJ45	63.0	17.0	3.0	5.0	1.2	-	-	3.3	2.3	-

The mineralogical composition of the BA, determined with X-Ray diffraction, reveals the existence of two peaks, see Figure 3. The first is quartz and the second is mullite. These two mineralogical phases are the same as those found for some BA studied in literature [28,29].

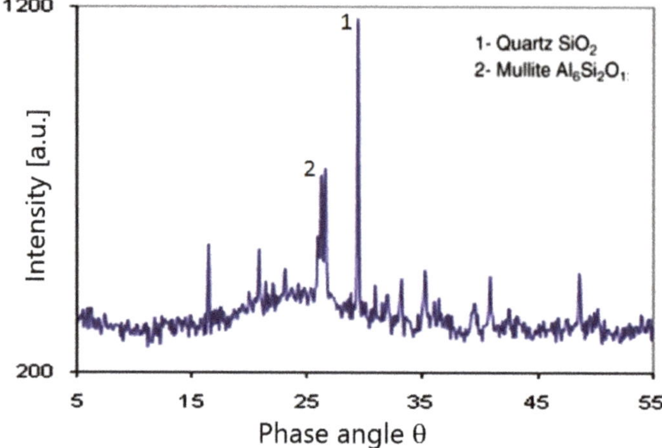

Figure 3. Mineralogical composition of bottom ash (BA) using X-ray diffraction.

3.2. Physical Characteristics of Bottom Ash

The visual observation of the BA shows that it has a gray color and a porous texture. Its absolute density is 1760 kg/m^3. This value is low compared to natural aggregates such as silica sand which has an absolute density of 2600 kg/m^3 [28], and BA tested in [9] with an absolute density of 2560 kg/m^3, but very close to the value for BA of 1880 kg/m^3 reported in [30]. In Table 2 from Reference [2] values ranging between 1390 and 2470 kg/m^3 are shown. The porous texture is the same as that observed for BA produced in other countries [2,28]. As discussed in Section 2.3, the cleanliness of the BA is verified by two tests: the sand equivalent (SE) and the methylene blue. The result of the first test is SE = 78%, which classifies the BA as clean sand with a low percentage of fine clay, which is ideal for construction work [19]. The second test yielded a methylene blue value of 0.5. This value is less than 1.5, which represents the threshold for silty sandy soils in sandy loam soils. It can be deduced from these two tests that the BA is clean and insensitive to water which encourages their use in the road domain.

The granulometric study of the BA (see Figure 4) shows that their particle size distribution (PSD) is very close to the sand. It also shows that their granulometric curve is located within the ASTM standard granular limits for BA used as foundation layer [31].

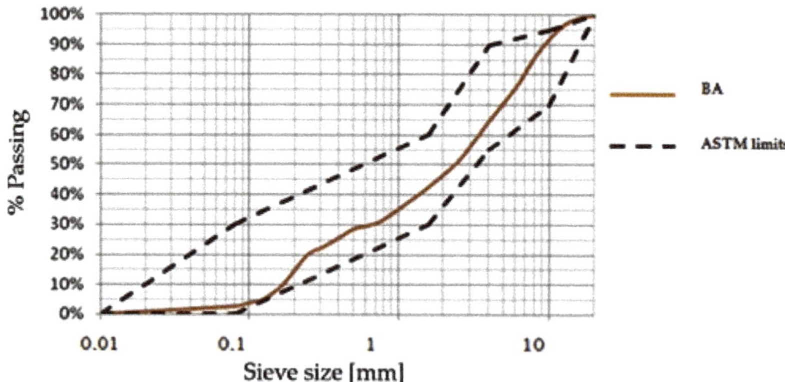

Figure 4. Granulometric curve of JLEC BA.

3.3. Compaction of Raw Bottom Ash

The Proctor test shows that the maximum density of the BA is 1.26 kN/m^3, and is obtained at a water content of 21.6%. This high value of the Proctor moisture content indicates a high porosity of the material. Moreover, this value of the optimum Proctor of the BA is very low compared to conventional materials such as clay and sand whose dry density is respectively between 16 and 21 kN/m^3. Hence the need to treat the BA in order to improve this characteristic.

The CBR_d value obtained is 35%, exceeding the min. standard value of 20%. The BA is then classified in the lift class AR3 [26]. On the other hand, the punching test performed on a BA sample immersed for four days gives a CBR_w value of 58%, which exceeds the CBR_d value of 35%. This increase in lift is significant and should be exploited in the road sector.

Finally, the recorded value of the swelling (G) was zero. This result shows that the BA does not represent any risk of swelling and promotes its use in the road construction.

3.4. Pozzolanic Activity of BA

The evaluation of the mechanical behavior is performed by determining the compressive strength of mortar prisms, see Figure 2a, prepared from a mixture of CPJ 45 cement, water and BA as a partial substitute for the standardized sand. The results show that the compressive strength of the prisms containing BA is, at any age, adversely influenced by the use of BA, see Figure 5. The decrease in resistance becomes very important for 50% of BA and especially at a young age. These results are in the same order of magnitude as those found by other researchers [8,29]. They also found that the decrease of the compressive strength of mortar prisms containing BA can be explained by the dominant role of porosity, which is inversely proportional to compressive strength. In fact, the substitution of the most resistant material (standardized sand) by the weaker and more porous material (BA) increases the fraction of the pores of the mortar which decreases the compressive strength.

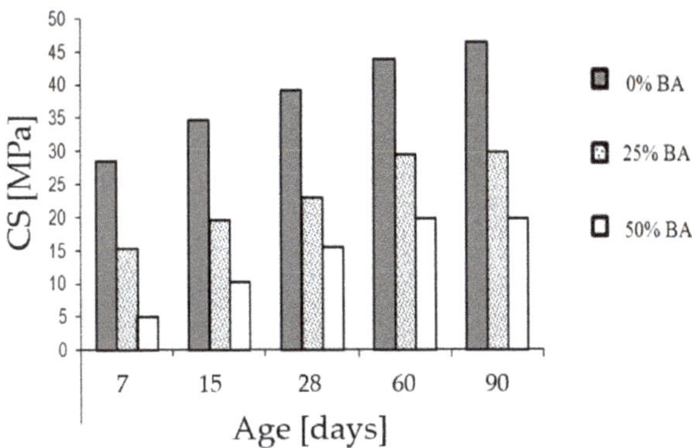

Figure 5. Compressive strength of mortar containing bottom ash.

As shown in Figure 5, an increase of the pozzolanic potential of the BA, especially at 60 and 90 days, becomes apparent. Therefore, during the treatment of the BA, the evolution of the compressive strength will be studied at 60 days instead of 28 days, because at this age the pozzolanic power of the BA is important and the compressive strength is significant. The composition of each mortar mixture and the values of their compressive strength at different ages (one sample per test) are given in Table 4.

Table 4. Composition of the mortar mixture and compressive strength at different ages.

Type	Composition (g)				CS (MPa) at Different Ages (days)				
Mortar	Cement	Water	Sand	BA	7	15	28	60	90
Reference	450	225	1350	0	28.5	34.7	39	43.8	46.4
25% BA	450	225	1012.9	337.5	15.1	19.4	22.8	29.3	29.7
50% BA	450	225	675	675	5	10.3	15.4	19.7	19.8

3.5. Property of the BA-Lime-Sand Mixture

3.5.1. Presentation of Experimental Results on Treated BA

The average results and standard deviation of all tests performed on lime and sand treated BA are given in Table 5. Each value represents the average of three tests. These results show that in addition to the specific weight of the BA, mechanical properties such as compressive strength, and the dry and wet CBR indices are positively influenced by the addition of lime and sand. This result may be due to the increased compactness thanks to the addition of the sand or the activation of a pozzolanic reaction after the addition of lime. We also note that, for each treated BA mixture (TBA), CBR values after submersion (CBR_w) are higher than those of dry CBR (CBR_d) which represents a lift gain that cannot be found for conventional materials. This result is very interesting in terms of lift gain and reveals the importance of the use of BA in the road domain. Moreover, the swelling remains very low, as we wish, which encourages us to use the treated BA without risk of volume instability. The dry density was determined from its wet density using Equation (6):

$$\gamma_d = \frac{\gamma_w}{1 + W} \tag{6}$$

Table 5. Average results of tests carried out on treated BA.

Test	Levels of Factors			Average Responses and Standard Deviation				
Number	Lime (%)	Water (%)	Sand (%)	CBR_d (%)	CBR_w (%)	γ_d (kN/m^3)	G (%)	CS_{60} (MPa)
1	1	21.6	5	79.0 (6.5)	98.3 (9.0)	11.7 (0.20)	0.06 (0.03)	1.10 (0.09)
2	4	21.6	5	82.9 (9.8)	107.0 (15.8)	11.5 (0.03)	0.15 (0.02)	2.36 (0.08)
3	1	22.4	5	58.6 (1.5)	73.0 (9.0)	11.0 (0.11)	0.03 (0.00)	1.63 (0.03)
4	4	22.4	5	75.2 (6.9)	86.0 (14.3)	11.5 (0.20)	0.13 (0.02)	2.84 (0.02)
5	1	21.6	20	74.0 (3.7)	90.7 (6.6)	11.6 (0.10)	0.12 (0.01)	1.65 (0.07)
6	4	21.6	20	80.4 (5.1)	72.7 (3.1)	11.3 (0.03)	0.14 (0.01)	1.74 (0.00)
7	1	22.4	20	74.5 (5.3)	78.0 (1.6)	11.5 (0.05)	0.04 (0.00)	1.87 (0.05)
8	4	22.4	20	66.3 (8.1)	85.7 (6.6)	11.9 (0.05)	0.11 (0.02)	2.75 (0.08)
9	0	22.0	12.5	68.7 (5.2)	36.3 (5.7)	10.7 (0.12)	0.02 (0.00)	0.43 (0.04)
10	5	22.0	12.5	79.2 (0.5)	83.3 (6.9)	11.8 (0.15)	0.12 (0.02)	2.61 (0.13)
11	2.5	21.4	12.5	66.3 (3.9)	77.3 (13.7)	11.0 (0.13)	0.11 (0.02)	2.02 (0.06)
12	2.5	22.6	12.5	59.4 (4.5)	65.3 (10.3)	11.7 (0.08)	0.02 (0.00)	2.48 (0.14)
13	2.5	22.0	0	70.0 (4.9)	58.3 (6.2)	10.6 (0.09)	0.02 (0.00)	1.97 (0.02)
14	2.5	22.0	25	75.0 (6.1)	59.0 (2.9)	12.1 (0.15)	0.01 (0.00)	1.87 (0.03)
15	2.5	22.0	12.5	62.8 (1.4)	69.0 (11.0)	11.4 (0.02)	0.12 (0.02)	2.08 (0.08)
16	2.5	22.0	12.5	62.5 (0.4)	64.3 (2.5)	11.3 (0.06)	0.12 (0.02)	2.06 (0.04)

3.5.2. Modeling Responses

For each response studied, we proceed to the modeling of the results using the Design-Expert software. This starts with the analysis of the variance, which allows the determination of the influencing factors and the elimination of the insignificant factors. Then a model combining each response to influential factors is developed. From this fact, a mathematical function connecting the response to the factors is proposed by the model. The modeling equations for the responses (CBR_d, CBR_w, γ_d, G and CS_{60}) are expressed in terms of the percentages of lime (L), water (W) and sand (S). We thus obtain the following Equation (7)–(11):

$$CS_{60}{}^2 = 1395.96 - 21.41 \times L - 126.8 \times W + 0.14 \times S + 1.08 \times L \times W - 0.06 \times L \times S - 0.08 \times L^2 + 2.88 \times W^2 \quad (7)$$

$$\gamma_d{}^3 = 12525.76 - 2892.09 \times L - 513.29 \times W - 442.31 \times S + 133.3 \times L \times W + 20.66 \times W \times S \quad (8)$$

$$CBR_d{}^{-0.85} = -0.04 + (13.42 \times L + 27.28 \times W + 2.78 \times S + 0.69 \times L \times S - 0.52 \times L2 - 0.19 \times S2) \times 10^{-4} \quad (9)$$

$$CBR_w{}^2 = 5397.02 + 588.98 \times L - 2246.85 \times W \quad (10)$$

$$G^{0.65} = 2.19 + 0.04 \times L - 0.1 \times W + 0.02 \times S - 6.68 \times 10^{-4} \times S^2 \quad (11)$$

3.5.3. Optimal Predicted Formulations

Taking into account the prerequisites of using BA as a foundation material, the models obtained make it possible for us to propose two optimal formulations, predicted from the actual test results.

The first optimal formulation is technical and takes into consideration the following requirements:

1. Maximize CS_{60}, CBR_w, and CBR_d.
2. Minimize swelling (G) and percentages of lime and added sand.

The first optimal formulation is obtained with 2% of lime, 22.4% of water and 5% of sand. The predicted results for this first formulation are:

- $CBR_d = 62\%$
- $CBR_w = 68\%$
- Swelling (G) = 0.05%
- $CS_{60} = 2.05$ MPa

The second variant is economical. It aims to reduce the cost by minimizing the processing materials while keeping acceptable mechanical properties and it takes into consideration the following requirements:

1. Minimize swelling (G)
2. Minimize the cost
3. CS60 > 1.5 MPa

The second optimal formulation is obtained with 0.3% of lime, 22.4% of water and 20% of sand. The predicted results for the second formulation are:

- $CBR_d = 74\%$
- $CBR_w = 61\%$
- Swelling (G) = 0.01%
- $CS_{60} = 1.5$ MPa

Both optimal formulations are acceptable although the compressive strength is less than 3 MPa. The CBR-values of both variants are sufficient for the proposed application and it is the major parameter taken into consideration for the dimensioning of this kind of structure. Moreover, the compressive strength parameter is studied in our case just to ensure the efficiency of the treatment. Higher compressive strength values are required for rigid structures (with concrete) which is not the case for our proposed pavement structure.

3.6. Use of Treated BA in Road Pavement

Characterization of the BA shows that they have very satisfactory characteristics: their granulometric curve is located inside the granular limits of BA valued in the roads. They are also clean and are insensitive to water. Moreover, their treatment with lime and sand has clearly improved their mechanical properties: compressive strength (CS_{60}), dry and wet CBR indices (CBR_d, CBR_w) and dry density γ_d all reached satisfactory levels after this treatment, while the swelling G remains very low. These results encourage us to use the BA in the foundation layer of a rural road, e.g. near the city of El Jadida in Morocco, thus minimizing transportation costs. The design method adopted is the classic method of the New Pavement Structure Catalog [19]. The parameters taken into consideration in the design of this route are the traffic which is class TPL3, the climate which is considered wet for the rural area of El Jadida, and the supporting ground which has a bearing capacity class of ST1. The conventional pavement structure proposed by the catalog, presented in Figure 6a, consists of 10 cm AC + 20 cm UGF2 + 20 cm UGB + SC, with:

- AC: Anti-Contamination layer;
- UGF2: Untreated Gravel for Foundation layer type 2;
- UGB: Untreated Gravel for the base layer;
- SC: Superficial Coating.

In the proposed pavement structure, we used the TBA (BA treated with lime and sand) as a foundation layer material instead of the UGF2. However, taking into account the modest CS of the TBA, we propose an increase in the thickness of the foundation layer, 30 cm of TBA instead of 20 cm of UGF2. The new proposed variant, see Figure 6b, consists of 10 cm AC + 30 cm TBA + 20 cm UGB + SC.

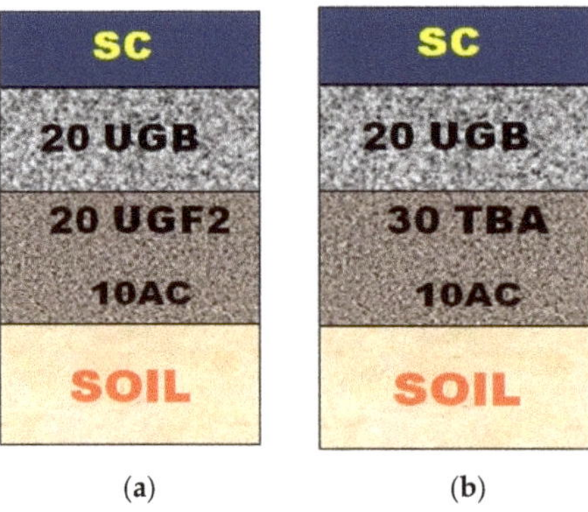

(a)   (b)

Figure 6. (a) Conventional pavement structure, (b) Pavement structure based on TBA.

4. Conclusions

The results obtained in this study show that, despite their low density, the BA studied has interesting properties that can be used in order to promote their value in road engineering, particularly in the foundation layer. The results of this study are summarized as follows:

1. The Jorf Lasfar bottom ash BA is class F. It has a significant pozzolanic power, which favors their treatment with a hydraulic binder.
2. The lime treatment of the BA significantly improves their properties, which are: the compressive strength (CS_{60}), the dry density, and the CBR indices. The best mechanical performances are obtained for a mixture with 4% of lime.
3. The use of calcarenite sand increases the dry density of the treated BA mixture (TBA) compared to that of the original BA. It is deduced that sand corrects the porous texture of BA by increasing their compactness. The maximum density is registered for 25% of sand.

4. According to the performance obtained after the treatment of BA with lime and sand, we propose their use as a road material in the rural roads of the city of El Jadida in Morocco. In the proposed pavement structure, we used the TBA as a foundation layer instead of the UGF2 (untreated granulate for foundation layer type 2). This valorization makes it possible, on the one hand, to provide a cheaper ecological rural road network and, on the other hand, to find a sustainable solution for landfilling the BA.

The results of this research constitute a first step forward in the field of valorization of BA in Moroccan pavements. However, the stakes remain very important if we take into account the millions of tons of this waste produced worldwide each year. Further research should include an analysis of leaching from the treated BA mixture and preferably a complete Life Cycle Assessment. If these studies are positive, the proposed pavement structure, containing BA, should be tested further, under real traffic conditions, in order to monitor its behavior and ensure its durability.

Author Contributions: methodology, M.M. and K.L.; investigation, S.E.M.E.A.; writing—original draft preparation, S.E.M.E.A.; writing—review and editing, C.V., R.M., A.M. and N.H.; supervision, M.M. and K.L.; funding acquisition, W.V.d.b.

Funding: This research was partly funded by the international office of the University of Antwerp and the VLIR-UOS South Programmes.

Acknowledgments: The authors would like to warmly thank Mhimra Youssef and Mohamed el Yousoufi who participated in this study as part of their end-of-study project during their third year at the Hassania Public Works School in Morocco.

Conflicts of Interest: The authors declare no conflict of interest.

References

1. Churchill, E.V.; Amirkhanian, S.N. Coal ash utilization in asphalt concrete mixtures. *J. Mater. Civ. Eng.* **1999**, *11*, 295–301. [CrossRef]
2. Singh, N.; Mithulraj, M.; Arya, S. Influence of coal bottom ash as fine aggregates replacement on various properties of concretes: A review. *Resour. Conserv. Recycl.* **2018**, *138*, 257–271. [CrossRef]
3. Lin, Y.; Hu, C.; Adhikari, S.; Wu, C.; Yu, M. Evaluation of Waste Express Bag as a Novel Bitumen Modifier. *Appl. Sci.* **2019**, *9*, 1242. [CrossRef]
4. Hassan, H.F. Recycling of municipal solid waste incinerator ash in hot-mix asphalt concrete. *Constr. Build. Mater.* **2005**, *19*, 91–98. [CrossRef]
5. Hasnain Saeed, M.; Shah, S.A.R.; Arshad, H.; Waqar, A.; Imam, M.A.H.; Sadiq, A.N.; Hafeez, S.; Mansoor, J.; Waseem, M. Sustainable Silicon Waste Material Utilization for Road Construction: An Application of Modified Binder for Marshall Stability Analysis. *Appl. Sci.* **2019**, *9*, 1803. [CrossRef]
6. El Moudni El Alami, S. Contribution à l'étude de valorisation des cendres à charbon et des scories d'aciéries en Génie Civil. Ph.D. Thesis, Faculté des sciences, Université Chouaib Doukkali, El Jadida, Morocco, June 2010.
7. Van den Heede, P.; Ringoot, N.; Beirnaert, A.; Van Brecht, A.; Van den Brande, E.; De Schutter, G.; De Belie, N. Sustainable High Quality Recycling of Aggregates from Waste-to-Energy, Treated in a Wet Bottom Ash Processing Installation, for Use in Concrete Products. *Materials* **2016**, *9*, 9. [CrossRef] [PubMed]
8. Singh, M.; Siddique, R. Effect of coal bottom ash as partial replacement of sand on properties of concrete. *Resour. Conserv. Recycl.* **2013**, *72*, 20–32. [CrossRef]
9. Tambara Júnior, L.U.D.; Cheriaf, M.; Rocha, J.C. Development of Alkaline-Activated Self-Leveling Hybrid Mortar Ash-Based Composites. *Materials* **2018**, *11*, 1829. [CrossRef] [PubMed]
10. Lee, K.J.; Kim, S.K.; Lee, L.H. Flowable Backfill Materials from Bottom Ash for Underground Pipeline. *Materials* **2014**, *7*, 3337–3352. [CrossRef] [PubMed]
11. Chindaprasirt, P.; Jaturapitakkul, C.; Chalee, W.; Rattanasak, U. Comparative study on the characteristics of fly ash and bottom ash geopolymers. *Waste Manag.* **2009**, *29*, 539–543. [CrossRef] [PubMed]

12. Antunes Boca Santa, R.A.; Bernardin, A.M.; Riella, H.G.; Kuhnen, N.C. Geopolymer synthetized from bottom coal ash and calcined paper sludge. *J. Clean. Prod.* **2013**, *57*, 302–307. [CrossRef]

13. Kalaw, M.E.; Culaba, A.; Hinode, H.; Kurniawan, W.; Gallardo, S.; Promentilla, M. Optimizing and Characterizing Geopolymers from Ternary Blend of Philippine Coal Fly Ash, Coal Bottom Ash and Rice Hull Ash. *Materials* **2016**, *9*, 580. [CrossRef] [PubMed]

14. El Moudni El Alami, S.; Monkade, M.; Lahlou, K. Valorisation du Mélange Cendres de Foyer de Jorf Lasfar-Scories de l'Aciérie Sonasid dans les Chaussées Routières. *Rev. Francoph. d'Ecologie Ind. Déchets Sci. Tech.* **2009**, *56*, 18–22.

15. Barbosa, R.; Lapa, N.; Lopes, H.; Gulyurtlu, I.; Mendes, B. Stabilization/solidification of fly ashes and concrete production from bottom and circulating ashes produced in a power plant working under mono and co-combustion conditions. *Waste Manag.* **2011**, *31*, 2009–2019. [CrossRef] [PubMed]

16. Cabrera, M.; Galvin, A.P.; Agrela, F.; Beltran, M.G.; Ayuso, J. Reduction of Leaching Impacts by Applying Biomass Bottom Ash and Recycled Mixed Aggregates in Structural Layers of Roads. *Materials* **2016**, *9*, 228. [CrossRef] [PubMed]

17. Hashemi, S.S.G.; Mahmud, H.B.; Ghuan, T.C.; Chin, A.B.; Kuenzel, C.; Ranjbar, N. Safe disposal of coal bottom ash by solidification and stabilization techniques. *Constr. Build. Mater.* **2019**, *197*, 705–715. [CrossRef]

18. Linder, R. *Les plans d'expériences: Un outil indispensable à l'expérimentateur*, 1st ed.; Presses de l'école nationale des Ponts et Chaussées: Paris, France, 2005; 328p.

19. Ministère de l'Equipement, des Transports et des logements (France). Catalogue des structures types de chaussées neuves; Direction des routes et de la circulation routière. 1998. 321p. Available online: http://lyceecherioux.fr/Mooc/Routes/structure_route/~{}gen/structure%20des%20routes. publi/auroraW/res/catalogue_des_structures_1998.pdf (accessed on 10 July 2019).

20. *Moroccan Standard NM 10.1.004 on Hydraulic Binders-Cement Composition, Specifications and Conformity Criteria*; CQAPC version Moroccan Industrial Standardisation Service: Rabat, Morocco, 5 June 2018.

21. *French standard, NF P18-598 on the equivalent of sand, "Évaluation des fines — Équivalent de sable"*; French Association for Standardization (AFNOR): Paris, France, October 1991.

22. *French standard, NF P94-068 on the methylene blue test, "Sols: reconnaissance et essais - Mesure de la capacité d'adsorption de bleu de méthylène d'un sol ou d'un matériau rocheux-Détermination de la valeur de bleu de méthylène d'un sol ou d'un matériau rocheux par l'essai à la tache"*; French Standardization Association (AFNOR): Paris, France, October 1998.

23. Dupain, R.; Lanchon, R.; Saint-Arroman, J.C.; Capliez, A. *Granulats, sols, ciments et bétons: Caractérisation des matériaux de génie civil par les essais de laboratoire*; Casteilla: Montigny Le Bretonneux, France, 2000.

24. *French standard, NF P18-560 on granulometric analysis by sieving, "Granulats-Analyse granulométrique par tamisage"*; French Association for Standardization (AFNOR): Paris, France, September 1990.

25. Standard Specification for Materials for Soil-Aggregate Subbase, Base, and Surface Courses. American Society for Testing and Materials. *Annual Book of ASTM Standards*; ASTM: West Conshohocken, PA, USA, 1994; Volume 04.08.

26. Laboratoire central des ponts et chaussées (France). *Traitement des sols à la chaux et-ou aux liants hydrauliques: Application à la réalisation des remblais et des couches de forme*; Technical Guide; LCPC: Paris, France, January 2000; Available online: http://www.chercheinfo.com/uploads/1958-0cc46bff31.pdf (accessed on 10 July 2019).

27. *French standard NF P94-078 on the CBR index after immersion, Immediate CBR index. Immediate Bearing Index, "Sols: reconnaissance et essais - Indice CBR après immersion. Indice CBR immédiat. Indice Portant Immédiat - Mesure sur échantillon compacté dans le moule CBR."*; French Association for Standardization (AFNOR): Paris, France, May 1997.

28. Hashemi, S.S.G.; Mahmud, H.B.; Djobo, J.N.Y.; Tan, C.G.; Ang, B.C.; Ranjbar, N. Microstructural characterization and mechanical properties of bottom ash mortar. *J. Clean. Prod.* **2018**, *170*, 797–804. [CrossRef]

29. Gimhan, P.G.S.; Disanayaka, J.P.B.; Nasvi, M.C.M. Geotechnical Engineering Properties of Fly Ash and Bottom Ash: Use as Civil Engineering Construction Material. *Eng. J. Inst. Eng. Sri Lanka* **2018**, *51*, 49–57. [CrossRef]

30. Lin, W.T.; Weng, T.L.; Cheng, A.; Chao, S.J.; Hsu, H.M. Properties of Controlled Low Strength Material with Circulating Fluidized Bed Combustion Ash and Recycled Aggregates. *Materials* **2018**, *11*, 715. [CrossRef] [PubMed]

31. Mhimra, Y.; El Youssoufi, M. *Valorisation des Cendres de Foyer de la Centrale Thermique de Jorf Lasfar (JLEC) dans le domaine routier*; Ecole Hassania des Travaux Publics: Casablanca, Morocco, June 2008.

Article

Slag Substitution as a Cementing Material in Concrete: Mechanical, Physical and Environmental Properties

María Eugenia Parron-Rubio [1], Francisca Perez-Garcia [2], Antonio Gonzalez-Herrera [2], Miguel José Oliveira [3] and Maria Dolores Rubio-Cintas [1,*]

[1]   Departamento de Ingeniería Industrial y Civil, Universidad de Cádiz, 11205 Algeciras, Spain
[2]   Departamento de Ingeniería Civil, Materiales y Fabricación, Universidad de Málaga, 29071 Málaga, Spain
[3]   Civil Engineering Department, University of Algarve, 8005-139 Faro, Portugal
*   Correspondence: mariadolores.rubio@uca.es

Received: 17 July 2019; Accepted: 30 August 2019; Published: 4 September 2019

Abstract: A circular economy is a current tenet that must be implemented in the field of construction. That would imply the study of the possibilities of the use of waste generated, for obtaining materials the used in construction as replacements for the raw material used. One of these possibilities is the substitution of the cement by slag, which contributes to the reduction of cement consumption, decreasing CO_2 emissions, while solving a waste management problem. In the present paper, different types of concrete made by cement substitution with different type of slags have been studied in order to evaluate the properties of these materials. Cement is replaced by slag from different steel mills, both blast furnace and ladle furnace slag. The percentages of slag substitution by cement are 30%, 40% and 50% by weight. Mechanical, physical and environmental properties have been evaluated. Compressive and flexural strength have been analysed as the main mechanical properties. As far as physical properties go, density and porosity tests were be reported and analysed, and from an environmental point of view, a leachate study was performed. It has been found that some kinds of slag (blast furnace slag) are very suitable as substitutes for cement, providing properties above those of the reference concrete, while other types (ladle furnace slag) could be valid for non-structural applications, contributing in both cases to a circular economy.

Keywords: concrete; slag; valorisation; cement; circular economy

---

1. Introduction

A circular economy is a currently accepted tenet, in which the traditional linear economy is transformed into a circular economy, where every activity is conceived as a cycle, where waste materials are considered as potential new resources, instead of by-products to discard.

In the field of construction, the challenge is to exploit the possibilities of the waste generated in the building industry as raw materials to be integrated in the same construction cycle.

One of the fields where this strategy is feasible is the incorporation of the slag generated during the steel production into concrete production. It has been used in many processes in the cement production and paving industries. It is interesting to focus attention on this problem and to study, thoroughly, all the possibilities that the steel by-product presents.

The substitution of the cement by slag provides two clear advantages; the first one is the use of a waste that must be managed in a landfill, and the second one, even more relevant, is the reduction in cement consumption, so the reduction of CO_2 emissions needed for its production.

Nowadays, there is already a lot of research that supports the adequacy of steel slag for the production of cementitious matrices [1–6]. Additionally, many studies in which aggregates are replaced by those types of by-products exist; e.g., blast furnace slag, copper slag, electric arc and fume dust have been used [2,7–16].

There are also some studies on the substitutions of cement by ground granulated blast furnace slag (GGBFS) [5,17–20], even getting up to 80% of the cement removed by this type of slag. Khatib et al. [21] replaced up to 80% of cement by GGBFS making different substitutions. Good results were obtained in the substitutions up to 60%, since compressive strengths similar to conventional concrete were obtained. After 28 and 90 days, the strength was increased. Nevertheless, worse results were obtained when replacing 80% of the slag, and in the first days of setting, the strength of the reference concrete was not reached.

Less attention has been paid to the substitution of cement by ladle furnace slag (LFS). In previous works, different types of slag have been studied and compared, with a maximum amount of 25% of cement replaced [22]. These studies provided promising results.

In this paper, different types of concrete have been elaborated on, in which the cement is replaced by slag from different steel mills, both blast furnace and ladle furnace slag.

The percentages of slag substitution by cement were 30%, 40% and 50% by weight. The substitution of cement was made in each mix by types of slag from different factories in Spain. According to different studies, it is known that the component with the highest influence over the durability of cementitious mixtures is SiO_2. In this work, we will focus on the relationship between the amount of this component in each slag and the mechanical properties.

Compressive and flexural strength were analysed as the main mechanical properties, making a comparison between all of them to evaluate which one provides the best characteristics.

Additionally, some of the physical and environmental properties evaluated were included in the present paper. For physical properties, density and porosity test were reported and analysed. For a sake of brevity, other tests made are omitted. From an environmental point of view, a leachate study of the material was carried out, since it was essential, considering that waste material was being put into service.

The paper is structured as follows. In Sections 2 and 3, the materials studied and the tests performed have been briefly described. A longer Section 4 is devoted to the results obtained, along with a broad discussion with a special attention to the analysis of the mechanical properties. Finally, conclusions are outlined in Section 5.

2. Materials

In this work, a 52.5 R Portland cement (PC) is used; this cement was chosen as it is free of any additives; that is to say, composed of clinker between 95%–100% and between 0%–5% of minor components, without other additives that change their composition. The substitutions were made for the different types of slag that are shown below:

➢ Slag 1 (GGBFS): Granulated blast furnace slag ground in ball mill.
➢ Slag 2 (LFS1): Ladle furnace slag (LFS).
➢ Slag 3 (LFS2): Ladle furnace slag (LFS).

LFS1 and LFS2 are ladle furnace slag with different origins and composition. Slag 1 (GGBFS) has a particle size <0.063 μm provided by the company, while the LFSs were sieved in the laboratory to obtain equal granulometry from them. This is an important fact to keep in mind, since it would be interesting to see what would happen if the LFS slags were also treated in the same way as the GGBFS, but the companies that provided us with that type of slag did not have the technology to do so. Therefore, it was decided to carry out the study with screened slags to study its results.

Three different concrete mixes were designed by substituting 30%, 40% and 50% of the weight of the cement with slag obtained from three different steel mills in Spain.

The most characteristic chemical values of these slags are shown in Table 1. These values were determined by X-ray fluorescence (XRF). This test was performed with the LFS slags once screened.

Table 1 shows the major components of the slags studied; the rest of secondary compounds are described in another paper which used the same slags [22].

Table 1. Cement and slag chemical composition.

Cement and Slag Origin/Chemical Composition	SiO_2	Al_2O_3	Fe_2O_3	CaO	MgO
	%	%	%	%	%
PC	16.6 ± 0.5	4.25 ± 0.5	3.02 ± 0.02	67.92 ± 0.5	1.43 ± 0.05
GGBFS	32.3 ± 0.5	10.7 ± 0.5	0.29 ± 0.02	47.14 ± 0.5	7.64 ± 0.05
LFS1	13.7 ± 0.5	9.1 ± 0.5	1.57 ± 0.02	55.18 ± 0.5	16.9 ± 0.05
LFS2	18.8 ± 0.5	12.5 ± 0.5	2.34 ± 0.02	54.9 ± 0.5	6.99 ± 0.05

The main components (CaO, SiO_2, Al_2O_3) of each of the slags are transcribed on a ternary diagram (Figure 1). Observe how the blast furnace slags are those that have better pozzolanic properties, by containing a higher percentage of SiO_2.

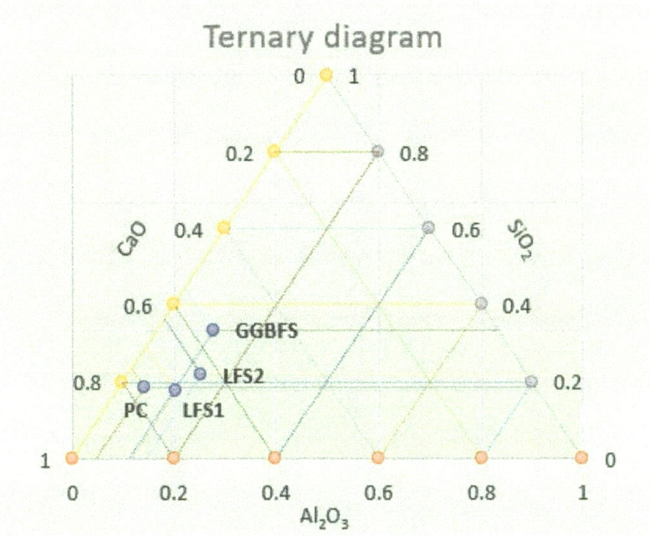

Figure 1. Ternary diagram indicating the compositions of Portland cement (PC), ground granulated blast furnace slag (GGBFS) and ladle furnace slag (LFS) in the system.

The different concrete mixtures were named as follows:

Mix 1 (MPC): Ordinary concrete without slag.
Mix 2 (MGGBFS): Concrete with 30%, 40% or 50% cement replaced with processed slag.
Mix 3 (MLFS1): Concrete with 30%, 40% or 50% cement replaced with unprocessed slag.
Mix 4 (MLFS2): Concrete with 30%, 40% or 50% cement replaced with stainless steel slag.

Table 2 shows the dosages and the percentages of substitution to be made in each mixture. The W/C (Cement water ratio) (ratio is 0.5; the tests are the continuation of the paper belonging to this research group, in which only substitutions were made up to 25% [22]. Additional details can be found in that reference.

Table 2. Concrete mixture proportion.

Mix	Water (w/b Ratio)	Dosage	Cement	Slag	Additive (Superpla-Sticizer)	Dosage	Fine Sand 0–2 mm	Sand 0–4 mm	Gravel 4–16 mm
MPC			100%	0%					
M30GGBFS			70%	30%					
M40GGBFS			60%	40%					
M50GGBFS			50%	50%					
M30LFS1	0.5	300 kg/m^3	70%	30%	3.9 kg/m^3	2033.8 kg/m^3	15%	35%	50%
M40LFS1			60%	40%					
M50LFS1			50%	50%					
M30LFS2			70%	30%					
M40LFS2			60%	40%					
M50LFS2			50%	50%					

3. Tests' Descriptions

Concrete mixes defined in the preceding section were subject to different tests. The main objective of these tests was to evaluate the effects on the mechanical characteristics (flexural and compressive strength), when cement is replaced by slag.

Concrete was made with the proportions shown in Table 2, where a 30%, 40% and 50% of the PC was substituted by the different slag according to Table 1, providing the different samples previously described.

The different mixture proportion was made according to the EN 12390-2 norm [23] for testing hardened concrete.

3.1. Physical Properties

The densities and porosities of the new materials were studied. They were obtained according to the EN 12390-7 norm [24].

A cubic specimen of 10 cm of edge were used for this test. Two specimens for each type of concrete were tested. The determination of the parameters was made for concrete of more than 28 days of age.

The formulation that was used to obtain the parameters is the following:

$$\text{Density} \quad D = \frac{P_s}{P_{sss} - P_{sum}} \tag{1}$$

$$\text{Porosity} \quad P = 100\frac{P_{sss} - P_s}{P_{sss} - P_{sum}} \tag{2}$$

The parameters are obtained in the following way:

- Psum: Weight obtained by the hydrostatic balance (submerged weight), placing the specimen inside. This test piece must be completely saturated with water
- Psss (saturated surface dry weight): Obtained by drying the surface water with a damp cloth.
- Ps (dry weight): It is obtained by drying the test pieces in the oven and checking every 24 hours that the mass loss is not less than 0.2%, at a temperature of 105 ± 5 °C. As indicated in the EN_12390-7 [24] standard, to carry out the test, the hydrostatic balance is used, to which a basket is attached where the test piece is introduced. In that way, one obtains the weight of the submerged sample.

3.2. Mechanical Properties

In order to obtain the compressive strength of the concrete specimens, we used cubes with edges of 10 cm and an automated press with a 2.000 kN capacity. The specimens were made according to the normative EN 12390-3 [25] and EN 12390-4 [26]; the fresh mixes were vibrated on a vibrating table and they were cured in a water bath 20 ± 2 °C. Then, they were tested at the ages of 1, 7, 28 and 90 days. For each of the mix proportions (Table 2), three different mixtures were made, and two specimens were tested at the different concrete ages (see reference [16] for details).

For the flexural strength, tests prismatic specimens were used with dimensions $4 \times 4 \times 16$ cm^3, made of the same kneaded as for the rest of the trial. That parameter was calculated using the uniform application of centred load. The same type of curing as the compressive test specimens was applied. This test was performed at 28 and 90 days. The prismatic test specimen was subjected to a bending moment by applying a load through upper and lower rollers, registering the maximum applied load calculating the flexural strength by EN-12390-5 [27]

3.3. Leachate

Finally, the mixtures are subjected to leachate tests. The cement substitution percentage chosen was 30% for every mixture. Additionally, we carried the test out using a 50% slag substitution percentage for GGBFS, since throughout the research, it showed similar behaviour to conventional concrete. Specimens were immersed in 1 litre of distilled water for 2 days. The treatment that was made to the water sample in the laboratory was to sieve the sample with a 0.45 μm filter, and acidify them to pH < 2. Once that process was done, it was introduced into the spectrometer.

4. Results and Discussion

The results obtained for the physical (density and porosity), mechanical (compressive and flexural) and environmental (leachate) tests for each of the mixtures and their different substitutions are shown below, making a comparison between the percentages of loss and gains of strength, and the differences between the varieties.

4.1. Density

In the differently manufactured mixtures, it was observed (Table 3) that the density varied very little with respect to the standard mixture, decreasing only in the mixtures made with the slag LFS1; therefore, this indicates that LFS1 aerates the mixture more. It was significant, and the porosity of the material also increased significantly. For the mixtures with the other two types of slags, there were no significant differences; therefore, replacing them in the mixtures would not pose any problem for this property.

Table 3. Density of the mixtures.

	Density (kg/m^3)			
	PC	**GGBFS**	**LFS1**	**LFS2**
30%	2500 ± 100	2490 ± 100	2450 ± 50	2530 ± 10
40%	2500 ± 100	2490 ± 100	2370 ± 30	2510 ± 10
50%	2500 ± 100	2490 ± 100	2370 ± 40	2510 ± 10

4.2. Porosity

The results of the porosity test are shown in the Table 4 and plotted in Figure 2.

Table 4. Porosity of the mixtures.

	Porosity (%)			
	PC	**GGBFS**	**LFS1**	**LFS2**
0%	1.81 ± 0.5	-	-	-
30%	-	1.51 ± 0.2	4.01 ± 0.5	0.91 ± 0.1
40%	-	1.27 ± 0.2	6.1 ± 0.5	0.78 ± 0.1
50%	-	1.53 ± 0.2	6.60 ± 0.5	-

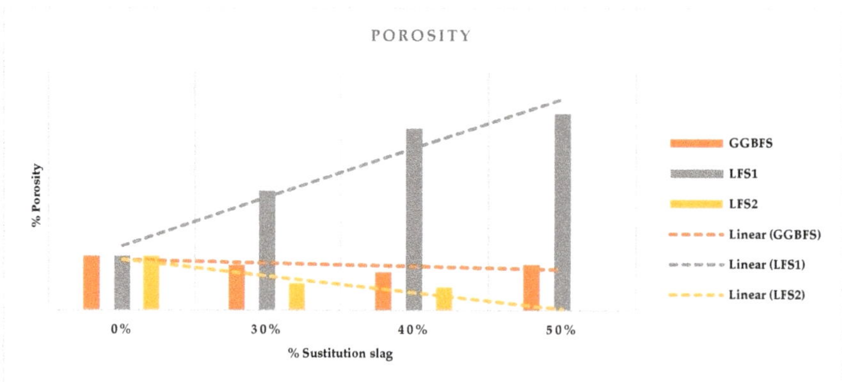

Figure 2. Porosity.

This property is tightly linked to the durability of the concrete. Taking as reference the PC mixture, for the mixtures with GGBFS, we see that it is practically constant for all the substitutions, even decreasing its porosity by 25% in the 50% substitution, with respect to conventional concrete. Regardless of the percentages of cement substitution, it was observed that the mixtures with slag LFS1 had a high porosity, since its porosity increased significantly. We saw in point Section 4.1, that for the mixture LFS1, the porosity increased up to 60%, so it was confirmed that this type of slag increases the incorporation of air into the concrete. On the contrary, it is observed that those made with slag LFS2, obtained a lower porosity, reaching up to a 43% improvement of results with respect to conventional concrete; this indicates that the links between particles that occur within the mixture are greater with this type of slag, without increasing its density by the same percentage; therefore, they provide better mixing conditions at the time of commissioning.

In short, the GGBFS and LFS2 mixtures' lower porosity means a better performance in the long run, as this makes it more difficult for external agents to lead to the deterioration of the material, which affects the steel frame, in the case of reinforced elements. The opposite occurs with the LFS1 mixtures; it has a higher porosity, which can lead to the material having a shorter shelf life thanks to external agents that can damage it.

4.3. Compressive Strength

Results of mixtures for slag substitution.

4.3.1. Slag GGBF

Figure 3 shows the compressive strength over time and how the GGBF slag influences concrete properties, from 1 day to 90 days of testing. These data are mean values for six samples per test (for the different concrete age). Results show that the compressive strength in the first days of hardening was lower than in conventional mixtures, becoming equal after 7 days and even increasing after

30 days. This is coincident with our previous results [16] and in accordance with the results reported by Wang [28,29].

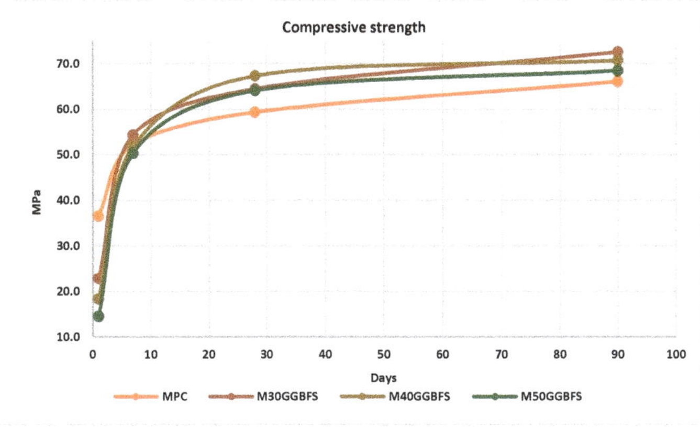

Figure 3. Compressive strength GGBFS.

Figure 4 shows the percentages of loss or increase in strength at day one and after 90 days. It is easy to visualize how, at one day there was up to 50% less strength than the conventional mixture. Nevertheless, after 90 days, it acquired up to 10% more compressive strength in the 30% replacement.

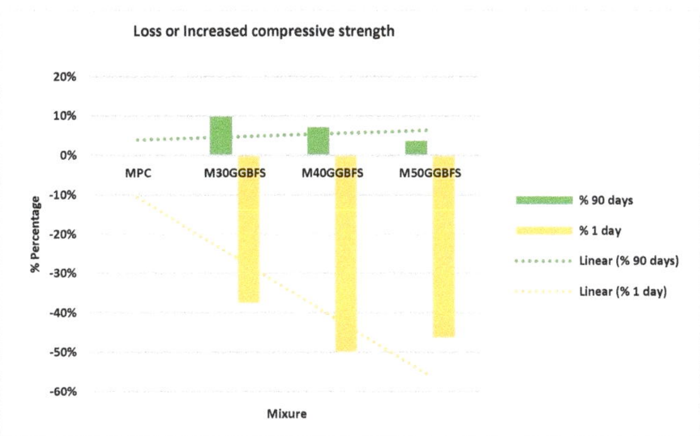

Figure 4. Percentage of loss or increase in compressive strength GGBFS.

4.3.2. Slag LFS1

Figure 5 shows the mean values obtained for the samples made with slag LFS1. This average has been made with six test specimens for each of the four mixtures.

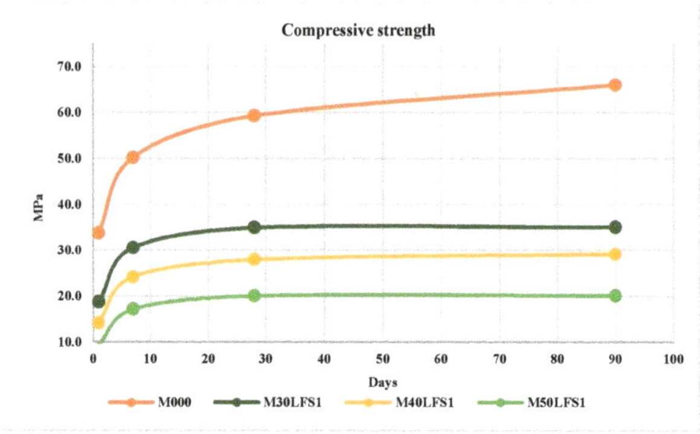

Figure 5. Compressive strength LFS1.

The results show that in the LFS1 slag there was a lesser strength for day one, but in this case the compressive strength decreased over time. The loss at 90 days with a 50% substitution of cement was of almost 70% of PC's strength, as shown in Figure 6. The loss of strength is maintained at both day one and after 90 days.

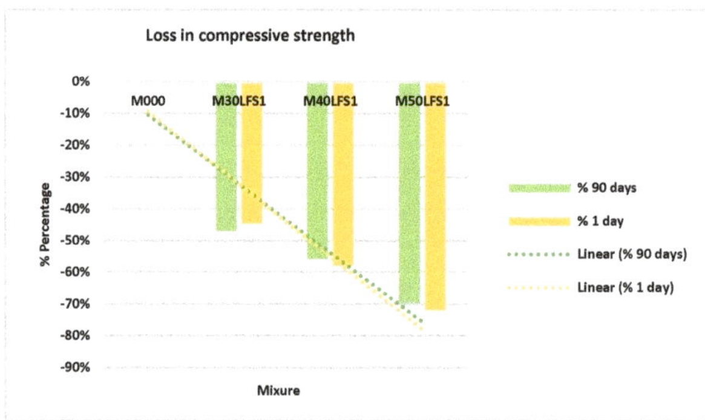

Figure 6. Percentage of loss in compressive strength LFS1.

4.3.3. Slag LFS2

Figure 7 shows the mean values for the specimens made of slag LFS2. Again, six specimens per mixture and age were tested.

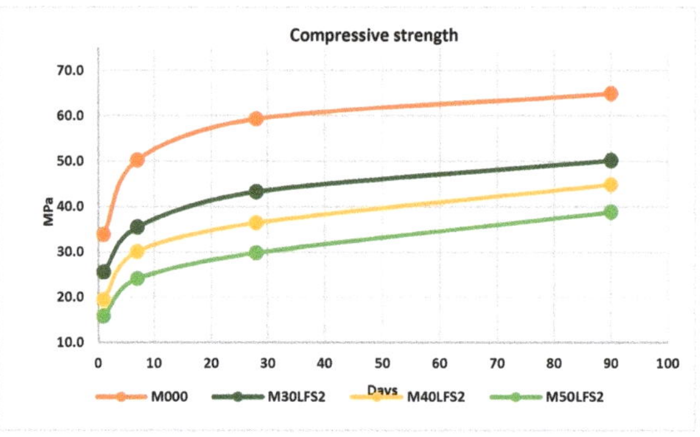

Figure 7. Compressive strength LFS2.

With the slag LFS2, it was observed, as in the mixtures with slag LFS1, that there was a decrease of the strength both on day one, and over time. At 90 days and with a 50% replacement of cement, the loss was 40% (Figure 8).

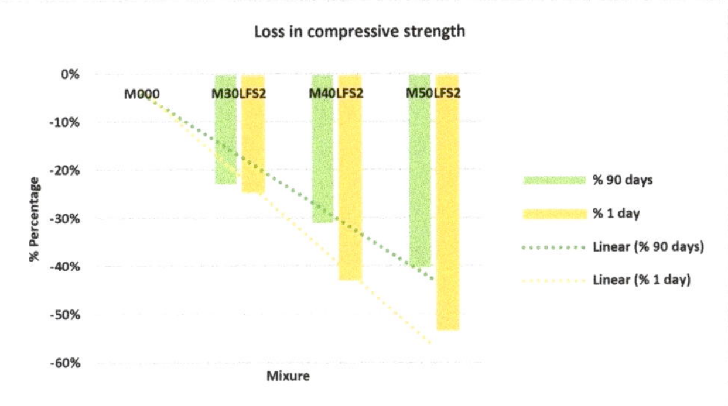

Figure 8. Percentage of loss in compressive strength LFS2.

The results lead us to think that, for high percentage substitutions, specimens with ladle furnace slag (LFS) substitutions have a higher strength loss compared with conventional concrete than those with blast furnace slag (GGBFS) substitutions. Particularly, the mixtures with LFS1 substitutions showed the worst behaviour at compressive strength tests, obtaining up to a 70% of compressive strength loss, when the cement substitution percentage was 50%.

4.3.4. Comparison between the Mixtures

In Figure 9, the comparison of compressive strength for 90 day is shown for the different mixtures. In order to complete the figure, results previously reported with a 25% cement replacement [16] have also been included in the Figure.

It is observed how GGBFS slag presents an increase in strength; it provides pozzolanic benefits to the mixture. On the contrary, what occurs in those made with LFS slags has already been observed by other researchers, like Manso [30,31]. In this figure, it can be seen that the proportion of loss of strength is not the same in the two cases being the substitution up to 30% of LFS2 admissible for concretes with minor strength needs.

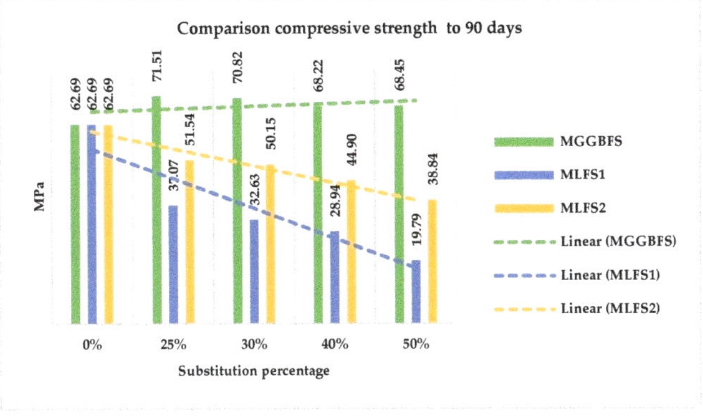

Figure 9. Comparison compressive strength to 90 days, for the three different slag.

We can conclude that the blast furnace slag (GGBFS) is a good substitute for cement in terms of compressive strength. On the other hand, there is such a loss of strength on the other two ladle furnace slags (in the best of cases, 23%), that rule out any possibility of using this concrete as structural material. Nevertheless, they could be acceptable, being able to withstand medium-environmental pressure for situations in which the strength needs are lesser.

Returning to the ternary diagram in Figure 1, it is clear that a higher amount of SiO_2, means better pozzolanic characteristics in the mixture, and in this case, it was observed that the mixtures with slag LFS1 were those that obtained the worst compressive strength, a cause not only of the greater porosity, and therefore greater amount of voids that weaken the mixture, but the slag also contained less SiO_2.

4.4. Flexure Strength Tests

A similar study has been made to evaluate the flexure strength. In this case, the results obtained for all the mixtures are shown in Figures 10 and 11, breaking the test pieces at 28 and 90 days, respectively.

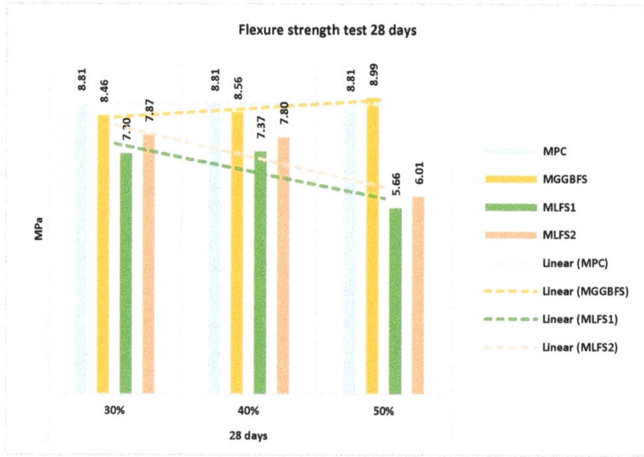

Figure 10. Flexure strength test 28 Days.

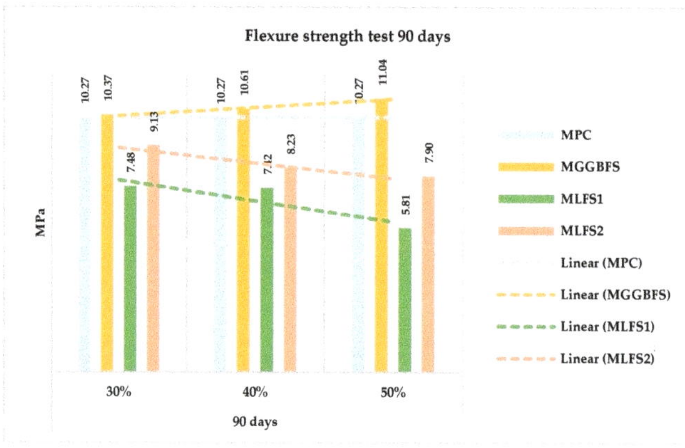

Figure 11. Flexure strength test at 90 days.

Following a similar trend to the previous section, the mixtures made with ladle furnace slag (LFS) obtained worse strength than the mixture without substitution. On the contrary, those made with blast furnace slag (GGBFS) were equal to and even improved the reference concrete.

This loss or increase in strength was observed not only at 28 days, but also at 90 days (Figure 10). It also shows that mixtures with substitutions LFS1 behave worse than LFS2; additionally, showing clear differences between them. This behaviour is due to the chemical composition of the slag.

The percentages of loss of flexure strength of each one of the cases are shown in Figure 12, where it is seen how the GGBFS contributes to the material, the same characteristics as the PC, increasing its strength by 4% with a substitution of 50%.

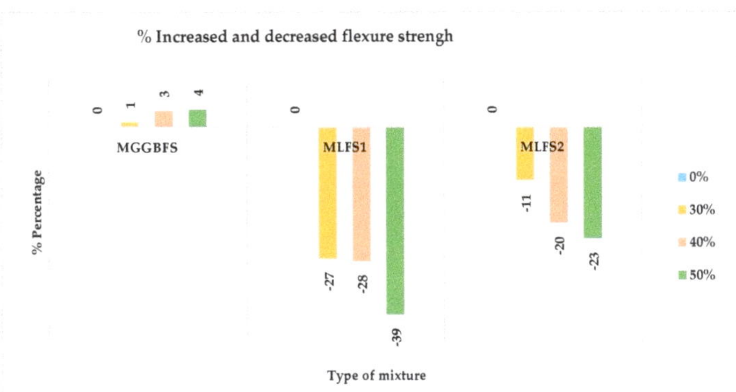

Figure 12. Percentage increased and decreased flexure strength.

4.5. Comparison between 90-Day Flexure and Compressive Strength

The significance of this study is highlighted in the comparison of the gain or loss of the strength against flexure and compression combined. We will focus on that point in this section.

In Figure 13, the comparison in percentage of the strength, both compressive and flexural, in each of the mixtures made, highlights that the losses of flexural strength of the mixtures with slag LFS are much lower than those of compressive strength. This decrease in strength is practically half in most cases. In the strength of the mixtures with GGBFS does not increase twice as much in the flexural strength, as would be expected by the previous results, but only half. Apart from substitutions of 50%, this percentage of 4% is maintained.

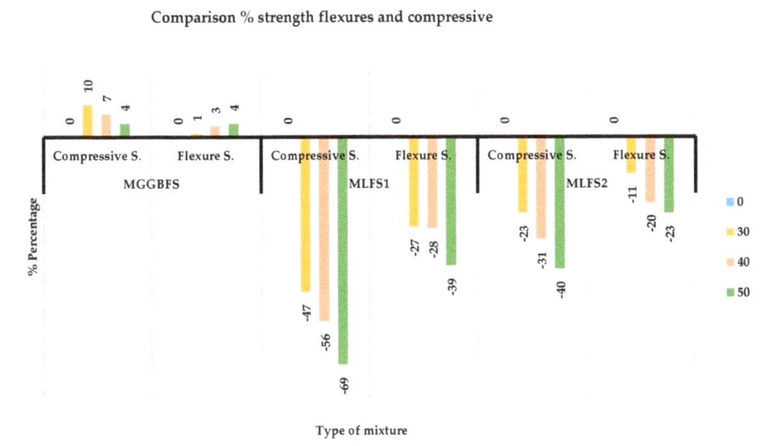

Figure 13. Comparison of flexural and compressive strength.

This indicates the feasibility of LFS slag in non-structural elements, valuing a waste as a by-product, reducing the production of cement that generates a large amount of CO_2 into the environment.

In all the investigations where the laws of mechanical behaviour (constitutive models) for materials are established, outcomes are considered a representative volumetric element of the same. It is assumed that the material behaves as a continuous medium; that is, it has the same elastic properties at each point.

4.6. Leachate

The results obtained are shown in Table 5.

Table 5. Leachate of the mixtures.

Chemical Element	PC	GGBFS (30%)	GGBFS (50%)	LFS1 (30%)	LFS2 (30%)	CFR 40/261.24 (mg/L)
[Mg] µg/L	20.2 ± 1.0	21.3 ± 0.2	28.1 ± 1.0	26.0 ± 1.2	31.3 ± 1.0	-
[Si] µg/L	8.60 ± 0.33	13.1 ± 1.0	13.3 ± 1	9.30 ± 0.40	7.90 ± 0.73	-
[Ti] µg/L	<0.100	<0.100	<0.100	<.0,100	<0.100	-
[Cr$_{total}$] µg/L	15.6 ± 0.6	1.10 ± 0.05	0.394 ± 0.013	0.673 ± 0.040	0.162 ± 0.023	5
[Mn] µg/L	0.100 ± 0.010	0.180 ± 0.010	0.190 ± 0.04	0.150 ± 0.020	0.101 ± 0.01	-
[Fe] µg/L	18.2 ± 0.7	17.0 ± 0.2	10.6 ± 0.03	1.82 ± 0.03	8.23 ± 0.20	-
[Ni] µg/L	0.270 ± 0.030	0.510 ± 0.020	0.280 ± 0.012	0.150 ± 0.013	0.270 ± 0.021	-
[Cu] µg/L	1.92 ± 0.10	4.3 ± 0.36	8.80 ± 0.033	1.50 ± 0.03	4.00 ± 0.04	-
[Zn] µg/L	6.60 ± 0.20	6.10 ± 0.12	3.90 ± 0.20	2.92 ± 0.10	4.40 ± 0.14	-
[As] µg/L	<0.200	<0.200	<0.200	<0.200	<0.200	5
[Cd] µg/L	<0.100	<0.100	<0.100	<0.100	<0.100	1
[Sn] µg/L	<0.100	<0.100	<0.100	<0.100	<0.100	-
[Pb] µg/L	0.543 ± 0.022	2.02 ± 0.02	0.230 ± 0.010	0.190 ± 0.001	0.333 ± 0.002	-

In general terms, it can be observed how most of the values obtained had a decrease in relation to conventional concrete. The most significant is to see how the amount of chromium in the mixtures with slag, compared with the conventional concrete mixture, decreased.

It is significant how one of the most harmful elements in the leachate is chromium, and this element decreases with respect to the master mix. This is interpreted as the encapsulation of the slag being diluted into the cementitious matrix absorbing this metal, without generating any environmental danger when it is used.

There are some values which are slightly above those obtained with the reference concrete (PC); however, all these leachate ranges fall within the values allowed by the Code of Federal Regulations (CFR) 40CFR/261.24.

According to those maximum values the encapsulation of the slag in the concrete not only does not leach contaminant, but also reduces the leachate of one of the more dangerous components that are measured in the CFR—the Cr, not exceeding the limit of 5 mg/L.

5. Conclusions

According to the results described, we can outline the following conclusions:

→ The results lead us to think that, for high percentage substitutions, specimens with ladle furnace slag (LFS) substitutions have a higher strength loss compared with conventional concrete than those with blast furnace slag (GGBFS) substitutions. Particularly, the mixture with LFS1 slag substitutions showed the worst behaviour in compressive strength tests, obtaining up to a 72% of compressive strength loss with a cement substitution percentage of 50%.

→ On the other hand, specimens with blast furnace slag substitution showed an increase in compressive strength of 10% at 90 days. Mixtures using this type of slag substitution showed a slower hardening process, with a compressive strength reduction at day one, but gaining a compressive strength similar to or even above the conventional concrete after 7 days.

→ The chemical characteristics of the slag influence the mixtures and strength. It was observed in this study, that for essential components such as SiO_2, the lower the percentage, the lower the strength. As it can be seen in the mixes made with LFS1 and LFS2 slags, the lack of that compound makes them work worse. GGBFSs are the best performers, having twice the amount of that compound, increasing its strength even with respect to conventional concrete. The good pozzolanic activity that contributes to those types of cementitious mixtures was verified.

→ Another characteristic result of this investigation is the difference of compressive and flexural strength among the different mixtures. LFS presents a loss of flexural strength that is the half of the loss of compressive strength. This suggests that they could be used in other fields of civil engineering with lower strength requirements.

→ Leaching test confirmed that slag does not cause damage to the environment. The results of the leaching tests of concrete mixtures with slag are similar to the results of traditional concrete. Therefore, slag encapsulation into the concrete seems to be a good strategy to manage this waste product, instead of it being deposited in landfills where it will pollute the environment by leaching.

As a final conclusion, it is clear that blast furnace slag (GGBFS) is suitable for the production of a sustainable concrete, and as a substitute for cement, since it has been proven to bring the same characteristics to the mixture as cement.

For the other two types of mixtures with slag (LFS), a non-structural application would be suitable. This would put a value on the residue, avoiding the consumption of raw material and reducing the landfill deposit.

Concrete production with slag is a clear example of circular economics, since the steel manufactured, necessary to build structures, generates waste that can be incorporated to the same building cycle.

Author Contributions: M.E.P.-R., F.P.-G., A.G.-H., M.J.O. and M.D.R.-C. conceived and designed the experiments; M.E.P.-R., M.D.R.-C. and F.P.-G. performed the experiments and analysed the data; and M.E.P.-R. and A.G.-H. wrote the paper.

Funding: The authors acknowledge the financial support provided to this work by the Center of Industrial Technological Development (CDTI) of the Ministry of Economy and Competitiveness as a part of the research project IDI-20160509, through the companies DRACE and GEOCISA.

Conflicts of Interest: The authors declare no conflict of interest.

References

1. Manso, J.M.; Polanco, J.A.; Losañez, M.; González, J.J. Durability of concrete made with EAF slag as aggregate. *Cem. Concr. Compos.* **2006**, *28*, 528–534. [CrossRef]
2. González, J.S.; Gayarre, F.L.; Pérez, C.L.-C.; Ros, P.S.; López, M.A.S. Influence of recycled brick aggregates on properties of structural concrete for manufacturing precast prestressed beams. *Constr. Build. Mater.* **2017**, *149*, 507–514. [CrossRef]
3. New Views on Effect of Recycled Aggregates on Concrete Compressive Strength. *ACI Mater. J.* **2013**, *110*, 687–696.
4. Tsakiridis, P.; Papadimitriou, G.; Tsivilis, S.; Koroneos, C.; Tsakiridis, P. Utilization of steel slag for Portland cement clinker production. *J. Hazard. Mater.* **2008**, *152*, 805–811. [CrossRef] [PubMed]
5. Çelik, E.; Nalbantoglu, Z. Effects of ground granulated blastfurnace slag (GGBS) on the swelling properties of lime-stabilized sulfate-bearing soils. *Eng. Geol.* **2013**, *163*, 20–25. [CrossRef]
6. Shi, C.; Hu, S. Cementitious properties of ladle slag fines under autoclave curing conditions. *Cem. Concr. Res.* **2003**, *33*, 1851–1856. [CrossRef]
7. Pellegrino, C.; Cavagnis, P.; Faleschini, F.; Brunelli, K. Properties of concretes with Black/Oxidizing Electric Arc Furnace slag aggregate. *Cem. Concr. Compos.* **2013**, *37*, 232–240. [CrossRef]

8. Vijayaraghavan, J.; Jude, A.B.; Thivya, J. Effect of copper slag, iron slag and recycled concrete aggregate on the mechanical properties of concrete. *Resour. Policy* **2017**, *53*, 219–225. [CrossRef]

9. Adegoloye, G.; Beaucour, A.-L.; Ortola, S.; Noumowe, A. Concretes made of EAF slag and AOD slag aggregates from stainless steel process: Mechanical properties and durability. *Constr. Build. Mater.* **2015**, *76*, 313–321. [CrossRef]

10. Al-Jabri, K.S.; Hisada, M.; Al-Saidy, A.H.; Al-Oraimi, S. Performance of high strength concrete made with copper slag as a fine aggregate. *Constr. Build. Mater.* **2009**, *23*, 2132–2140. [CrossRef]

11. Al-Jabri, K.; Taha, R.; Al-Hashmi, A.; Al-Harthy, A. Effect of copper slag and cement by-pass dust addition on mechanical properties of concrete. *Constr. Build. Mater.* **2006**, *20*, 322–331. [CrossRef]

12. Rubio-Cintas, M.; Barnett, S.; Perez-García, F.; Parron-Rubio, M. Mechanical-strength characteristics of concrete made with stainless steel industry wastes as binders. *Constr. Build. Mater.* **2019**, *204*, 675–683. [CrossRef]

13. López Boadella, Í.; López Gayarre, F.; Suárez González, J.; Gómez-Soberón, J.; López-Colina Pérez, C.; Serrano López, M.; de Brito, J.; López Boadella, Í.; López Gayarre, F.; Suárez González, J.; et al. The Influence of Granite Cutting Waste on The Properties of Ultra-High Performance Concrete. *Materials* **2019**, *12*, 634. [CrossRef] [PubMed]

14. Juan-Valdés, A.; García-González, J.; Rodríguez-Robles, D.; Guerra-Romero, M.; López Gayarre, F.; De Belie, N.; Morán-del Pozo, J.; Juan-Valdés, A.; García-González, J.; Rodríguez-Robles, D.; et al. Paving with Precast Concrete Made with Recycled Mixed Ceramic Aggregates: A Viable Technical Option for the Valorization of Construction and Demolition Wastes (CDW). *Materials* **2018**, *12*, 24. [CrossRef] [PubMed]

15. Panda, B.; Paul, S.C.; Hui, L.J.; Tay, Y.W.D.; Tan, M.J. Additive manufacturing of geopolymer for sustainable built environment. *J. Clean. Prod.* **2017**, *167*, 281–288. [CrossRef]

16. Amario, M.; Pepe, M.; Martinelli, E.; Filho, R.D.T. Rheological Behavior at Fresh State of Structural Recycled Aggregate Concrete. In Proceedings of the fib symposium 2017, Maastricht, The Netherlands, 12–14 June 2017; pp. 215–223.

17. Öner, A.; Akyuz, S. An experimental study on optimum usage of GGBS for the compressive strength of concrete. *Cem. Concr. Compos.* **2007**, *29*, 505–514. [CrossRef]

18. Wang, H.; Sun, X.; Wang, J.; Monteiro, P.J. Permeability of Concrete with Recycled Concrete Aggregate and Pozzolanic Materials under Stress. *Materials* **2016**, *9*, 252. [CrossRef]

19. Pal, S.; Mukherjee, A.; Pathak, S. Investigation of hydraulic activity of ground granulated blast furnace slag in concrete. *Cem. Concr. Res.* **2003**, *33*, 1481–1486. [CrossRef]

20. Perez-Garcia, F.; Parron-Rubio, M.E.; Garcia-Manrique, J.M.; Rubio-Cintas, M.D. Study of the Suitability of Different Types of Slag and Its Influence on the Quality of Green Grouts Obtained by Partial Replacement of Cement. *Materials* **2019**, *12*, 1166. [CrossRef]

21. Khatib, J.; Hibbert, J.; Khatib, J. Selected engineering properties of concrete incorporating slag and metakaolin. *Constr. Build. Mater.* **2005**, *19*, 460–472. [CrossRef]

22. Parron-Rubio, M.E.; Pérez-García, F.; Gonzalez-Herrera, A.; Rubio-Cintas, M.D. Concrete Properties Comparison When Substituting a 25% Cement with Slag from Different Provenances. *Materials* **2018**, *11*, 1029. [CrossRef] [PubMed]

23. *Testing Hardened Concrete—Part 2: Making and Curing Specimens for Strength Tests*; EN 12390-2: 2009; European Committee for Standardization: Brussels, Belgium, 2009.

24. *Testing Hardened Concrete. Density of Hardened Concrete*; EN 12390-7:2009; European Committee for Standardization: Brussels, Belgium, 2009.

25. *Testing Hardened Concrete. Part 3: Compressive Strength of Test Specimens*; EN 12390-3: 2001; European Committee for Standardization: Brussels, Belgium, 2001.

26. *Testing of Hardened Concrete. Part 4: Compressive Strength Specification for Testing Machines*; EN 12390-4: 2009; European Committee for Standardization: Brussels, Belgium, 2009.

27. *Testing Hardened Concrete. Part 5: Flexural Strength of Test Specimens*; EN 12390-5:2009; European Committee for Standardization: Brussels, Belgium, 2009.

28. Wang, Q.; Yan, P.; Mi, G. Effect of blended steel slag–GBFS mineral admixture on hydration and strength of cement. *Constr. Build. Mater.* **2012**, *35*, 8–14. [CrossRef]

29. Wang, Q.; Yang, J.; Yan, P. Cementitious properties of super-fine steel slag. *Powder Technol.* **2013**, *245*, 35–39. [CrossRef]

30. Rodriguez, Á.; Manso, J.M.; Aragón, Á.; Gonzalez, J.J. Strength and workability of masonry mortars manufactured with ladle furnace slag. *Resour. Conserv. Recycl.* **2009**, *53*, 645–651.
31. Manso, J.M.; Ortega-López, V.; Polanco, J.A.; Setién, J. The use of ladle furnace slag in soil stabilization. *Constr. Build. Mater.* **2013**, *40*, 126–134. [CrossRef]

Article

Estimation of Recycled Concrete Aggregate's Water Permeability Coefficient as Earth Construction Material with the Application of an Analytical Method

Wojciech Sas, Justyna Dzięcioł * and Andrzej Głuchowski

Water Centre Laboratory, Faculty of Civil and Environmental Engineering, Warsaw University of Life Sciences, 02-787 Warsaw, Poland
* Correspondence: justyna_dzieciol@sggw.pl

Received: 15 July 2019; Accepted: 3 September 2019; Published: 10 September 2019

Abstract: Creating models based on empirical data and their statistical measurements have been used for a long time in the economic sciences. Increasingly, these methods are used in the technical sciences, such as construction and geotechnical engineering. This allows for reducing the costs of geotechnical research at the design stage. This article presents the research carried out on Recycled Concrete Aggregate (RCA) material with is reclaimed crushed concrete rubble. Permeability tests were carried out using the constant head method. Tests were conducted on blends of RCA with the following particle size ranges: 0.02–16 mm, 0.05–16 mm, 0.1–16 mm, and 0.2–16 mm. The gradients used during the tests were between 0.2 to 0.83, which corresponds to gradients encountered in earth construction and are below the critical gradient. Directly from the tests, the flux velocity for the range of tested gradients were calculated based on filtered water volume measurements. The values of the permeability coefficient (k) were then recalculated. Finally, statistical methods were used to determine which physical parameters of the tested material affect the permeability coefficient. The physical parameters selected from the statistical analysis were used to create a model describing the phenomenon. The model can be used to determine the permeability coefficient for a mixed RCA material. The article ends with conclusions and proposals concerning the use of models and the limits of their applicability.

Keywords: statistical analysis; estimation; permeability; constant head method; estimation coefficient of permeability; recycled concrete aggregate

---

1. Introduction

The construction industry was one of the most strongly affected industries during the 2008 global financial crisis in Europe. However, local construction markets have been steadily strengthening since then, which is visible mainly in the countries of Western and Northern Europe. It is forecasted that the total growth of Northwestern and Southern Europe markets will average 2.5% per annum for 2018–2022 [1]. For the construction industry in Central and Eastern Europe, which after the economic crisis is trying to catch up with the rest of Europe, the expected higher market growth is estimated at an average of 4.4% for 2018–2022 [1,2].

Despite these optimistic forecasts, the industry faces numerous problems. The most important are the falling profitability of investments and the increase in employee remuneration costs. This is the reason why it is so important to look for solutions that help reduce the cost of construction projects. One of the options is to reduce material costs by using recycled materials [2,3].

Aggregates are materials commonly used in civil engineering. Natural aggregates are mainly used as materials in earth constructions, such as dams, embankments, or other earth structures, e.g., road bases; therefore, the demand for this material is very high [4]. Natural aggregates represent approximately 88% of the demand in the market [4]. Problems with the sustainable development of the natural aggregates market, and with waste management as well, forces engineers to use anthropogenic aggregates in earth structures. The mechanical and physical characteristics of recycled aggregates vary from natural aggregates and need to be further investigated. Better knowledge of their response to different kind of loadings or the permeability characteristics will allow for wider application of this kind of material by designers and engineers [5].

2. Literature Review

The proper assessment of soil properties and their parameters is very important in geotechnical modeling. A well-chosen model allows one to save time and reduce the costs of construction investment [6]. However, for the model to provide reliable results, several conditions must be met. The first is to have a large database to assess and recognize the case [7]. Another important requirement is to determine the appropriate physical parameters that should be included in the model. It is also important to determine the nature of the function distribution for empirical research. The last stage, after creating the model, is to determine the parameters of its application. This is due to every model being created on the basis of empirical research that has the character of partial research [7,8].

The central pillar of statistical inference is the analysis of source data from empirical research, based on theorems of probability theory. Properly carried out inference leads to the creation of a model that maps the distribution of empirical data [9,10]. Generalization of the results of statistical observation to the whole phenomenon is carried out using statistical estimation or verification of the hypothesis. The model is created after estimation of the study population parameters' distributions, based on the observed results. By analyzing the interrelationships between different material properties, the hypotheses and the correctness of the initial assumptions are verified [10].

Proper planning of laboratory tests is the key to obtaining a reliable model. The selection of the research sample, followed by statistical analysis, should be based on the knowledge of the phenomenon and the previous research reports [11,12].

The calculation of the probability assumes randomness of the analyzed sample. Often, especially in the scientific research process, randomness is disturbed by study restrictions or is even impossible to achieve due to the limited ability to test the material properties. This is due to measurement errors during the test, e.g., destruction of the test material. The above-mentioned limitations are strictly connected with statistical sample construction and can be considered directed or expert. It leads to the need for representative sample estimation for a given study or phenomenon [13].

The credibility of the model is important from the points of view of the entrepreneur and the researcher. It limits cost and time expenditure. It is also important to evaluate the parameters for the possible substitution of laboratory tests on the analytical method of estimation. This article presents a way to replace laboratory tests of the permeability coefficient by estimating with the use of statistical tools [14–16].

Reclaimed concrete is widely used in earthworks, especially in road substructures. This allows for its recycling and the reduction of the number of landfills and concrete debris, which can be reused after mechanical crushing. It also reduces the need for natural aggregates in earth construction. Crushed concrete is classified as a material suitable for auxiliary foundations, basic foundations, and cut-off layers [17,18]. A special form of this kind of material application is used as a filtration layer for the construction of levees and dams.

Recycled concrete aggregates (RCAs) are obtained as a result of the crushing process, excluding brick and soft materials, which are then used to obtain residual concrete with a grain size of 0 to 63 mm [19]. RCAs can be used as aggregate for earth construction. They are mainly used in road engineering, in which their geotechnical parameters (strength and deformation susceptibility) have already been recognized [19,20].

The determination of the angle of internal friction (φ) for RCAs with gradation sandy gravel (saGr) has already been studied by Sas et al. [21] and Soból et al. [19], who confirmed the results of the test presented by O'Mahony [22]. This is important that RCAs are characterized by a complicated structure and can give the effect of cohesion in non-adhesive soil, which improves the mechanical properties of RCAs [23]. When utilizing RCAs in road engineering constructions and in designing dams and shafts, the permeability coefficient is of particular importance. In the case of natural aggregates, in order to reduce investment costs via replacement of laboratory methods, predictions regarding the hydraulic conductivity based on porosity or grain size distribution are calculated [24]. This is possible because natural aggregates are characterized by the presence of rounded grains, composed mainly of quartz origin with a low roughness. RCA aggregates are rougher and have an irregular shape, which has a significant impact on the water flow and the surface area [25].

The significant difference when comparing RCAs with natural aggregates is the remains of hydrated cement on the surface of aggregate grains. As a result, this property leads to a lower specific density of the grains, differentiation in aggregate quality, and a higher absorption ability [26,27].

The porosity of both types of materials are different. In the case of natural aggregates, it is about 3%, and for RCA, it is about 15% [28]. On the surface of RCA aggregates, there is a residue of cement mortar, which affects the ability of the aggregate to absorb water [29,30]. During one day of carrying, the water absorption by RCA increased by 2.56% [30], which is important for the permeability parameters of this material.

Recycled aggregates used in earth constructions are usually cheaper than the natural aggregates. Re-use of construction waste is an environmentally friendly solution. The growing popularity of this material contributes to a better understanding of the properties of this material. Understanding its limitations and recommendations regarding its use already at the stage of the design, then improves the construction works [31].

3. Materials and Methods

3.1. Material

The material used for the tests came from the demolition site of a building and was crushed using an impact crusher. The class of concrete strength was evaluated based on laboratory tests. The results of the tests showed that the tested RCA was concrete with a strength of class from C16/20 to C30/35.

The aggregates were 99% composed from broken cement concrete, the rest being of the bulk mass, e.g., glass and brick ($\Sigma(Rb, Rg, X) \leq 1\%$ m/m), in accordance with the standard EN 933-11:2009 [32–34], and contained no asphalt or tar elements.

Recycling concrete aggregate is regarded as an environmental safety material to use in road pavement and road construction. Maia et al. [35] studied many articles about the chemical analysis of RCA from the last few years. After a review on leaching tests, the obtained information has shown that critical compounds of RCAs are chromium, sulfate, antimony, and selenium. The concentration of these elements should be periodically monitored but the threat of their elevated concentration is very low due to the fact that these elements exist naturally in soil. Leaching of these elements from RCA and concentration in the soil are rare and usually does not exceed existing standards of acceptability. Rodrigues P. et al. [36] more in our research take were notice of supervening and concentration heavy metals and anion in RCA. She suggested that materials should be tested for heavy metals and anions before re-use. Using the laboratory leaching method, one can estimate the actual amount of elements that are eluted into the soil. This allows one to evaluate the ecotoxicity of the material before using it [36–39]. The norm for other elements, such as cadmium (Cd), mercury (Hg), and lead (Pb), are not exceeded in RCA material [36,40].

The pH value is of great importance for the intensity of element leaching. The higher the acidity, the greater the leaching [41]. This is important information from the point of view of practical application.

The material was fractioned to the appropriate fractions according to Galvín et al. [39], and then the RCAs were divided into four blends: 0.02–16 mm, 0.05–16 mm, 0.1–16 mm, and 0.2–16 mm. Each blend was composed of obtained fractions. The grain gradation curve was adopted with respect to the Polish technical standard and was placed between the upper and lower grain grading limits. The resulting mixtures are suitable for earth structures such as dams and embankments [33].

A series of tests were carried out on the obtained blends to determine their physical properties. According to Eurocode, the tested material was classified on the basis of sieve analysis as sandy gravel (saGr) [34]. Figure 1 presents the graining curves for the tested blends, where the particle size distribution from 0 mm to 16 mm is in the standard range for aggregates used as an auxiliary base and improved substrate in road engineering and in earth structures [33]. The coefficient of curvature (C_C) and coefficient of uniformity (C_U) were calculated in order to classify the shape of the grading curve of tested blends. The C_U value was in the range from 15.0 to 17.33 and C_C was in the range from 0.42 to 0.64. The soil was therefore classified as medium-graded according to the Eurocode [34]. For the four blends, the Proctor tests were performed in order to establish the optimum moisture content of the RCAs. The tests were conducted with the energy density of the compaction equal to 0.59 J/cm^3 and the results are presented in Figure 2.

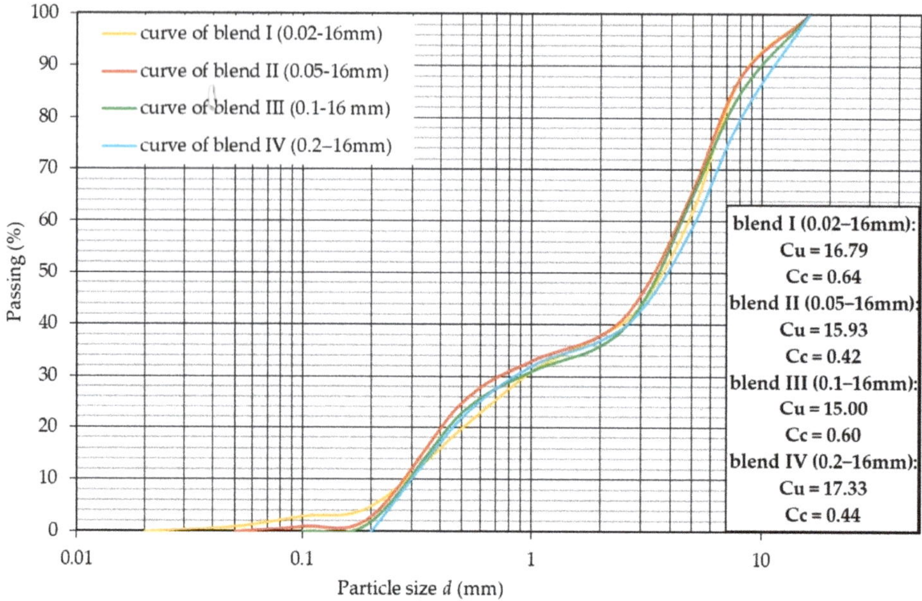

Figure 1. Grain size distribution curve of each tested RCA blends.

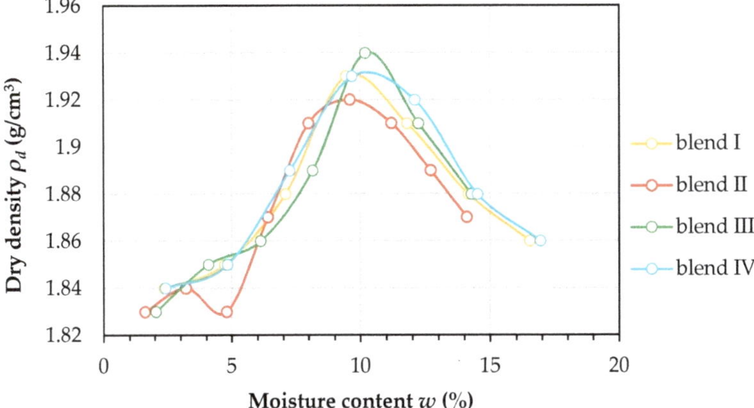

Figure 2. Results of the Proctor tests for RCA blends with an energy density of compaction Ec equal to 0.59 J/cm^3.

Results of the Proctor tests indicated that the RCA blends had an optimum moisture contents between 9.4% and 10.2% (9.44%, 9.6%, 10.2%, and 9.6%, respectively, for blends I, II, III, and IV). The maximum dry density ρ_d max was between 1.92 and 1.94 g/cm^3. The RCAs were compacted in the optimum moisture content in a permeability mold to achieve the maximum dry density. These kinds of conditions represent the state in which the subbase soils would be in. The quality of soil samples in permeability tests was measured with the use of the compaction index *IS*, which is the quotient of the soil dry density in the permeability mold $\rho_{d,m}$ to the maximum dry density for tested blend $\rho_{d,max}$ ($IS = \rho_{d,m}/\rho_{d,max}$). The RCAs sample quality was qualified based on this quotient. In this study, the samples had an IS parameter in the range of 0.98 to 1.01. Samples below an *IS* equal to 0.98 were excluded from this study.

3.2. Permeability Test

In poor and very poor permeability soils, filtration can only be initiated after a certain hydraulic gradient (i_0) has occurred. This means that the graph $v = f(i)$ does not come out at the beginning of the hydraulic gradient axis, and even in the initial period, shows curvature caused by a gradual increase in permeability. This filtration is called pre-linear filtration (Figure 3). In order to initiate the movement of the pore water, the threshold stress (τ_0) must be overcome by the tangential stress (τ) [42–44].

In this study, the constant head method test was used to estimate the RCAs' permeability characteristics. The method is characterized by simplicity and unchanging test conditions, and the constant head method alone is one of the most reliable techniques for measuring permeability in non-cohesive soil [45].

The permeameter construction (Figure 4) consists of internal and external cylinders made of stainless steel. Dimensions of the inner cylinder were: height h = 0.17 m, diameter d = 0.205 m, and the dimensions of the outer cylinder were h = 0.27 m and d = 0.19 m. Cylinders were connected by a permeability cylinder with a perforated bottom, also made of stainless steel, where the sample was placed. Then, after installing the sample, a perforated cover was placed on top. The permeability cylinder was attached to the inner cylinder by means of four screws and a rubber ring to ensure that there was no unexpected water leakage from the external cylinder. The principle of operation of this device is based on communication vessels that allow the flow of water from the external cylinder to the inner cylinder through the soil sample. The hydraulic gradient was simply determined by the difference between the outer and inner table heights. In practice, the internal water table is stationary and the hydraulic gradient is caused by the variable height of the external water table. The tests were carried out when both the internal and external water table were in a fixed position. Measurements of outflow water were repeated five times for each test point [31,45,46].

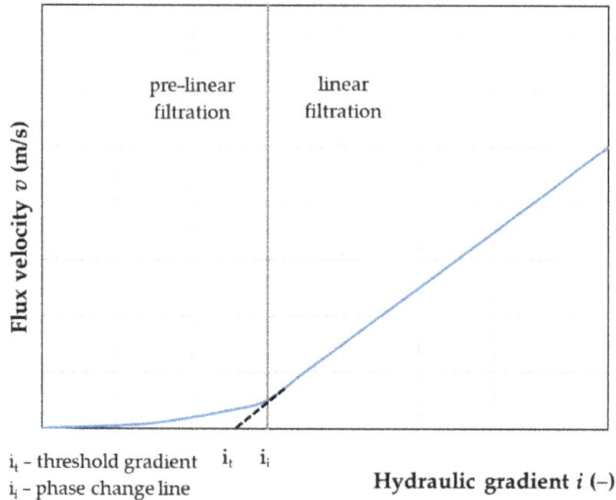

Figure 3. Pre-linear and linear filtration process.

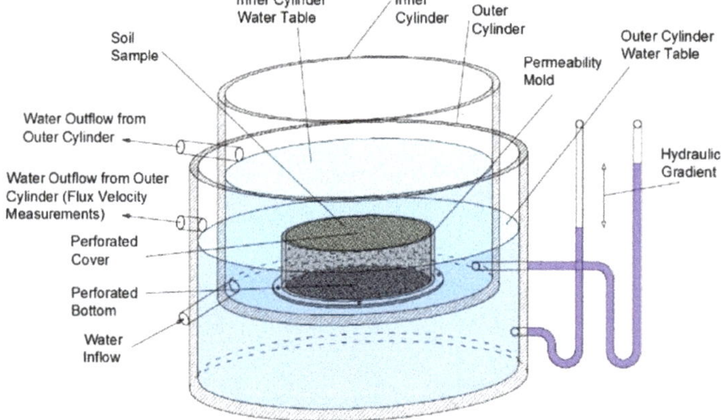

Figure 4. The permeameter scheme.

3.3. Estimation Theory

The theory of estimation concerns the inference of the correct probability distribution of the general population on the basis of independent variables from the tested sample. Using the knowledge about the distribution of classes in the test sample, an inference is made to the general sample. Parametric estimation occurs when the elements of the class of possible distributions of the general population differ only in the values of parameters. Non-parametric estimation is used in more complex cases when the differences in elements of the class of possible distributions of the general population concern not only the values of parameters, but also the form of the distribution function [47,48].

There are two parts of the theory of estimation: point estimation and interval estimation. The point estimation finds the function sampling and its value is taken as the best estimate of the value of the parameter for the overall sample. The interval estimation on the basis of the sample determines a numerical interval, which contains the value of a parameter of the general population, taking into account the assumed probability [48].

An estimator, with the general form given in Equation (1), is a statistic that serves to estimate the value of the distribution parameter with a function in the sample. It can be used to estimate an unknown parameter in a population.

$$\hat{Q}_k = f(X_1, X_2, X_3 \ldots, X_n) \tag{1}$$

In the theory of estimation, there are three main features, which should be met by any good estimator to be useful for the created model. It should be unbiased, consistent, and effective [47,49]. The unbiased estimator is characterized by the realization of a random variable X comprising an N-elemental sample, which meets the condition:

$$E(\hat{Q}_k) = q_k \tag{2}$$

where: $\hat{Q}_k$ is an estimator, and q_k is a parameter for a random variable X.

Estimators meeting this relationship are called unloaded estimators and the load of the estimator $\Delta\hat{Q}_k = E(\hat{Q}_k) - q_k$ is equal to zero.

The estimator is consistent if an unlimited increase in the sample size occurs. The estimation $\hat{Q}_k$ of the parameter q_k strives for a true value with a probability of one.

$$P(\hat{Q}_k \rightarrow q_k) = 1 \text{ for } N \rightarrow \infty \tag{3}$$

The effective estimator shall be that which shows a smaller dispersion of the values obtained from all possible samples. The measure of spread is, therefore, the variance, and the smaller the variance, the better the unloaded estimator [49,50].

4. Test Results

The aim was to create a model that allows for determining the permeability coefficient, k, based on the physical parameters of the tested material. Analytical methods using statistical tools are widely used in scientific research, saving time and money. In this case, Statistica®(version 13, TIBCO Software Palo Alto, CA 94304 USA) was used as a statistical analysis tool. Preparation of the model was preceded by the collection of an appropriate database. The research hypotheses were formulated and preceded by an in-depth analysis of the phenomenon with the reference of parameters to each other. It is good practice to divide the sample into control groups and research groups, which allows for independent verification of the model.

In the case of the studied phenomenon of permeability, the soil and test properties taken into account in the further analysis were grain size, gradient, porosity, specific density, and optimum moisture content. The values of gradients taken in this study were 0.2, 0.3, 0.4, 0.5, 0.58, 0.67, 0.75, and 0.83. The distribution of uniformity for the hydraulic gradients and granulometric compositions of the tested samples should also be emphasized. There were 20 independent tests for each granulometric compositions and for each gradient. The total sample was N = 640.

Initially, a preliminary statistical analysis of the conducted tests was carried out, in which the following descriptive statistics were evaluated for the parameters: mean, standard deviation, and standard error (Table 1).

Table 1. Average test against a fixed reference value.

Variable	Description	Mean	Standard Deviation	Standard Error	Test t
k	Coefficient of permeability	0.000120	0.000078	0.000003	38.958
v	Flux velocity	0.000072	0.000061	0.000002	29.890
ρ	Specific density	1.931437	0.019769	0.000781	2471.586
n	Porosity	0.374610	0.083305	0.003293	113.762
e	Void ratio	0.611371	0.190529	0.007531	81.177
w	Moisture content	0.105425	0.009157	0.000362	291.260
d_5	Particle size when passing 5%	0.230000	0.018723	0.000740	310.774
d_{10}	Particle size when passing 10%	0.287500	0.013001	0.000514	559.457
d_{30}	Particle size when passing 30%	0.837500	0.086200	0.003407	245.793
d_{60}	Particle size when passing 60%	4.675000	0.334739	0.013232	353.318
d_{90}	Particle size when passing 90%	9.350000	0.753915	0.029801	313.747

Then a correlation analysis of the coefficient of permeability in terms of all physical parameters was carried out. Correlation results are presented in Table 2.

Table 2. Correlation table for coefficient of permeability k (m/s).

Variable	Value of Correlation for Coefficient of Permeability k (m/s)
i *	0.504794
ρ	−0.754960
n	−0.112145
e	−0.137592
w	−0.737912
d_5	0.772737
d_{10}	0.569209
d_{30}	−0.140365
d_{60}	0.408963
d_{90}	0.701564

* i—Hydraulic gradient.

The parameters with the highest degree of correlation are marked in Table 2 in yellow. Correlations of material physical properties with the coefficient of permeability were the highest for granulometric coefficients, d_5 and d_{90}, specific density ρ_d, and optimal water content w. All analyzed correlation coefficients were statistically significant with $p < 0.05$.

The next stage was to examine the distribution of the flow velocity in relation to the hydraulic gradient for different blends included in the study. This was a dependence characteristic of the studied phenomenon. After examining the characteristics of the flow velocity distribution in relation to the gradient, it was found that the dynamics of the flow velocity changed with the gradient. For gradients above 0.3, it was linear. With gradients from 0.2 to 0.3, dynamics were slowed down and the changes had a pre-linear character (Figure 5), which means pre-linear filtration. Such a procedure allows for formulating a hypothesis on the possible need to create not only one common model to determine the permeability coefficient at any gradient but to separate equations into two phases of flow velocity. This had a significant impact on the determination of models for the permeability coefficient, and in particular, on the determination of independent variables.

The best convergence effect of the expected value in relation to the tested one was obtained using the non-linear estimation method. A series of tests were carried out based on previously selected parameters (variables) that were best correlated with the permeability coefficient. In the case of the attribute pairs with a high correlation coefficient (above $R^2 = 0.6$), having a similar influence on the permeability coefficient, one of the attributes was omitted. This was to improve the reliability of the regression model and to ensure the stability of the parameter estimation of this model.

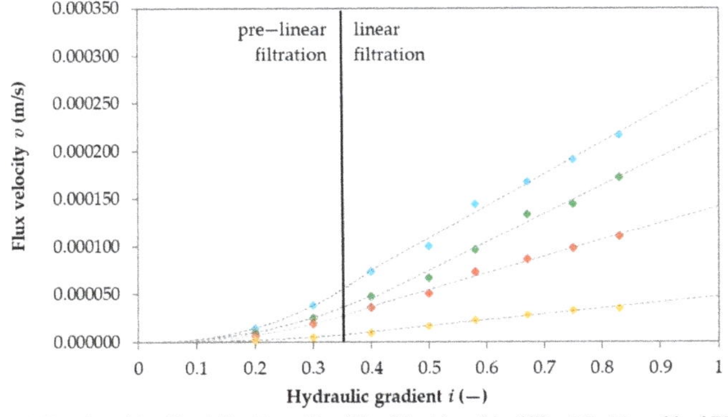

Figure 5. Results of average velocities of flow to a given gradient for different grain sizes.

The highest variance coefficient with $R^2 = 0.615$, was obtained using the following:

$$k = (\rho_d \times (-0.00122)) - ((-0.0022955) \times (d_5 \times d_{90}))^{0.09942} \tag{4}$$

The assessment of fixed parameters for the model is included in Table 3.

Table 3. Evaluation of fixed parameters for model 1.

Independent Variable's Label	Independent Variable's Value	Standard Error
a1	−0.001220	0.000231
b1	−0.002296	0.000464
c1	0.099420	0.029056

To verify the hypothesis of the effect significance for the two phases of flux velocity in relation to the hydraulic gradient, further statistical analysis was performed. An analysis of the normality of the residue distribution (Figure 6a) and the distribution of the observed values in relation to the predicted ones (Figure 6b) were prepared. Both analyses supported the belief that it is reasonable to adjust the model to include both pre-linear and linear phases. In order to maintain the universality of the designated model, it was decided to adjust only the independent variables to it, while retaining the dependent variables used in model 1 (ρ_d, d_5, d_{90}).

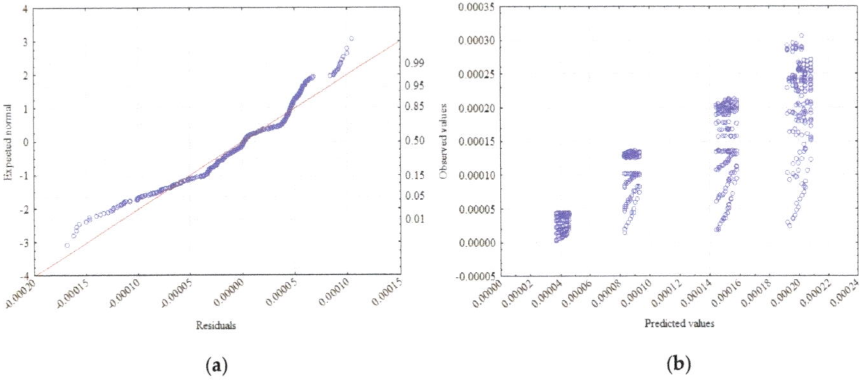

Figure 6. Normality diagram of residuals (**a**) and values observed against predicted (**b**).

For the pre—linear phase (gradients of 0.2–0.3), after the determination of independent variables, the explained variance was obtained at the level of $R^2 = 0.622$. The values of the independent variables, together with the description of their standard errors, are described in Table 4.

Table 4. Evaluation of fixed parameters for model 2.

Independent Variable's Label	Independent Variable's Value	Standard Error
a2	0.001734	0.000319
b2	−0.002594	0.000455
c2	−0.000040	0.000030

The new formula (Equation (5)) is characterized by a better matching of observed values in relation to the calculated values (Figure 7a,b), but also by a better distribution of residual values.

$$k = 0.00173 + \left(\rho_d^{0.0259}\right) - \left((d_5 \times d_{90})^{0.00004}\right) \tag{5}$$

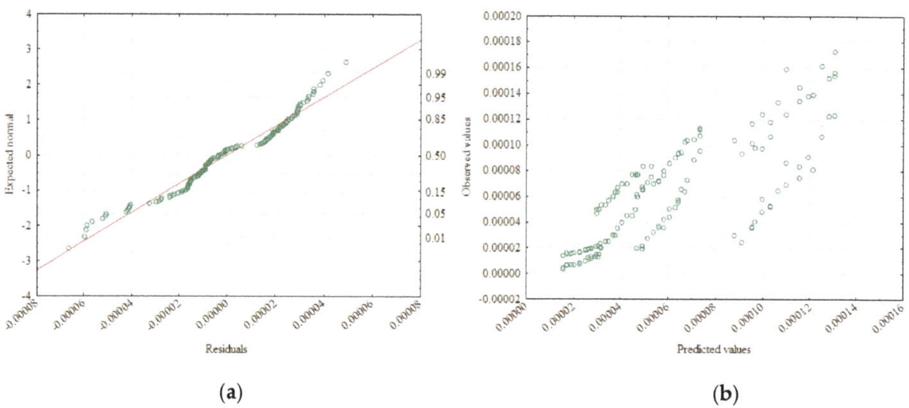

(a)

(b)

Figure 7. Normality diagram of residuals (**a**) and values observed against predictions (**b**).

A further step was to determine the independent variables for gradients 0.4–0.83 for the linear phase, where the explained variance was $R^2 = 0.883$. Table 5 shows the determined independent variables, which also contains parameter values together with standard errors for the determined variables.

Table 5. Evaluation of fixed parameters for model 3.

Independent Variable's Label	Independent Variable's Value	Standard Error
a2	−0.001169	0.000150
b2	0.00165	0.000256
c2	−0.163	0.058118
d2	0.00185	0.000307

The correctness of the model (Equation (6)) was supported by the diagram of the normality of residuals distribution (Figure 8a) and the distribution of observed values in relation to predicted values (Figure 8b).

$$k = (-0.001169 \times \rho_d) + \left(\frac{0.00165 \times d_5}{d_{90}^{-0.163}}\right) + 0.00185 \tag{6}$$

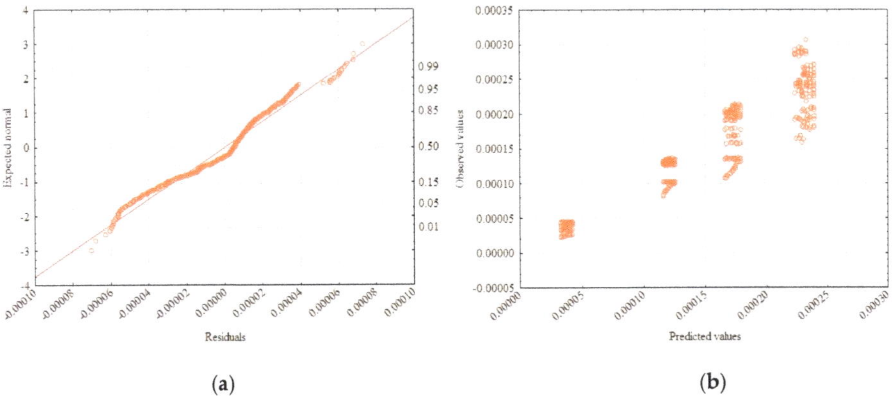

(a)   (b)

Figure 8. Normality diagram of residuals (**a**) and values observed against predicted (**b**).

5. Result Discussion and Conclusions

Determination of new independent variables and the application of pre-linear and linear phases allowed for estimating the calculated flow velocity results. This was not ensured by the first solution but fulfilled the assumptions of the hypothesis presented in the article. At the same time, the importance of accounting for the division into phases when determining the flow velocity was confirmed. This is best illustrated in Figure 9a, which shows the residual normality distributions for all models. By dividing the phases, a significantly better result of the explained variance for the linear phase was achieved. A better matched model for the pre-linear phase was also found, resulting in an increased accuracy of the resulting model. This is illustrated in Figure 9b, which compares the values observed and calculated for all models.

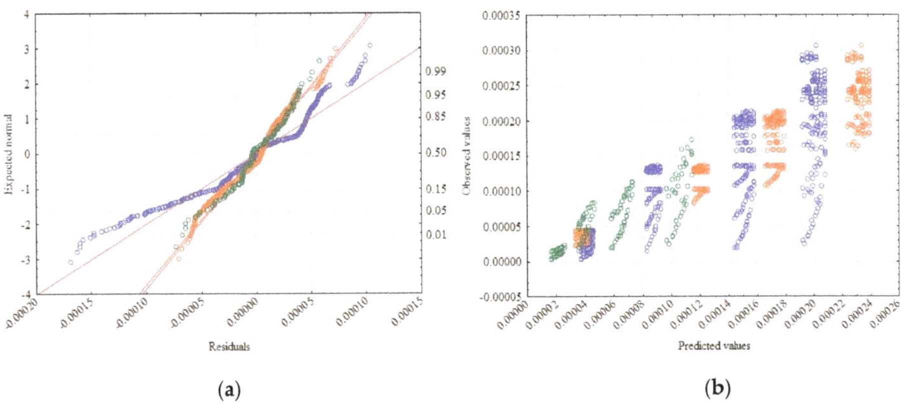

(a)   (b)

Figure 9. Normality diagram of residuals (model 1—blue color, model 2—red color, and model 3—green color) (**a**) and values observed against predictions (model 1—blue color, model 2—green color, and model 3—orange color) (**b**).

The parameters considered when creating the models were used to create the limits of applicability of the model by means of basic descriptive statistics. For each parameter (variable) used, the mean, standard deviation, and minimum and maximum value were calculated. The results are presented in Table 6.

Table 6. Parameters of applicability of the formula.

Variable	Mean (m/s)	Std. Dev.	Minimum (m/s)	Maximum (m/s)
k	0.000120	0.000078	0.000004	0.00031
ρ_d	1.931438	0.019769	1.895000	1.95600
d_5	0.230000	0.018723	0.200000	0.25000
d_{90}	9.350000	0.753915	8.600000	10.20000

The presented data defines the limits of applicability of the designated models. The conclusions are summarized below:

- The RCA material tested for blends with 0.02–16 mm, 0.05–16 mm, 0.1–16 mm, and 0.2–16 mm grains was characterized by good permeability from 3.1×10^{-4} to 4×10^{-6} m/s. Reported coefficients of permeability by Azram and Cameron [51] of RCAs with gradation in the range of 0–20 mm with 6 to 7% fine particles ($d < 0.0063$ mm) were in the range of 2×10^{-7} to 2×10^{-8} m/s. Arulrajach et al. [52], based on constant head test experiments, estimated the coefficient of permeability in range of 2.04×10^{-3} to 3.3×10^{-8} m/s, which indicates the existence of non-Darcian flow. Tests performed by McCulloch et al. [53] were conducted for RCAs with 4% fines and with fractions of 0–50 mm. The reported coefficients of permeability were in the range of 1×10^{-4} to 3×10^{-4} m/s. Tests on RCAs with poorly graded RCAs with fractions of 6–12 mm and with no fine content have proven a high water permeability of such material with a coefficient of permeability of approximately 1×10^{-3} m/s [53]. Tests performed on an RCA blend with gradation of 0–50 mm and with 5% fine contents have shown that the coefficient of permeability calculated based on a constant head permeability test is equal to 1.06×10^{-6} m/s [54]. As can be seen, the coefficient of permeability value strongly depends on the fines content. Test results presented in this article corresponds with the test results presented by other studies.
- The specific density, optimal moisture content, and particle sizes d_5 and d_{90} had a significant influence on the determination of the permeability coefficient.
- Regarding RCAs, a relationship between the flow velocity and the hydraulic gradient showed the existence of two phases, namely pre-linear and linear.
- Models created for individual phases gave greater confidence in determining the permeability coefficient.
- The models were created on the basis of the same set of variables, which facilitated their application and implementation in practice.
- Each of the models was examined in terms of the discrepancy with the observed value in relation to the forecasted results.
- For each of the models, the limits of its applicability were estimated.

The solutions presented here indicate the possibility that their use in calculating the flow rate was particularly advantageous due to the lack of such characteristics for this type of materials. To properly design a geotechnical structure, it is important to have information about geotechnical parameters and the solution presented here provides the basis to receive them.

Author Contributions: Conceptualization, W.S. and J.D.; methodology, W.S., J.D. and A.G.; formal analysis, J.D.; investigation, W.S., J.D. and A.G.; writing—original draft preparation, J.D.; writing—review and editing, W.S., J.D. and A.G.; visualization, J.D.; supervision, W.S.; project administration, W.S. and J.D.; funding acquisition, W.S.

Funding: This research received no external funding.

Conflicts of Interest: The authors declare no conflict of interest.

References

1. Global Construction Outlook 2022. Report ID: 4496591. Available online: https://www.researchandmarkets.com/reports/4496591/global-construction-outlook-2022 (accessed on 3 September 2019).

2. Podvezko, V.; Kildienė, S.; Zavadskas, K.E. Assessing the performance of the construction sectors in the Baltic states and Poland. *Panoeconomicus* **2017**, *64*, 493–512. [CrossRef]

3. Wilkinson, S.; Richards, A.Y.; Sapeciay, Z.; Costello, S.B. Improving construction sector resilience. *Int. J. Disaster Resil. Built Environ.* **2016**, *7*, 173–185. [CrossRef]

4. *European Aggregates Association A Sustainable Industry for a Sustainable Europe*; UEPG: Brussels, Belgium, 2017.

5. Cardoso, R.; Silva, R.V.; de Brito, J.; Dhir, R. Use of recycled aggregates from construction and demolition waste in geotechnical applications: A literature review. *Waste Manag.* **2016**, *49*, 131–145. [CrossRef] [PubMed]

6. Yilmaz, M.; BAKIŞ, A. Sustainability in construction sector. *Procedia Soc. Behav. Sci.* **2015**, *195*, 2253–2262. [CrossRef]

7. Michel, L.; Angela, B. *Quantitative and Statistical Data in Education*; Aix-Marseille University: Marseille, France, 2018; ISBN 9781786302281.

8. Cowan, G. Statistical models with uncertain error parameters. *Eur. Phys. J. C* **2019**, *79*, 133. [CrossRef]

9. Gnedenko, B.V. *Theory of Probability*; Routledge: London, UK, 2018.

10. Rao, J.N.K. Interplay between sample survey theory and practice: An appraisal. *Surv. Methodol.* **2005**, *31*, 117–138.

11. Kish, L. *Survey Sampling*; John Wily & Sons: New York, NY, USA, 1995.

12. Silverman, B.W. *Density Estimation for Statistics and Data Analysis*; Routledge: London, UK, 2018.

13. Hilbe, J.; Robinson, A. *Methods of Statistical Model Estimation*; Chapman and Hall/CRC: Boca Raton, FL, USA, 2013.

14. Li, X.R. *Probability, Random Signals, and Statistics*; CRC press: Boca Raton, FL, USA, 2017.

15. Brus, D.J. Statistical sampling approaches for soil monitoring. *Eur. J. Soil Sci.* **2014**, *65*, 779–791. [CrossRef]

16. Speak, A.; Escobedo, F.J.; Russo, A.; Zerbe, S. Comparing convenience and probability sampling for urban ecology applications. *J. Appl. Ecol.* **2018**, *55*, 2332–2342. [CrossRef]

17. Petkovic, G.; Engelsen, C.J.; Haoya, A.O.; Breedveld, G. Environmental impact from the use of recycled materials in road construction: Method for decision making in Norway. *Resour. Conserv. Recycl.* **2004**, *42*, 249–264. [CrossRef]

18. Poon, C.S.; Chan, D. Feasible use of recycled concrete aggregates and crushed clay brick as unbound road sub-base. *Constr. Build. Mater.* **2006**, *20*, 578–585. [CrossRef]

19. Herrador, R.; Pérez, P.; Garach, L.; Ordóñez, J. Use of recycled construction and demolition waste aggregate for road course surfacing. *J. Transp. Eng.* **2011**, *138*, 182–190. [CrossRef]

20. Gabryś, K.; Sas, W.; Soból, E.; Głuchowski, A. Application of bender elements technique in testing of anthropogenic soil—recycled concrete aggregate and its mixture with rubber chips. *Appl. Sci.* **2017**, *7*, 741. [CrossRef]

21. Sas, W.; Głuchowski, A.; Gabryś, K.; Soból, E.; Szymański, A. Deformation Behavior of Recycled Concrete Aggregate during Cyclic and Dynamic Loading Laboratory Tests. *Materials* **2016**, *9*, 780. [CrossRef] [PubMed]

22. O'Mahony, M.M. An analysis of the shear strength of recycled aggregates. *Mater. Struct.* **1997**, *30*, 599–606. [CrossRef]

23. Arm, M. Self-cementing properties of crushed demolished concrete in unbound layers: Results from triaxial tests and field tests. *Waste Manag.* **2001**, *21*, 235–239. [CrossRef]

24. Chapuis, R.P. Predicting the saturated hydraulic conductivity of soils: A review. *Bull. Eng. Geol. Environ.* **2012**, *71*, 401–434. [CrossRef]

25. Deshpande, Y.S.; Hiller, J.E. Pore characterization of manufactured aggregates: Recycled concrete aggregates and lightweight aggregates. *Mater. Struct.* **2011**, *45*, 67–79. [CrossRef]

26. Gee, K.K. *Use of recycled concrete Pavement as Aggregate in Hydraulic-Cement. 42 Concrete Pavement*; FHWA Publication: Washington, DC, USA, 2007.

27. Paranavithana, S.; Mohajerani, A. Effects of Recycled Concrete Aggregates on Properties of Asphalt Concrete. *Resour. Conserv. Recycl.* **2006**, *48*, 1–12. [CrossRef]

28. Gómez-Soberón, J.M. Porosity of recycled concrete with substitution of recycled concrete aggregate: An experimental study. *Cement Concr. Res.* **2002**, *32*, 1301–1311. [CrossRef]

29. Tam, V.W.Y.; Gao, X.F.; Tam, C.M.; Chan, C.H. New Approach in Measuring Water Absorption of Recycled Aggregates. *Constr. Build. Mater.* **2008**, *22*, 364–369. [CrossRef]

30. ASTM D698-12e1. Standard Test Methods for Laboratory Compaction Characteristics of Soil Using Standard Effort (12 400 ft-lbf/ft3 (600 kN-m/m3)). Annual Book of ASTM Standards, USA. 2008. Available online: https://www.techstreet.com/standards/astm-d5254-d5254m-92-2010-e1?product_id=1751704 (accessed on 8 January 2010).

31. Głuchowski, A.; Sas, W.; Dzięcioł, J.; Soból, E.; Szymański, A. Permeability and Leaching Properties of Recycled Concrete Aggregate as an Emerging Material in Civil Engineering. *Appl. Sci.* **2019**, *9*, 81. [CrossRef]

32. *EN 933-11:2009/AC:2009 Tests for geometrical properties of aggregates—Part 11: Classification test for the constituents of coarse recycled aggregate*; Spanish Association for Standardisation and Certification (AENOR): Madrid, Spain, 2009.

33. *WT-4 Unbound Mix for National Roads*; Technical Specifications; Directive No 102 of Polish General Director of National Roads and Motorways; Polish General Director of National Roads and Motorways: Warsaw, Poland, 2010.

34. ISO 17892-4:2016 Geotechnical Investigation and Testing—Laboratory Testing of Soil—Part 4: Determination of Particle Size Distribution. Available online: http://sklep.pkn.pl/pn-en-iso-17892-4-2017-01e.html (accessed on 18 January 2017).

35. Maia, M.B.; De Brito, J.; Martins, I.M.; Silvestre, J.D. Toxicity of recycled concrete aggregates: Review on leaching tests. *Open Constr. Build. Technol. J.* **2018**, *12*, 187–196. [CrossRef]

36. Rodrigues, P.; Silvestre, J.; Flores-Colen, I.; Viegas, C.; de Brito, J.; Kurad, R.; Demertzi, M. Methodology for the assessment of the ecotoxicological potential of construction materials. *Materials* **2017**, *10*, 649. [CrossRef] [PubMed]

37. Bestgen, J.O.; Cetin, B.; Tanyu, B.F. Effects of extraction methods and factors on leaching of metals from recycled concrete aggregates. *Environ. Sci. Pollut. Res.* **2016**, *23*, 12983–13002. [CrossRef] [PubMed]

38. Galvín, A.P.; Ayuso, J.; Agrela, F.; Barbudo, A.; Jiménez, J.R. Analysis of leaching procedures for environmental risk assessment of recycled aggregate use in unpaved roads. *Constr. Build. Mater.* **2013**, *40*, 1207–1214. [CrossRef]

39. Galvín, A.P.; Ayuso, J.; García, I.; Jiménez, J.R.; Gutiérrez, F. The effect of compaction on the leaching and pollutant emission time of recycled aggregates from construction and demolition waste. *J. Clean. Prod.* **2014**, *83*, 294–304. [CrossRef]

40. Saca, N.; Dimache, A.; Radu, L.R.; Iancu, I. Leaching behavior of some demolition wastes. *J. Mater. Cycles Waste Manag.* **2017**, *19*, 623–630. [CrossRef]

41. Butera, S.; Hyks, J.; Christensen, T.H.; Astrup, T.F. Construction and demolition waste: Comparison of standard up-flow column and down-flow lysimeter leaching tests. *Waste Manag.* **2015**, *43*, 386–397. [CrossRef]

42. Kowalski, J. *Hydrogeologia z podstawami geologii*; Wydaw AR: Wrocław, Poland, 1998. (in Polish)

43. Hansbo, S. Consolidation equation valid for both Darcian and non-Darcian flow. *Geotechnique* **2011**, *51*, 51–54. [CrossRef]

44. Hansbo, S. Deviation from Darcy's law observed in one-dimensional consolidation. *Geotechnique* **2003**, *53*, 601–605. [CrossRef]

45. Sas, W.; Dzieciol, J. Determination of the fi ltration rate for anthropogenic soil from the recycled concrete aggregate by analytical methods. *Sci. Rev. Eng. Environ. Sci.* **2018**, *27*, 80.

46. Batezini, R.; Balbo, J.T. Study on the hydraulic conductivity by constant and falling head methods for pervious concrete. *Rev. IBRACON Estrut. e Mater.* **2015**, *8*, 248–259. [CrossRef]

47. Allen, A.O. *Probability, Statistics, and Queueing Theory*; Academic press: London, UK, 2014.

48. Gatti, P.L. *Probability Theory and Mathematical Statistics for Engineers*; CRC Press: New York, NY, USA, 2004.

49. Giri, N.C. *Introduction to Probability and Statistics*; Routledge: New York, NY, USA, 1993.

50. WYWIAŁ, J. Estimation of population mean on the basis of non-simple sample when nonresponse error is present. *Stat. Transit.* **2001**, *5*, 443–450.

51. Azram, A.M.; Cameron, D.A. Geotechnical properties of blends of Recycled Clay Marsony and Recycled Concrete Aggregates in Unbound Pavement Construction. *J. Mater. Civ. Eng.* **2013**, *25*, 788–798. [CrossRef]

52. Arulrajah, A.; Piratheepan, J.; Ali, M.M.Y.; Bo, M.W. Geotechnical properties of recycled concrete aggregate in pavement sub-base applications. *Geotech. Test. J.* **2012**, *35*, 1–9. [CrossRef]

53. McCulloch, T.; Kang, D.; Shamet, R.; Lee, S.J.; Nam, B.H. Long-Term performance of recycled concrete aggregate for subsurface drainage. *J. Perform. Constr. Facil.* **2017**, *04017015*, 1–8. [CrossRef]
54. Bennert, T.; Maher, A. *The Use of Recycled Concrete Aggregate in a Dense Graded Aggregate Base Course*; Rutgers: Piscataway, NJ, USA, 2008.

Article

Properties of Mortar with Recycled Aggregates, and Polyacrylonitrile Microfibers Synthesized by Electrospinning

Manuel J. Chinchillas-Chinchillas [1,*]**, Manuel J. Pellegrini-Cervantes** [1,*]**, Andrés Castro-Beltrán** [1]**, Margarita Rodríguez-Rodríguez** [1,*]**, Víctor M. Orozco-Carmona** [2] **and Héctor J. Peinado-Guevara** [3]

[1] Faculty of Engineering Mochis, Autonomous University of Sinaloa, Fuente de Poseidón y Ángel Flores s/n, Col. Jiquilpan, Module B2, Los Mochis, Sinaloa 81210, Mexico; andres.castro@uas.edu.mx
[2] Corrosion Department, Advanced Materials Research Center, Chihuahua 31136, Mexico; victor.orozco@cimav.edu.mx
[3] Faculty of Economic and Administrative Sciences, Autonomous University of Sinaloa, San Joachín, Guasave 81101, Sinaloa, Mexico; hpeinado75@hotmail.com
* Correspondence: Manuel.chinchillas.fim@uas.edu.mx (M.J.C.-C.); Manuel.pellegrini@uas.edu.mx (M.J.P.-C.); maguirr_75@hotmail.com (M.R.-R.); Tel.: +52-6683978139 (M.J.C.-C.); +52-6681036979 (M.J.P.-C.); +52-6681370370 (M.R.-R.)

Received: 12 October 2019; Accepted: 19 November 2019; Published: 22 November 2019

Abstract: Currently it is necessary to find alternatives towards a sustainable construction, in order to optimize the management of natural resources. Thus, using recycled fine aggregate (RFA) is a viable recycling option for the production of new cementitious materials. In addition, the use of polymeric microfibers would cause an increase in the properties of these materials. In this work, mortars were studied with 25% of RFA and an addition of polyacrylonitrile PAN microfibers of 0.05% in cement weight. The microfibers were obtained by the electrospinning method, which had an average diameter of 1.024 µm and were separated by means of a homogenizer to be added to the mortar. Cementing materials under study were evaluated for compressive strength, flexural strength, total porosity, effective porosity and capillary absorption, resistance to water penetration, sorptivity and carbonation. The results showed that using 25% of RFA causes decreases mechanical properties and durability, but adding PAN microfibers in 0.05% caused an increase of 2.9% and 30.8% of compressive strength and flexural strength respectively (with respect to the reference sample); a decrease in total porosity of 5.8% and effective porosity of 7.4%; and significant decreases in capillary absorption (approximately 23.3%), resistance to water penetration (25%) and carbonation (14.3% after 28 days of exposure). The results showed that the use of PAN microfibers in recycled mortars allowed it to increase the mechanical properties (because they increase the tensile strength), helped to fill pores or cavities and this causes them to be mortars with greater durability. Therefore, the use of PAN microfibers as a reinforcement in recycled cementitious materials would be a viable option to increase their applications.

Keywords: recycled aggregate; polyacrylonitrile microfibers; electrospinning; durability; carbonation

---

1. Introduction

Construction industry is characterized by a significant demand of energy and raw materials such as clay, wood, metal, water, petrous materials, etc. [1]. In the construction sector, the aggregate is the predominant material from the volume point of view, substituting the natural fine aggregate (NFA) with recycled fine aggregates (RFA) contributes to the sustainability and conservation of natural resources [2–4]. Growing environmental concerns and landfill shortages, overexploitation of rivers and hills ecosystems, increased transportation, and rising landfill costs are the driving forces that promote the recycling of concrete demolition waste to be used in a new substitute material [5–7]. Mortar made with recycled aggregates is no longer just a field of research, it is already a practical reality and it has been used for some years in several countries [8]. The amount of construction and demolition waste deposited in landfills differs from one country to another, in Hong Kong, approximately 20 million tons of waste were produced in 2004, only 12% of the waste was disposed of in landfills and 88% was used as filling material. The generation of construction and demolition waste in the European Union (EU) reached 850 million tons in 2008, which represents 31% of total waste generation in the EU [9]. The countries that recycle more construction material in the European Union are Holland, Denmark and Belgium, generating and using waste of 90%, 81% and 87%, respectively [10]. In most research, it is mentioned that the use of a certain percentage of recycled coarse aggregate (RCA) does not damage the durability and the resistance of the cementitious materials, but the use of recycled fine aggregate (RFA) is restricted or even prohibited due to its unsatisfactory properties [11,12]. However, several studies suggest that its use is not very unfavorable and that results similar to the ones obtained with natural fine aggregates with low substitutions can be obtained [13]. According to some studies, replacing less than 20%–30% RFA does not cause a significant decrease in the physical and mechanical properties of the mortar [14,15]. Therefore, to contribute to sustainability, it is necessary to reuse these materials and to develop new mortars with less effects on the resistance and durability properties of the structures made with cementitious materials.

Mechanical properties represent the capacity of mortar to support stress [16] and they are related to the durability properties, which are indicators of the deterioration suffered by mortar due to some external effect, and it depends on the ability of a fluid to penetrate the microstructure of mortar allowing the introduction of molecules (carbon dioxide and chloride ions) that react and destroy the chemical stability [11]. In order to improve these properties, diverse researchers use additives (superplasticizers, accelerators, etc.), pozzolans (silica fume, fly ash, granulated slag, etc.), nanoparticles (silica, titanium, calcium carbonate, etc.), pieces of polymeric materials (rubber, PET bottles, polypropylene, etc.) and fibers of different nature (fiberglass, steel fibers, polymer fibers, etc.) [17–26]. In the last decades, fiber reinforced mortar has been used in applications such as pavements, coatings, repair of hydraulic structures patches, thin sheets, prefabrications, projected mortar, reinforcement of slabs, wall cladding, bridge decks, earthquakes, resistant structures, etc. [27]. Currently, it is well established that fiber incorporation improves the engineering performance of cementitious materials including a better resistance to cracking, an increase on ductility and toughness, as well as an improvement on fatigue and impact resistance [28]. Steel fibers are very expensive and produce corrosion [29], glass fibers are very fragile and do not transmit elasticity to the mortar, [30], therefore, polymer fibers are the best option, they are produced at a low cost, they have high elastic properties and they are easy to place in the mortar mixture, improving its dispersion [31], in addition to being the least studied in the construction engineering area. Some of the fibers used as reinforcement of cementitious materials are polypropylene fibers [32], polyethylene terephthalate (PET) fibers [33], polyvinyl alcohol (PVA) fibers [34], among others, but few studies have been done with polyacrylonitrile (PAN) reinforcing fibers [35]. PAN is a polymer widely used in the industry, and has been widely used for the synthesis of microfibers because it has good mechanical properties, easy spinning, chemical stability and high durability [36]. A very efficient method to synthesize PAN microfibers is electrospinning, it is a simple procedure and microfibers are obtained without defects and with diameters ranging from nanometer to micrometer [37,38]. This research studies the effect of the addition of PAN fibers at 0.05% (in cement weight) in the mechanical properties and in the durability in mortars made with 25% of recycled fine aggregates (replacing the natural aggregate), evaluating the resistance to compression, to bending and to water penetration, total porosity, effective porosity, resistance to, capillary absorption, sorptivity and carbonation, since the mechanical properties and durability are the most important measures to be taken into account to observe the useful life of mortar for future applications.

2. Materials and Methods

2.1. Materials

For the preparation of mortars, Portland Cement type III (30R) Cemex® brand (Monterrey, México) was used, which achieves a compressive strength of 30 MPa at 28 days (NMX-C-061) [39], distilled water, NFA and RFA was used, which was obtained by crushing waste concrete by means of a ball mill. To obtain PAN microfibers, PAN (Sigma-Aldrich, Estado de México, México) with a molecular weight of 150,000 g/mol^{-1} and dimethylformamide (DMF) of 99.85% purity was used. Figure 1 shows the granulometric curves of the aggregates, following ASTM C-33 [40]. In Table 1, the properties of the materials used in the manufacture of the mortars are shown.

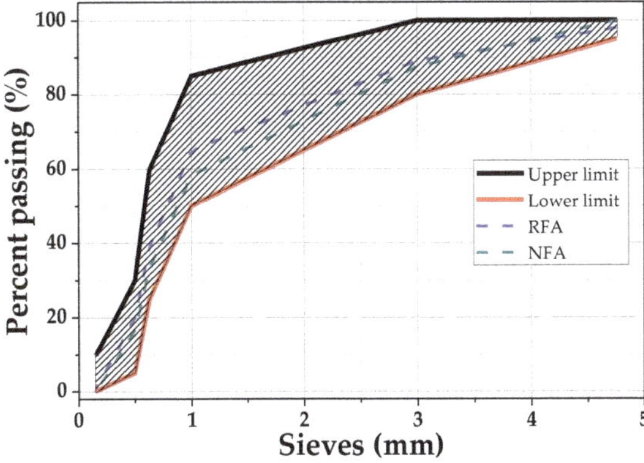

Figure 1. Granulometry of NFA and RFA following the ASTM C-33 standard.

Table 1. Properties of the materials used.

Properties	NFA	RFA	PAN Microfibers	Cement
Diameter range (µm)	-	-	0.6–1.3	-
Average microfiber diameter (µm)	-	-	1.024	
Modulus of elasticity (GPa)	-	-	2.5	-
Hardness (GPa)	-	-	0.48	-
PAN Molecular weight (g/mol)	-	-	150,000	-
PAN concentration (% by weight of DMF)	-	-	12	-
Specific gravity (kg/m^3)	2.56	2.15	-	3.150
Moisture (%)	6.15	3.04	-	-
Water absorption (%)	4.16	13.63	-	-
Fineness modulus	2.88	3.04	-	-
Type	Natural	Recycled	Polymeric	30R

2.2. Production of Recycled Fine Aggregate

The concrete of origin used for the production of the RFA was obtained from a quality control laboratory of a concrete company, which had been assigned to a road paving project with an average resistance of 23 MPa, the test cylinders were stored and free from contaminants. It was necessary to screen the material by means of mechanical crushing using a jaw crusher Maneklal Global Exports model JS-0804 (Los Mochis, México) to minimize the size of the samples. This process was carried out several times until an approximate size of 3 cm was obtained. The last step was to place the pieces of crushed concrete in a ball mill with a Falcon engine model 7518081PA56C (Culiacán, México) with 8–12 mm steel balls for 30 min until obtaining fine material with a maximum size of 0.6 cm, this fine aggregate was subjected to two washing times to eliminate possible contaminating materials. All the material was sieved to comply with the granulometry following the ASTM C-33 standard [40].

2.3. Production of PAN Microfiber by Electrospinning

The details of the electrospinning equipment can be observed in the Chinchillas et al. 2019 investigation. The electrospinning parameters were: 17 cm separation between the tip of the needle and the collector plate, 16 kV applied voltage and 1 mL/h flow. The polymeric solution of PAN, was elaborated by adding 12% in PAN weight and 88% of DMF in a vial and stirring for 12 h in room temperature, until a completely homogeneous solution was achieved. This solution was put into a 3 mL syringe and installed in the infusion pump of equipment, the process was started and the fibers were deposited in an aluminum container.

Dispersion of Microfibers in Aqueous Medium

To separate the PAN microfibers, a Branson ultrasonic tip model 450A (Chihuahua, México) was used for the homogenization process. First, the water from each mixture was put into a 1000 mL vial. Then 0.05% of the fibers were added and the ultrasound tip was placed into the vial for 3 hours with an amplitude of 45% to 50% until dispersion was achieved.

2.4. Design of Mortar Mix

In Table 2, the four mortar mixtures are described, showing the amount of materials needed to make the specimens, each mixture corresponds to a mortar volume of 0.000256 m^3. A water/cement ratio (w/c) of 0.58 was used, after verifying that it was the ideal ratio to obtain a mortar with a plastic consistency, the method used is described in ASTM C1437 [41] (by this procedure, it is possible to obtain a percentage of the workability of the mixture and know if it has a dry, plastic or wet workability [42]). The aggregate/cement ratio (ag/c) used was 2.75, following ASTM C109 [43]. It is worth mentioning that the RFA was previously saturated with water to avoid the absorption of mixing water [44].

The nomenclature used in the mixtures is classified as follows: the first three letters correspond to the origin of the fine aggregate (RFA and NFA), in the case of the RFA it refers to the mixture with a 25% replacement and the mixture with NFA refers to the 100% of natural aggregate used. The samples with the acronym PAN, refer to the addition of microfibers of PAN in 0.05% with respect to the weight of the cement.

Table 2. Mixture design.

Mixtures	% of RFA	% of NFA	% of Microfibers	Water (l)	Cement (kg)	NFA (kg)	RFA (kg)	PAN Microfibers (kg)	Mix Flow (%)
RFA	25	75	0	2.847	4.908	10.124	3.374	-	116.6
RFA/PAN	25	75	0.05	2.847	4.908	10.124	3.374	0.00245	115.7
NFA	-	100	0	2.847	4.908	13.499	-	-	117.2
NFA/PAN	-	100	0.05	2.847	4.908	13.499	-	0.00245	115.5

2.5. Preparation of Specimens and Test Procedures

The morphology and average size of PAN microfibers were observed by scanning electron microscope (SEM), FEI Nova Nano-SEM brand (Chihuahua, México) at 5 kV and at a working distance of 10 mm. The vibration of the polymer bonds was analyzed by infrared spectroscopy (FT-IR) Bruker Alpha II brand (Los Mochis, México) in a range of 4000–500 cm^{-1}. The thermogravimetric analysis and differential calorimetry scanning (TGA/DSC) was carried out by using SDT Q600 TA INSTRUMENTS (Los Mochis, México), from 0 to 700 °C with a heating rate of 10 °C/min in nitrogen environment.

Table 3 shows the summary of the tests, the shapes and sizes of the specimens, the curing time, and the norms followed in this research. In all the tests, three specimens were made for each group of samples, and the mixing procedure was following the ASTM C192 standard [45]. For durability tests, it was necessary to cut the cylinders using a Schneider cutter model 55-CO210 (Los Mochis, México) and in accelerated carbonation test, the penetration of CO_2 was measured through a 1% phenolphthalein solution in water and alcohol and the parameters of the carbonation chamber were 4% of CO_2, 56% relative humidity and a temperature of 25 °C.

Table 3. Tests, shape and size of the specimens, curing time and standards applied.

Test	Form of the Specimen (type)	Dimensions of the Specimen (cm)	Curing Time (days)	Standard/Reference
Flexural strength	Prisms	$4 \times 4 \times 16$	28	ASTM C 348 [46]
Compressive strength	Post-test pieces of mortar bending	4×4	28	ASTM C 349 [47]
Total porosity	Cubes	5×5	28	[48]
Effective porosity	Cylinders	7.5 in diameter and 8 in height	28	ASTM C 1585 [49]
Resistance to water penetration and capillary absorption	Cubes	4×4	-	[35]
Accelerated carbonation	Cylinders	7.5 in diameter and 2 in height	7, 15 and 28	[50]

3. Results and Discussion

3.1. Microfibers of PAN by Electrospinning

Figure 2 shows the characterizations that were made to the PAN microfibers obtained by electrospinning and the results after separation by means of the homogenization process. The FT-IR study was carried out in order to know the purity of PAN microfibers, when analyzing the vibration of the polymer links. With this study it is possible to determine if there is a certain amount of the solvent used in the manufacture of microfibers, or if there is any vibration of contaminating molecules. At 3445 and 2940 cm^{-1}, the vibrations of the bonds O–H and C–H respectively are observed, due to water absorption [51]. The stretching vibrations of C≡N, C–C and C–H representative of the PAN were observed at 2442, 1630 and 1450 cm^{-1} [52,53]. These results confirm that the structure of the microfibers is PAN. You could also observe some bands at approximately 1368, 1256 and 1088 cm^{-1}, which are characteristic of DMF, a solvent used in the synthesis of PAN microfibers [54,55]. On the other hand, the thermogravimetric analysis is shown in Figure 2b. PAN microfibers showed a total weight loss of 87.3% at 700 °C. The chemical decomposition of the polymer begins at approximately 250 °C, and an exothermic peak is observed at 288 °C in the DSC analysis [56]. This confirms that changes in molecular structure occur due to degradation of the material [57]. On the other hand, through SEM images, it is possible to observe the morphology of the microfibers and using "ImageJ" software (version number: 1.51, Wayne Rasband, National Institutes of Health, Bethesda, MD, USA), it is possible to calculate the average diameter. In Figure 2c, PAN microfibers are continuous, and smooth and defect-free morphology is shown, confirming that the parameters used for its fabrication were adequate. The diameters obtained were from 600 nm to 1.3 μm, presenting an average diameter of 1.024 μm (Figure 2d). Recent research has reported PAN microfibers with 1 μm diameter, using the same parameters [58]. Microfibers obtained by electrospinning are separated from each other, but continuous, i.e., it is not possible to control the length after the electrospinning process. So, it is not possible to incorporate them into the cement mortar, which is why a separation and cutting process through a homogenization process is needed. In Figure 2e,f, the cut and separated microfibers can be observed where no change in the microfiber shape is watched, but they are no longer continuous microfibers, now they are shorter (approximate sizes of 15–30 μm), comparable to some studies reported in literature, where microfibers within micrometer scale length are used in cementitious materials [59]. These results confirm that the homogenization process is useful for the separation and cutting of electro-spun microfibers.

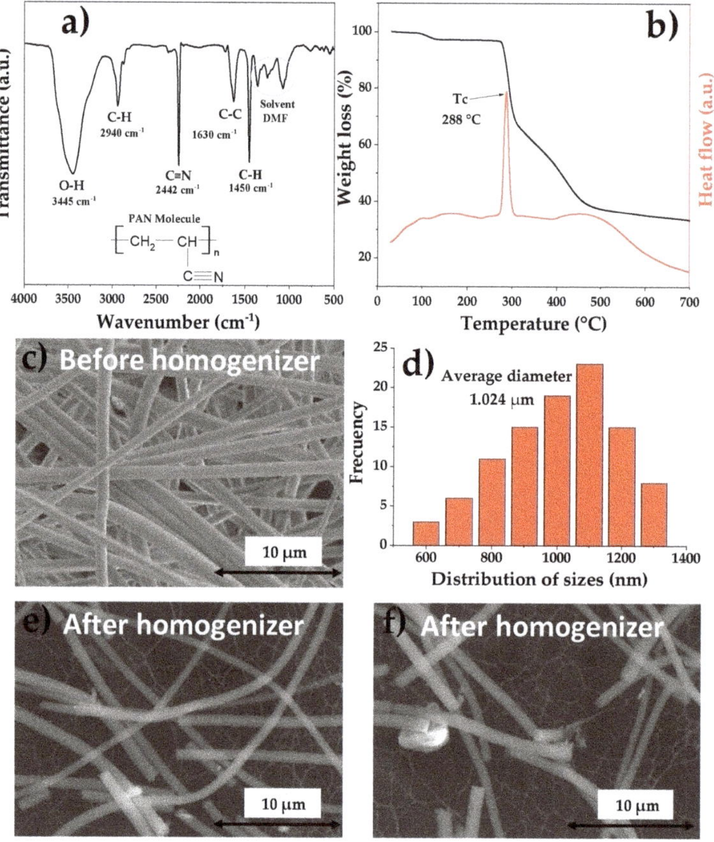

Figure 2. Characterization and separation of PAN microfibers: (**a**) FT-IR, (**b**) TGA/differential calorimetry scanning (DSC), (**c**) SEM, (**d**) average diameter and (**e**,**f**) microfibers after separation.

3.2. Mortar

3.2.1. Compressive and Flexural Strength

The use of recycled aggregates for the manufacture of mortars causes a decrease in their mechanical properties, generating materials with low density and high porosity [60]. Figure 3a shows the results of compressive strength. It can be seen that the mixture with RFA reached a compressive strength of 17.2 MPa, it is the sample with lower mechanical properties of the study, this is attributed to the use of recycled materials (high porosity) [7]. On the other hand, the incorporation of PAN microfibers in recycled mortars caused a slight increase of 2.9%, since microfibers help to withstand internal stresses in cementitious materials [61]. In addition, the NFA mortars reached a value of 17.9 MPa, a result similar to other mortar works with the same ratio w/c = 0.58 [62]. By incorporating PAN microfibers into natural mortars, it increased by 11% compared to the RFA sample. Similar results are observed in Figure 3b (flexural strength), where the RFA mixture was the one that reached a lower value (1.3 MPa). As well as using 0.05% of PAN microfibers in recycled and natural mortars caused an increase of 30.8% and 53.8% respectively. It should be noted that the flexural strength results are more outstanding than the compressive strength, because microfibers have a greater effect on tensile stresses, which are more abundant in flexion [63]. These results confirm that the use of recycled aggregates in the manufacture of mortars causes decreases in mechanical properties due to their high porosity and confirms that the addition of PAN microfibers in small percentages (0.05%) causes significant increases in the mechanical properties of mortars, due to the ability of microfibers to withstand stresses, retain crack propagation and decrease their porosity [64].

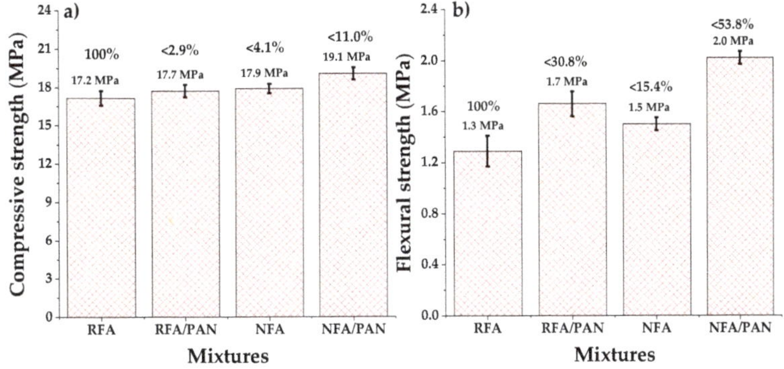

Figure 3. Mechanical properties of mortars: (**a**) compressive strength and (**b**) flexural strength.

3.2.2. Total and Effective Porosity

Total porosity refers to all the pores found throughout the mortar structure and the effective porosity to the interconnected pores in the cement matrix [65]. Figure 4 shows the result of the total porosity and the effective porosity of the mortars evaluated in this investigation. As expected, the RFA mixtures were the ones that reached the highest values, 25.9% and 21.7% of total porosity and effective porosity respectively. This is due to the fact that the recycled aggregate contains old high-porosity paste in its structure [66]. It was also observed that adding PAN microfibers in the mortar with recycled aggregate (RFA/PAN) caused a decrease in its porosities of 5.8% and 7.4% for total and effective porosity respectively (referring to the RFA mortar). On the other hand, the NFA mixture had similar values to other investigations for mortars with this ratio w/c = 0.58 [48,67]. Finally, the NFA/PAN mortars had a total porosity of 21.4% and an effective porosity of 18.7%, comparing this sample with RFA, the addition of microfibers caused a decrease of 17.4% and 13.8% respectively. As it is known, the use of recycled aggregates causes the cementitious materials to increase their porosities (due to the nature of the aggregate), but the outstanding thing is that with the use of PAN microfibers the mortars decreased their porosities, this is attributed to the ability of the fibers to fill cavities or function as a barrier, preventing the penetration of water into the structure of the material [31].

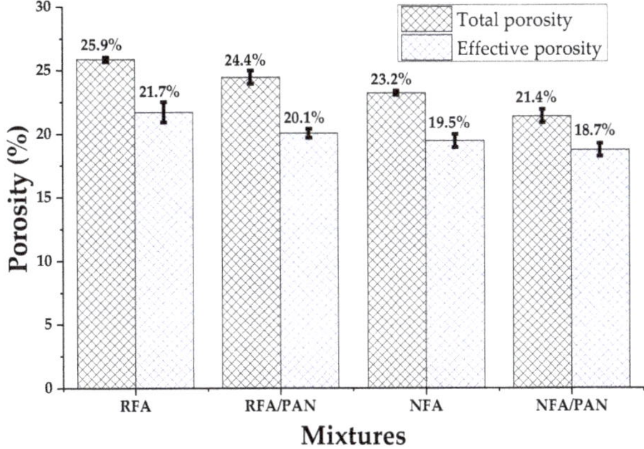

Figure 4. Test of total porosity and effective porosity of the mortar.

3.2.3. Capillary Absorption and Resistance to Water Penetration

The absorption and resistance to water penetration are important indicators of the durability of hardened mortar. Reducing water absorption and increasing its resistance to water penetration greatly improve the long-term performance and the useful life of mortar in aggressive environments [68]. Figure 5a shows the results of the resistance to water penetration, and Figure 5b the capillary absorption of the mixtures. It is observed that the RFA mortar had a lower resistance to water penetration and a greater capillary absorption compared to other samples. This is because this type of aggregate contains old mortar adhered to its structure, being a highly porous and very absorbent material [69]. However, it is also observed that adding PAN microfibers to the RFA and NFA caused an increase in resistance to water penetration and a decrease in capillary absorption. This is because microfibers serve as a filler material or as a barrier between the cementitious matrix and the aggregates present in mortar [70]. These results show that the addition of PAN microfibers helped to provide a denser cementitious matrix and consequently increased its durability.

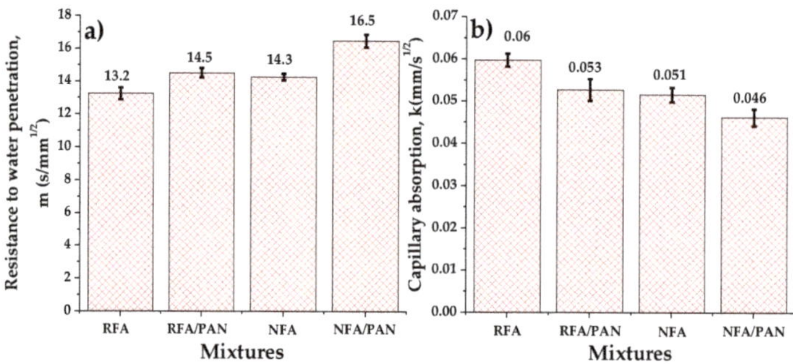

Figure 5. (**a**) Resistance to water penetration and (**b**) capillary absorption of the mortar mixtures.

3.2.4. Sorptivity

Sorptivity is a property of the materials that measures the tendency of a porous material to absorb and transmit water through capillarity [71,72]. Figure 6 shows the results of the absorption capacity of the mortars studied, highlighting that the samples with recycled aggregates were those that had higher absorption values due to the high water retention capacity of the recycled aggregate and its porous structure [73]. It can also be seen that the use of PAN microfibers causes a decrease in the specificity of sorptivity of cementitious materials. This result, in addition to those shown above, demonstrates that PAN microfibers were effectively preventing the flow of water in the mortar microstructure helping to reduce its porosity.

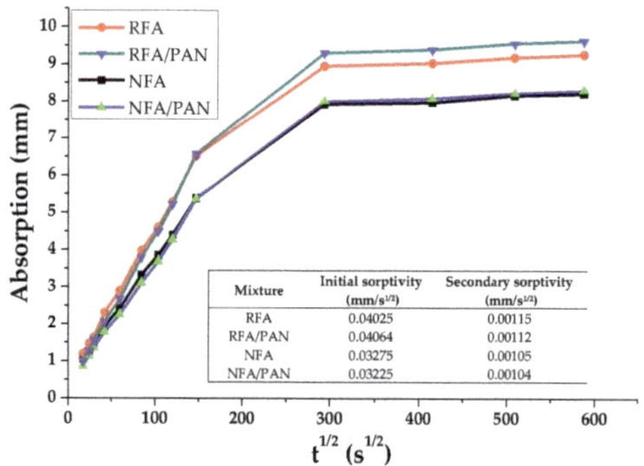

Figure 6. Sorptivity.

3.2.5. Carbonation

Carbonation is one of the most important studies of durability in cementitious materials and is caused by the reaction that occurs in CO_2 and moisture inside the pores of cementitious materials [74].

The results of CO_2 penetration in mortars made with RFA and PAN microfibers are shown in Figure 7. In which a clear trend was observed and it matched with the results shown above. At 7 days of exposure, the RFA mixture was the one with the highest CO_2 penetration (2.1 mm), due to its porous nature, on the other hand, the RFA/PAN mixture slightly decreased the penetration of CO_2 to 2 mm, because PAN microfibers cause a lower porosity reflecting a lower carbonation. On the other hand, the NFA mixture had a CO_2 penetration of 1.82 mm and by adding PAN microfibers helped reduce carbonation to 1.75 mm. The same trend was observable in the results at 15 and 28 days of exposure, demonstrating that PAN microfibers helped to counteract the decreases that were generated with the use of RFA in mortars [75].

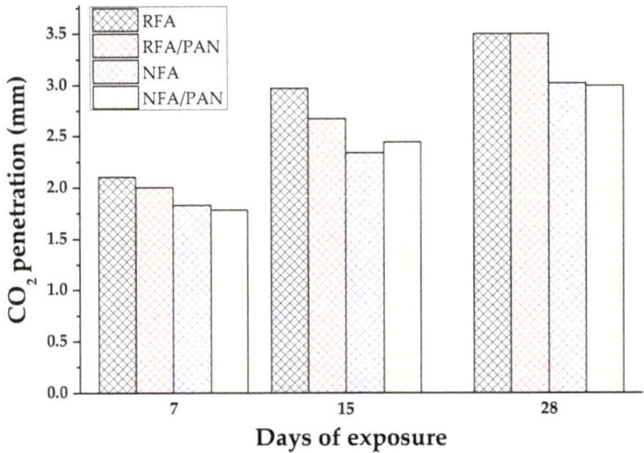

Figure 7. Carbonation results at different days of exposure (7, 15 and 28 days).

4. Conclusions

The formation of cementitious materials with the use of recycled aggregates opens up a world of possibilities and applications towards sustainable construction. The use of RFA in cementitious materials caused a decrease in the mechanical and durability properties of mortars, due to the large amount of old adhered paste, its high porosity, its high absorption and because it is less resistant than NFAs. However, this research showed an alternative to counteract this problem. The use of PAN microfibers caused an increase in the mechanical properties of compression and flexural strength of 2.9% and 30.8% respectively, it also caused a porosity decrease by 5.8% and 7.4% with respect to the total and effective porosity. By decreasing the porosity of the mortars with the use of PAN microfibers, the capillary absorption decreased by 23.3% and the resistance to water penetration increased by 25%. Finally, carbonation decreased 14.3%. It should be noted that PAN microfibers have not been reported in the literature as a reinforcement of recycled mortars. These results were due to the fact that PAN microfibers provide mortar with tensile strength, help fill cavities or pores within the mortar microstructure and prevent the spread of cracks. Therefore, the use of PAN microfibers as a reinforcement for recycled mortars opens up a world of possibilities for use in future constructions and could be an alternative for sustainable construction.

Author Contributions: Conceptualization, M.J.P.-C. and A.C.-B.; Data curation, M.J.C.-C. and M.R.-R.; Formal analysis, M.J.C.-C. and M.R.-R.; Investigation, M.J.C.-C. and H.J.P.-G.; Methodology, M.J.C.-C., M.J.P.-C. and A.C.-B.; Resources, V.M.O.-C. and H.J.P.-G.; Software, V.M.O.-C.; Supervision, M.J.P.-C. and V.M.O.-C.; Validation, M.J.C.-C. and V.M.O.-C.; Visualization, H.J.P.-G.; Writing—original draft, M.J.P.-C. and M.R.-R.; Writing—review and editing, A.C.-B.

Funding: This research received no external funding.

Acknowledgments: The authors thank CONACYT for its doctoral scholarship program, the Center for Research in Advanced Materials (CIMAV) and the Mochis School of Engineering (UAS).

Conflicts of Interest: The authors declare no conflict of interest.

References

1. Koenders, E.A.B.; Pepe, M.; Martinelli, E. Compressive Strength and Hydration Processes of Concrete with Recycled Aggregates. *Cem. Concr. Res.* **2014**, *56*, 203–212. [CrossRef]
2. Vázquez, E.; Barra, M.; Aponte, D.; Jiménez, C.; Valls, S. Improvement of the Durability of Concrete with Recycled Aggregates in Chloride Exposed Environment. *Constr. Build. Mater.* **2014**, *67*, 61–67. [CrossRef]
3. Sas, W.; Dzięcioł, J.; Głuchowski, A. Estimation of Recycled Concrete Aggregate's Water Permeability Coefficient as Earth Construction Material with the Application of an Analytical Method. *Materials* **2019**, *12*, 2920. [CrossRef] [PubMed]
4. Wang, F.; Sun, C.; Ding, X.; Kang, T.; Nie, X. Experimental Study on the Vegetation Growing Recycled Concrete and Synergistic Effect with Plant Roots. *Materials* **2019**, *12*, 1855. [CrossRef] [PubMed]
5. Muñoz-Ruiperez, C.; Rodríguez, A.; Gutiérrez-González, S.; Calderón, V. Lightweight Masonry Mortars Made with Expanded Clay and Recycled Aggregates. *Constr. Build. Mater.* **2016**, *118*, 139–145. [CrossRef]
6. Fang, S.-E.; Hong, H.-S.; Zhang, P.-H. Mechanical Property Tests and Strength Formulas of Basalt Fiber Reinforced Recycled Aggregate Concrete. *Materials* **2018**, *11*, 1851. [CrossRef]
7. Chinchillas-Chinchillas, M.J.; Rosas-Casarez, C.A.; Arredondo-Rea, S.P.; Gómez-Soberón, J.M.; Corral-Higuera, R. SEM Image Analysis in Permeable Recycled Concretes with Silica Fume. A Quantitative Comparison of Porosity and the ITZ. *Materials* **2019**, *12*, 2201. [CrossRef]
8. Silva, R.V.; De Brito, J.; Dhir, R.K. Properties and Composition of Recycled Aggregates from Construction and Demolition Waste Suitable for Concrete Production. *Constr. Build. Mater.* **2014**, *65*, 201–217. [CrossRef]
9. Zhao, Z. Re-Use of Fine Recycled Concrete Aggregates for the Manufacture of Mortars. Ph.D. Thesis, Université Lille 1, Lille, France, 14 February 2014.
10. Akça, K.R.; Çakır, Ö.; İpek, M. Properties of Polypropylene Fiber Reinforced Concrete Using Recycled Aggregates. *Constr. Build. Mater.* **2015**, *98*, 620–630. [CrossRef]
11. Evangelista, L.; de Brito, J. Durability Performance of Concrete Made with Fine Recycled Concrete Aggregates. *Cem. Concr. Compos.* **2010**, *32*, 9–14. [CrossRef]
12. Zhao, Z.; Courard, L.; Michel, F.; Remond, S.; Damidot, D. Influence of Granular Fraction and Origin of Recycled Concrete Aggregates on Their Properties. *Eur. J. Environ. Civ. Eng.* **2018**, *22*, 1457–1467. [CrossRef]
13. Zaharieva, R.; Buyle-Bodin, F.; Skoczylas, F.; Wirquin, E. Assessment of the Surface Permeation Properties of Recycled Aggregate Concrete. *Cem. Concr. Compos.* **2003**, *25*, 223–232. [CrossRef]
14. Thomas, C.; Setién, J.; Polanco, J.A.; Alaejos, P.; Sánchez De Juan, M. Durability of Recycled Aggregate Concrete. *Constr. Build. Mater.* **2013**, *40*, 1054–1065. [CrossRef]
15. Xiao, J.; Li, W.; Fan, Y.; Huang, X. An Overview of Study on Recycled Aggregate Concrete in China (1996–2011). *Constr. Build. Mater.* **2012**, *31*, 364–383. [CrossRef]
16. Güneyisi, E.; Gesoğlu, M.; Akoi, A.O.M.; Mermerdaş, K. Combined Effect of Steel Fiber and Metakaolin Incorporation on Mechanical Properties of Concrete. *Compos. Part B Eng.* **2014**, *56*, 83–91. [CrossRef]
17. Matias, D.; De Brito, J.; Rosa, A.; Pedro, D. Mechanical Properties of Concrete Produced with Recycled Coarse Aggregates—Influence of the Use of Superplasticizers. *Constr. Build. Mater.* **2013**, *44*, 101–109. [CrossRef]
18. Chen, Y.Y.; Chen, S.Y.; Yang, C.J.; Chen, H.T. Effects of Insulation Materials on Mass Concrete with Pozzolans. *Constr. Build. Mater.* **2017**, *137*, 261–271. [CrossRef]
19. Sabet, F.A.; Libre, N.A.; Shekarchi, M. Mechanical and Durability Properties of Self Consolidating High Performance Concrete Incorporating Natural Zeolite, Silica Fume and Fly Ash. *Constr. Build. Mater.* **2013**, *44*, 175–184. [CrossRef]
20. Shaikh, F.U.A.; Supit, S.W.M. *Compressive Strength and Durability of High-Volume Fly Ash Concrete Reinforced with Calcium Carbonate Nanoparticles*; Elsevier Ltd: Amsterdam, The Netherlands, 2015.
21. Park, S.J.; Chase, G.G.; Jeong, K.U.; Kim, H.Y. Mechanical Properties of Titania Nanofiber Mats Fabricated by Electrospinning of Sol-Gel Precursor. *J. Sol-Gel Sci. Technol.* **2010**, *54*, 188–194. [CrossRef]

22. Youssf, O.; Elgawady, M.A.; Mills, J.E.; Ma, X. An Experimental Investigation of Crumb Rubber Concrete Confined by Fibre Reinforced Polymer Tubes. *Constr. Build. Mater.* **2014**, *53*, 522–532. [CrossRef]

23. Fraternali, F.; Farina, I.; Polzone, C.; Pagliuca, E.; Feo, L. On the Use of R-PET Strips for the Reinforcement of Cement Mortars. *Compos. Part B Eng.* **2013**, *46*, 207–210. [CrossRef]

24. Avdeeva, A. Reinforcement of Concrete Structures by Fiberglass Rods 1 Introduction 2 Materials and Methods. *MATEC Web Conf.* **2016**, *6*, 1–5.

25. Hawileh, R.A.; Rasheed, H.A.; Abdalla, J.A.; Al-Tamimi, A.K. Behavior of Reinforced Concrete Beams Strengthened with Externally Bonded Hybrid Fiber Reinforced Polymer Systems. *Mater. Des.* **2014**, *53*, 972–982. [CrossRef]

26. Gao, D.; Zhang, L.; Nokken, M.; Zhao, J. Mixture Proportion Design Method of Steel Fiber Reinforced Recycled Coarse Aggregate Concrete. *Materials* **2019**, *12*, 375. [CrossRef]

27. Chuah, S.; Pan, Z.; Sanjayan, J.G.; Wang, C.M.; Duan, W.H. Nano Reinforced Cement and Concrete Composites and New Perspective from Graphene Oxide. *Constr. Build. Mater.* **2014**, *73*, 113–124. [CrossRef]

28. Carneiro, J.A.; Lima, P.R.L.; Leite, M.B.; Toledo Filho, R.D. Compressive Stress-Strain Behavior of Steel Fiber Reinforced-Recycled Aggregate Concrete. *Cem. Concr. Compos.* **2014**, *46*, 65–72. [CrossRef]

29. Behfarnia, K.; Behravan, A. Application of High Performance Polypropylene Fibers in Concrete Lining of Water Tunnels. *Mater. Des.* **2014**, *55*, 274–279. [CrossRef]

30. Khan, M.; Ali, M. Use of Glass and Nylon Fibers in Concrete for Controlling Early Age Micro Cracking in Bridge Decks. *Constr. Build. Mater.* **2016**, *125*, 800–808. [CrossRef]

31. Afroughsabet, V.; Ozbakkaloglu, T. Mechanical and Durability Properties of High-Strength Concrete Containing Steel and Polypropylene Fibers. *Constr. Build. Mater.* **2015**, *94*, 73–82. [CrossRef]

32. Caggiano, A.; Gambarelli, S.; Martinelli, E.; Nisticò, N.; Pepe, M. Experimental Characterization of the Post-Cracking Response in Hybrid Steel/polypropylene Fiber-Reinforced Concrete. *Constr. Build. Mater.* **2016**, *125*, 1035–1043. [CrossRef]

33. Foti, D. Use of Recycled Waste Pet Bottles Fibers for the Reinforcement of Concrete. *Compos. Struct.* **2013**, *96*, 396–404. [CrossRef]

34. Noushini, A.; Samali, B.; Vessalas, K. Effect of Polyvinyl Alcohol (PVA) Fibre on Dynamic and Material Properties of Fibre Reinforced Concrete. *Constr. Build. Mater.* **2013**, *49*, 374–383. [CrossRef]

35. Chinchillas-Chinchillas, M.J.; Orozco-Carmona, V.M.; Gaxiola, A.; Alvarado-Beltrán, C.G.; Pellegrini-Cervantes, M.J.; Baldenebro-López, F.J.; Castro-Beltrán, A. Evaluation of the Mechanical Properties, Durability and Drying Shrinkage of the Mortar Reinforced with Polyacrylonitrile Microfibers. *Constr. Build. Mater.* **2019**, *210*, 32–39. [CrossRef]

36. Jayawickramage, R.A.P.; Balkus, K.J., Jr.; Ferraris, J.P. Binder Free Carbon Nanofiber Electrodes Derived from Polyacrylonitrile-Lignin Blends for High Performance Supercapacitors. *Nanotechnology* **2019**, *30*, 355402. [CrossRef] [PubMed]

37. Jiang, S.; Chen, Y.; Duan, G.; Mei, C.; Greiner, A.; Agarwal, S. Electrospun Nanofiber Reinforced Composites: A Review. *Polym. Chem.* **2018**, *9*, 2685–2720. [CrossRef]

38. Zhang, Y.; Li, J.; Li, W.; Kang, D. Synthesis of One-Dimensional Mesoporous Ag Nanoparticles-Modified TiO$_2$ Nanofibers by Electrospinning for Lithium Ion Batteries. *Materials* **2019**, *12*, 2630. [CrossRef]

39. *Norm NMX-C-061-ONNCCE-2001: Construction Industry, Cement, Determination of the Compressive Strength of Hydraulic Cements*; National Organization for Norms and Certification for Construction: Mexico City, Mexico, 2001.

40. ASTM C33. *Standard Specification for Concrete Aggregates*; ASTM International: West Conshohocken, PA, USA, 2003; Volume 4, pp. 1–11.

41. ASTM C1437-99. *Standard Test Method for Flow of Hydraulic Cement Mortar*; ASTM International: West Conshohocken, PA, USA, 1999.

42. De Guzman, D.S. *Tecnologia Del Concreto Y Del Mortero*; Biblioteca de la construcción; Bhandar Ed.: Santafé de Bogotá, Colombia, 2001.

43. ASTM C109/C109M. *Standard Test Method for Compressive Strength of Hydraulic Cement Mortars*; ASTM International: West Conshohocken, PA, USA, 2005.

44. Mendivil-Escalante, J.M.; Gómez-Soberón, J.M.; Almaral-Sánchez, J.L.; Cabrera-Covarrubias, F.G. Metamorphosis in the Porosity of Recycled Concretes through the Use of a Recycled Polyethylene Terephthalate (PET) Additive. Correlations between the Porous Network and Concrete Properties. *Materials* **2017**, *10*, 176. [CrossRef]

45. Test, C.C.; Content, A.; Rooms, M.; Concrete, P. *Standard Practice for Making and Curing Concrete Test Specimens in the Field*; ASTM International: West Conshohocken, PA, USA, 2002; Volume 4, pp. 1–8.

46. ASTM C348. *Standard Test Method for Flexural Strength of Hydraulic-Cement Mortars*; ASTM International: West Conshohocken, PA, USA, 2008.

47. ASTM C349. *Standard Test Method for Compressive Strength of Soil-Cement Using Portions of Beams Broken in Flexure (Modified Cube Method)*; ASTM International: West Conshohocken, PA, USA, 2013.

48. Pellegrini, M.J.; Barrios, C.P.; Nuñez, R.E.; Arredondo, S.P.; Baldenebro, F.J.; Rodríguez, M.; Ceballos, L.G.; Castro, A.; Fajardo, G.; Almeraya, F.; et al. Performance of Chlorides Penetration and Corrosion Resistance of Mortars with Replacements of Rice Husk Ash and Nano-SiO2. *Int. J. Electrochem. Sci.* **2015**, *10*, 332–346.

49. ASTM C1585-04. *Standard Test Method for Measurement of Rate of Absorption of Water by Hydraulic Cement Concretes*; ASTM International: West Conshohocken, PA, USA, 2004.

50. ASTM, C.G. *Standard Test Method for Corrosion Potentials of Uncoated Reinforcing Steel in Concrete*; ASTM International: West Conshohocken, PA, USA, 2003.

51. Shi, M.; Tang, C.; Yang, X.; Zhou, J.; Jia, F.; Han, Y.; Li, Z. Superhydrophobic Silica Aerogels Reinforced with Polyacrylonitrile Fibers for Adsorbing Oil from Water and Oil Mixtures. *RSC Adv.* **2017**, *7*, 4039–4045. [CrossRef]

52. Zhao, W.; Lu, Y.; Wang, J.; Chen, Q.; Zhou, L.; Jiang, J.; Chen, L. Improving Crosslinking of Stabilized Polyacrylonitrile Fibers and Mechanical Properties of Carbon Fibers by Irradiating with γ-Ray. *Polym. Degrad. Stab.* **2016**, *133*, 16–26. [CrossRef]

53. Huang, F.; Xu, Y.; Liao, S.; Yang, D.; Hsieh, Y.-L.; Wei, Q. Preparation of Amidoxime Polyacrylonitrile Chelating Nanofibers and Their Application for Adsorption of Metal Ions. *Materials* **2013**, *6*, 969–980. [CrossRef] [PubMed]

54. Pang, T.; Zhou, Z.; Li, D.; Liu, H.; Zhang, Z.; Qi, L.; Song, C.; Gao, G.; Lv, Y. Crystal Structure and Reversible Photochromism of Pb (II)-N, N-Dimethylformamide Modified Keggin-Type Polyoxometalates. *Cryst. Res. Technol.* **2019**, *54*. [CrossRef]

55. Zhang, C.; Ren, Z.; Yin, Z.; Jiang, L.; Fang, S. Experimental FTIR and Simulation Studies on H-Bonds of Model Polyurethane in Solutions. I: In Dimethylformamide (DMF). *Spectrochim. Acta Part. A Mol. Biomol. Spectrosc.* **2011**, *81*, 598–603. [CrossRef] [PubMed]

56. Hameed, N.; Sharp, J.; Nunna, S.; Creighton, C.; Magniez, K.; Jyotishkumar, P.; Salim, N.V.; Fox, B. Structural Transformation of Polyacrylonitrile Fibers during Stabilization and Low Temperature Carbonization. *Polym. Degrad. Stab.* **2016**, *128*, 39–45. [CrossRef]

57. Qiao, M.; Kong, H.; Ding, X.; Zhang, L.; Yu, M. Effect of Graphene Oxide Coatings on the Structure of Polyacrylonitrile Fibers during Pre-Oxidation Process. *RSC Adv.* **2019**, *9*, 28146–28152. [CrossRef]

58. Chinchillas-Chinchillas, M.J.; Orozco-Carmona, V.M.; Alvarado-Beltrán, C.G.; Almaral-Sánchez, J.L.; Sepulveda-Guzman, S.; Jasso-Ramos, L.E.; Castro-Beltrán, A. Synthesis of Recycled Poly(ethylene terephthalate)/Polyacrylonitrile/Styrene Composite Nanofibers by Electrospinning and Their Mechanical Properties Evaluation. *J. Polym. Environ.* **2019**, *27*, 659–669. [CrossRef]

59. Sassani, A.; Arabzadeh, A.; Ceylan, H.; Kim, S.; Gopalakrishnan, K.; Taylor, P.C.; Nahvi, A. Polyurethane-Carbon Microfiber Composite Coating for Electrical Heating of Concrete Pavement Surfaces. *Heliyon* **2019**, *5*, e02359. [CrossRef]

60. Duan, Z.H.; Poon, C.S. Properties of Recycled Aggregate Concrete Made with Recycled Aggregates with Different Amounts of Old Adhered Mortars. *Mater. Des.* **2014**, *58*, 19–29. [CrossRef]

61. Mousavi, S.M.; Ranjbar, M.M.; Madandoust, R. Combined Effects of Steel Fibers and Water to Cementitious Materials Ratio on the Fracture Behavior and Brittleness of High Strength Concrete. *Eng. Fract. Mech.* **2019**, *216*. [CrossRef]

62. Singh, S.B.; Munjal, P.; Thammishetti, N. Role of Water/cement Ratio on Strength Development of Cement Mortar. *J. Build. Eng.* **2015**, *4*, 94–100. [CrossRef]

63. Kwan, A.K.H.; Dong, C.X.; Ho, J.C.M. Axial and Lateral Stress-Strain Model for FRP Confined Concrete. *Eng. Struct.* **2015**, *99*, 285–295. [CrossRef]

64. Hou, D.; Li, D.; Hua, P.; Jiang, J.; Zhang, G. Statistical Modelling of Compressive Strength Controlled by Porosity and Pore Size Distribution for Cementitious Materials. *Cem. Concr. Compos.* **2019**, *96*, 11–20. [CrossRef]

65. Zhong, R.; Wille, K. Material Design and Characterization of High Performance Pervious Concrete. *Constr. Build. Mater.* **2015**, *98*, 51–60. [CrossRef]

66. Katz, A.; Kulisch, D. Performance of Mortars Containing Recycled Fine Aggregate from Construction and Demolition Waste. *Mater. Struct. Constr.* **2017**, *50*, 199. [CrossRef]

67. Pellegrini Cervantes, M.J.; Barrios Durstewitz, C.P.; Núñez Jaquez, R.E.; Almeraya Calderón, F.; Rodríguez Rodríguez, M.; Fajardo-San-Miguel, G.; Martinez-Villafañe, A. Accelerated Corrosion Test in Mortars of Plastic Consistency with Replacement of Rice Husk Ash and Nano-SiO$_2$. *Int. J. Electrochem. Sci.* **2015**, *10*, 8630–8643.

68. Sanawung, W.; Cheewaket, T.; Tangchirapat, W.; Jaturapitakkul, C. Influence of Palm Oil Fuel Ash and W/B Ratios on Compressive Strength, Water Permeability, and Chloride Resistance of Concrete. *Adv. Mater. Sci. Eng.* **2017**, *2017*. [CrossRef]

69. Jiménez, L.F.; Moreno, E.I. Durability Indicators in High Absorption Recycled Aggregate Concrete. *Adv. Mater. Sci. Eng.* **2015**, *2015*, 1–8. [CrossRef]

70. Sobczak, L.; Brüggemann, O.; Putz, R.F. Polyolefin Composites with Natural Fibers and Wood-modification of the Fiber/filler–matrix Interaction. *J. Appl. Polym. Sci.* **2013**, *127*, 1–17. [CrossRef]

71. Hall, C. Water Sorptivity of Mortars and Concretes: A Review. *Mag. Concr. Res.* **1989**, *41*, 51–61. [CrossRef]

72. Awasthi, A.V.K.M.S.; Goyal, S.K.N.O.Y. Recycled Aggregate from C&D Waste Modified by Dry Processing and Used as A Partial Replacement of Coarse Aggregate in Concrete. *J. Mater. Sci. Surf. Eng.* **2017**, *5*, 671–678.

73. Pan, G.; Zhan, M.; Fu, M.; Wang, Y.; Lu, X. Effect of CO$_2$ Curing on Demolition Recycled Fine Aggregates Enhanced by Calcium Hydroxide Pre-Soaking. *Constr. Build. Mater.* **2017**, *154*, 810–818. [CrossRef]

74. Paul, S.C.; Panda, B.; Huang, Y.; Garg, A.; Peng, X. An Empirical Model Design for Evaluation and Estimation of Carbonation Depth in Concrete. *Measurement* **2018**, *124*, 205–210. [CrossRef]

75. Almeida, A.E.F.S.; Tonoli, G.H.D.; Santos, S.F.; Savastano, H. Improved Durability of Vegetable Fiber Reinforced Cement Composite Subject to Accelerated Carbonation at Early Age. *Cem. Concr. Compos.* **2013**, *42*, 49–58. [CrossRef]

Article

Numerical Studies on Damage Behavior of Recycled Aggregate Concrete Based on a 3D Model

Yao Wang [1,2,*], Huawei Zhao [1,*], Minyao Xu [1], Chunyang Wu [1], Jiajia Fu [1,3], Lili Gao [1] and Mahmoud M. A. Kamel [2,4]

1   Department of Architecture and Engineering, Yancheng Polytechnic College, Jiangsu 224005, China; miaoyun@njtech.edu.cn (M.X.); zhhpan@just.edu.cn (C.W.); njtfjj@njtech.edu.cn (J.F.); weimin@khu.ac.kr (L.G.)
2   Department of Architecture and Engineering, Beijing University of Technology, Beijing 100124, China; mahmoud.kamel@fagoum.edu.eg
3   School of Civil Engineering, Nanjing Technology University, Nanjing 211816, China
4   Department of Civil Engineering, Faculty of Engineering, Fayoum University, Fayoum 63514, Egypt
*   Correspondence: yaowang@emails.bjut.edu.cn (Y.W.); zhaohuawei@ycit.edu.cn (H.Z.); Tel.: +86-188-1104-9095 (Y.W.); +86-158-9518-0056 (H.Z.)

Received: 16 November 2019; Accepted: 9 January 2020; Published: 12 January 2020

Abstract: This paper develops a 3D base force element method (BFEM) based on the potential energy principle. According to the BFEM, the stiffness matrix and node displacement of any eight-node hexahedral element are derived as a uniform expression. Moreover, this expression is explicitly expressed without a Gaussian integral. A 3D random numerical model of recycled aggregate concrete (RAC) is established. The randomness of aggregate was obtained by using the Monte Carlo random method. The effects of the recycled aggregate substitution and adhered mortar percentage on the elastic modulus and compressive strength are explored under uniaxial compression loading. In addition, the failure pattern is also studied. The obtained data show that the 3D BFEM is an efficient method to explore the failure mechanism of heterogeneous materials. The 3D random RAC model is feasible for characterizing the mesostructure of RAC. Both the substitution of recycled aggregate and the percentage of adhering mortar have a non-negligible influence on the mechanical properties of RAC. As the weak points in the specimen, the old interfacial transition zone (ITZ) and adhered mortar are the major factors that lead to the weakened properties of RAC. The first crack always appears in these weak zones, and then, due to the increase and transfer of stress, approximately two-to-three continuous cracks are formed in the 45° direction of the specimen.

Keywords: 3D BFEM; recycled aggregate concrete; numerical simulation; failure pattern

---

1. Introduction

In the last several years, recycled aggregate concrete (RAC) has become a popularly used construction material that can effectively alleviate the shortage of natural resources. As a kind of green building material, RAC has attracted many researchers to explore its mechanical performance [1–5]. Compared to natural aggregate concrete (NAC), RAC has a highly heterogeneous internal composite, and its mechanical behavior is related to the mesostructure of its components. The literature has revealed that the substitution of recycled aggregate, water/cement ratio, aggregate content, the percentage of adhered mortar, air content, etc. play a significant role in the mechanical properties of RAC [6,7]. At the mesostructural level, the component of RAC identified to be a five-phase system including recycled aggregate, adhered mortar, new cement mortar, an old interfacial transition zone (ITZ), and a new ITZ. There are two ways to explore the mechanical properties of engineering materials, namely macroscopic experimental tests and mesoscopic numerical simulations. Since the concept of the finite element method (FEM) was proposed by Clough in the 1960s [8], FEM has become an effective and accurate approach for assisting macroscopic experiments [9–13]. In addition, this method reduces the consumption of natural resources, time, and testing costs.

At the mesoscopic level, lots of work about the modelling of aggregate have been conducted by many researchers. For the simulation of concrete, a numerical concrete concept was proposed by Zaitsev and Wittmann, and three random 2D structures including spherical geometry, polygons and arbitrary polygonal were first generated based on meso-mechanics [14,15]. Subsequently, different mesoscale structures were proposed to simulate the concrete fracture process according to the FEM. For instance, Peng et al. [16] established a model of a circular aggregate model to explore the mechanical properties of concrete according to the Walraven formula [17] and the Monte Carlo random sampling principle. Additionally, based on the Monte Carlo random sampling principle, a particle model was established to represent the fragile aggregates by Bazant et al. [18]. Wang et al. [7] proposed a convex aggregate to model crushed stone based on a round aggregate. Wriggers et al. [19] and Chen et al. [20] proposed a 3D geometrical model for NAC according to the random mesostructure of natural aggregates in a specimen.

These natural aggregate models have provided effective reference methods to simulate recycled aggregate. Some researchers conducted a series of numerical research studies, and some conclusions have been obtained about RAC. Xiao et al. [11] designed a nine recycled aggregate model to study the effect of the relative elastic modulus of ITZs of cement mortar on the damage crack of RAC under uniaxial compression and tensile loading; the obtained data showed that the relative elastic modulus had a major effect on the stress–strain curves and RAC strength. Sun et al. [21] presented a 3D FEM model to research the effect of recycled aggregate substitution on shear strength by using the ABAQUS/Standard module software. The data showed that the shear stress was reduced by up to 13.8% by ranging the substitution from 0% to 100%. Chen et al. [22] designed four levels of recycled aggregate substitution to explore the damage mechanism of RAC under uniaxial compression loading. The data showed that the compressive strength reduced as the replacement ratio increased. Jayasuriya et al. [23] presented four different adhered mortar percentages to analyze the effect of adhering mortar on compressive strength, and their numerical data showed that the compressive strength was reduced up to 9% as changing the adhering mortar from 2% to 50%. Wang et al. [7] established two types of aggregate shapes to explore the effect of recycled aggregate replacement ratios on mechanical properties, and their simulation results demonstrated that the elastic modulus reduced by up to 16%~25%, and the compressive strength reduced by up to 12%~15% as the replacement ratios increased from 0% to 100%. Due to the exceedingly complicated stiffness matrix and multiple degrees of freedom per each element in the 3D level, a structure and mesh topology was rarely generated [19]. Most numerical simulations of RAC mechanical properties have been concentrated on the 2D level and scaled up by a thickness of the fictitious slice, and only a little consideration has been focused on 3D models of RAC.

As mentioned previously, the FEM has been proven to be an efficient approach to explore the mechanical properties of materials. In recent years, according to the potential energy principle, a new FEM concept and a new 2D FEM method were presented by Gao [24] and Peng et al. [25], named base force and the base force element method (BFEM), respectively. Based on the BFEM, the element stiffness matrix is conveyed by an explicit tensor formulation for an element with an arbitrary shape in any coordinate system. Moreover, the Gaussian integrals are not used in the calculations and deriving processes of the element stiffness matrix.

For this paper, the 2D BFEM was developed into a 3D BFEM. In addition, a hexahedron element was established. A 3D numerical RAC model was established according to the 3D Fuller grading curve and the Monte Carlo random sampling method. The recycled aggregate was assumed to be a spherical particle. Several numerical models of RAC with five substitutions of recycled aggregate and six different percentages of adhered mortar were designed. These 3D RAC models were subjected to uniaxial compression loading that was controlled by displacement. The effects of the recycled aggregate replacement ratio and the adhered mortar content on the compressive strength and elastic modulus were investigated. Additionally, the failure pattern of a 3D RAC model was also calculated.

2. Basic Formula of 3D Base Force Theory

For a 3D region of a solid medium in the Lagrangian coordinate system, $x^i (i = 1, 2, 3)$ denotes the coordinate axes, and P, Q denote the initial/after position vector. The base vectors of material points can be defined as:

$$P_i = \frac{\partial P}{\partial x^i}, Q_i = \frac{\partial Q}{\partial x^i} \tag{1}$$

The displacement gradient can be expressed by the base vectors, as follows:

$$u_i = \frac{\partial u}{\partial x^i} = P_i - Q_i = \frac{\partial P}{\partial x^i} - \frac{\partial Q}{\partial x^i} \tag{2}$$

To express the stress state of the point Q, assuming a current configuration of this solid medium in the Cartesian coordinate system x^i, define a parallel hexahedron element and let dx^1, dx^2, dx^3 denote the element edges, as illustrated in

The force applied on the front surface of the element is marked as dT^i, so let:

$$T^i = \frac{1}{dx^{i+1} dx^{i-1}} dT^i \quad dx^i \to 0 \tag{3}$$

In Equation (3), the indexes are promised $3 + 1 = 1$ and $1 - 1 = 3$, where $T^i (i = 1, 2, 3)$ is the base force acting on the point Q in the 3D coordinate system x^i. Figure 1.

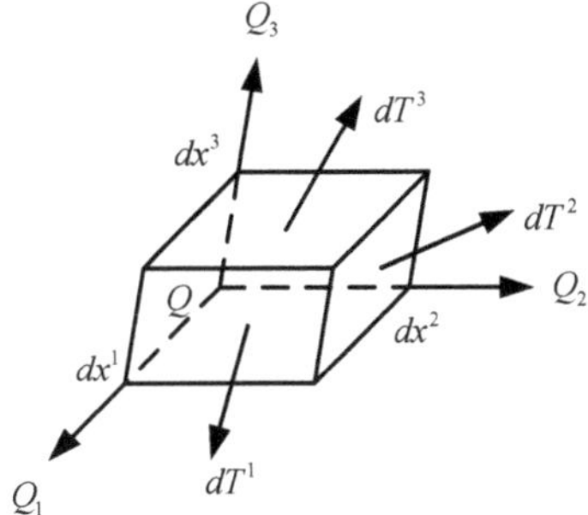

Figure 1. Base forces.

To describe $T^i (i = 1, 2, 3)$, define an arbitrary plane with normal n. The plane and coordinate axes x^i intersect at dx^1, dx^2, dx^3, as shown in Figure 2. According to the equilibrium condition, the stress vector that acts on the surface can be obtained as:

$$\sigma^n dS = \frac{1}{2} dx^1 dx^2 dx^3 \left(\frac{1}{dx^1} T^1 + \frac{1}{dx^2} T^2 + \frac{1}{dx^3} T^3 \right) \tag{4}$$

Let V_Q denote the current base volume of x^i system, and

$$V_Q = (Q_1, Q_2, Q_3) = Q_1 \cdot (Q_2 \times Q_3). \tag{5}$$

Then,

$$\sigma^n = \frac{1}{V_Q} T^i \frac{\partial n}{\partial x^i} \tag{6}$$

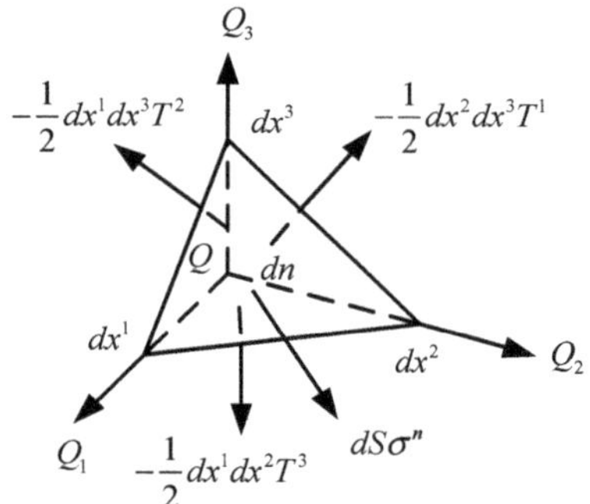

Figure 2. Forces on a tetrahedron.

The key point to note here is that:

$$\frac{\partial n}{\partial x^i} = \boldsymbol{Q}_i \cdot \boldsymbol{n} = n_i \tag{7}$$

Equation (6) can be derived as:

$$\sigma^n = \frac{n_i}{V_Q} \boldsymbol{T}^i \tag{8}$$

Then, for a small deformation case, the strain ε can be obtained as:

$$\varepsilon = \frac{1}{2}\left(\boldsymbol{u}_i \otimes \boldsymbol{P}^i + \boldsymbol{P}^i \otimes \boldsymbol{u}_i\right) \tag{9}$$

The relationship between the base forces and the various stress tensors can be obtained according to the base force.

As for the Cauchy stress, σ is:

$$\sigma = \frac{1}{V_Q} \boldsymbol{T}^i \otimes \boldsymbol{Q}_i \tag{10}$$

As for the Piola stress, τ is:

$$\tau = \frac{1}{V_P} \boldsymbol{T}^i \otimes \boldsymbol{P}_i \tag{11}$$

As for the Kirchhoff stress, Σ is:

$$\Sigma = \boldsymbol{P}_i \otimes \boldsymbol{Q}^i \frac{1}{V_P} \boldsymbol{T}^i \otimes \boldsymbol{P}_i \tag{12}$$

The equilibrium equation is the balance of the stress, inertial force and volume forces of the structure. For static problems, the equilibrium equation can be expressed by the base force, as follows:

$$\frac{\partial}{\partial x^i} \boldsymbol{T}^i + \rho_0 V_P f = 0, \tag{13}$$

and the geometric equation can be obtained according to the displacement gradient:

$$\varepsilon = \frac{1}{2}\left(\boldsymbol{u}_i \cdot \boldsymbol{P}_j + \boldsymbol{P}_i \cdot \boldsymbol{u}_j\right)\boldsymbol{P}^i \otimes \boldsymbol{P}^j \tag{14}$$

Similarly, based on the base force, the physical equation is:

$$\boldsymbol{T}^i = \rho_0 V_P \frac{\partial W}{\partial \boldsymbol{Q}_i} \tag{15}$$

in which W is the strain energy of the element.

3. BFEM Model of Hexahedron Element

Following the BFEM, a hexahedron element considering boundary conditions can be presented, as depicted in Figure 3. $A, B, \ldots, G$ are the vertices of the element, $u_{Ij}(I = A, B, \ldots G; j = x, y, z)$ denote the component of the displacement of the point I on the coordinate axes J, and $\alpha, \beta, \gamma \ldots$ are the six areas of the model.

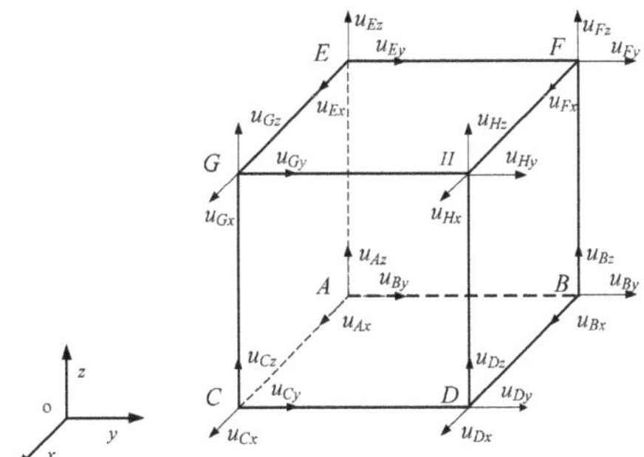

Figure 3. A hexahedron element.

The hexahedron elements contact each other through the faces in the model. A relationship can be established between the displacements of the points and the displacements of the faces. Take any plane in the hexahedron as a typical face and let it be represented by α, as depicted in Figure 4. Connect the centroid point and the midpoint of the four sides; therefore, the quadrilateral is divided into four parts. Let S_α express the area of α and $S_{\alpha I}(I = A, B, C \ldots)$ denote the area of the separated part.

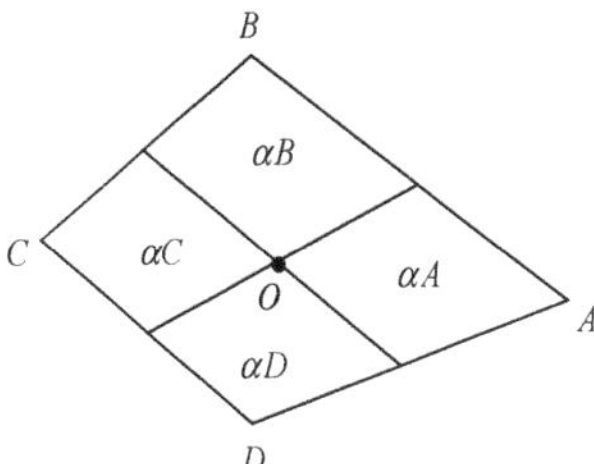

Figure 4. A typical face.

It is hypothesized that in the process of deformation, the shape and line segments are always kept flat and straight, respectively. Hence, the deformation of the centroid can be obtained as:

$$u_\alpha = \frac{1}{S_\alpha}(S_{\alpha A}u_A + S_{\alpha B}u_B + \cdots)$$
(16)

3.1. Strain Tensor

Assume that the volume V_Q of the hexahedron element is small complete and the actual strain ε can be replaced by the average strain $\bar{\varepsilon}$. In addition, the average strain $\bar{\varepsilon}$ can be obtained as:

$$\bar{\varepsilon} = \frac{1}{V}\int_V \varepsilon dV$$
(17)

Then, by substituting Equation (9) into Equation (17), one can obtain:

$$\bar{\varepsilon} = \frac{1}{2V} \int_V (\boldsymbol{u}_\alpha \otimes \boldsymbol{P}^\alpha + \boldsymbol{P}^\alpha \otimes \boldsymbol{u}_\alpha)dV \tag{18}$$

According to Gauss theorem, Equation (18) can be replaced by:

$$\bar{\varepsilon} = \frac{1}{2V}\left(\boldsymbol{u}_I \otimes \boldsymbol{m}^I + \boldsymbol{m}^I \otimes \boldsymbol{u}_I\right) \tag{19}$$

Equation (19) implies the summation rule, so $\boldsymbol{m}^I$ is:

$$\boldsymbol{m}^I = S_{\alpha I}\boldsymbol{n}_\alpha + S_{\beta I}\boldsymbol{n}_\beta + S_{\gamma I}\boldsymbol{n}_\gamma + \ldots \tag{20}$$

The expressions of the hexahedron element are:

$$\begin{aligned}
\boldsymbol{m}^I &= S_{\alpha I}\boldsymbol{n}_\alpha + S_{\beta I}\boldsymbol{n}_\beta + S_{\gamma I}\boldsymbol{n}_\gamma \\
&= S_{\alpha I}\left(n_{\alpha x}\boldsymbol{e}_x + n_{\alpha y}\boldsymbol{e}_y + n_{\alpha z}\boldsymbol{e}_z\right) + S_{\beta I}\left(n_{\beta x}\boldsymbol{e}_x + n_{\beta y}\boldsymbol{e}_y + n_{\beta z}\boldsymbol{e}_z\right) \\
&\quad + S_{\gamma I}\left(n_{\gamma x}\boldsymbol{e}_x + n_{\gamma y}\boldsymbol{e}_y + n_{\gamma z}\boldsymbol{e}_z\right) \\
&= \left(S_{\alpha I}n_{\alpha x} + S_{\beta I}n_{\beta x} + S_{\gamma I}n_{\gamma x}\right)\boldsymbol{e}_x + \left(S_{\alpha I}n_{\alpha y} + S_{\beta I}n_{\beta y} + S_{\gamma I}n_{\gamma y}\right)\boldsymbol{e}_y \\
&\quad + \left(S_{\alpha I}n_{\alpha z} + S_{\beta I}n_{\beta z} + S_{\gamma I}n_{\gamma z}\right)\boldsymbol{e}_z
\end{aligned} \tag{21}$$

Then, by substituting Equation (21) into Equation (19) and by letting x, y, z represent the Cartesian coordinate system, the following can be obtained:

$$\begin{aligned}
\bar{\varepsilon} = \frac{1}{2V}\sum_{I=1}^{n}\Big(&2u_{Ix}m_x^I\boldsymbol{e}_x \otimes \boldsymbol{e}_x + 2u_{Iy}m_y^I\boldsymbol{e}_y \otimes \boldsymbol{e}_y \\
&+2u_{Iz}m_z^I\boldsymbol{e}_z \otimes \boldsymbol{e}_z + \left(u_{Ix}m_y^I + u_{Iy}m_x^I\right)\boldsymbol{e}_y \otimes \boldsymbol{e}_x \\
&+\left(u_{Ix}m_y^I + u_{Iy}m_x^I\right)\boldsymbol{e}_x \otimes \boldsymbol{e}_y + \left(u_{Ix}m_z^I + u_{Iz}m_x^I\right)\boldsymbol{e}_x \otimes \boldsymbol{e}_z \\
&+\left(u_{Ix}m_z^I + u_{Iz}m_x^I\right)\boldsymbol{e}_z \otimes \boldsymbol{e}_x + \left(u_{Iy}m_z^I + u_{Iz}m_y^I\right)\boldsymbol{e}_y \otimes \boldsymbol{e}_z \\
&+\left(u_{Iy}m_z^I + u_{Iz}m_y^I\right)\boldsymbol{e}_z \otimes \boldsymbol{e}_y\Big)
\end{aligned} \tag{22}$$

or

$$\begin{aligned}
&\bar{\varepsilon}_x = \frac{1}{V}\sum_{I=1}^{n}\left(u_{Ix}m_x^I\right) && \bar{\varepsilon}_y = \frac{1}{V}\sum_{I=1}^{n}\left(u_{Iy}m_y^I\right) \\
&\bar{\varepsilon}_z = \frac{1}{V}\sum_{I=1}^{n}\left(u_{Iz}m_z^I\right) && \bar{\gamma}_{xz} = \frac{1}{V}\sum_{I=1}^{n}\left(u_{Ix}m_z^I + u_{Iz}m_x^I\right) \\
&\bar{\gamma}_{xy} = \frac{1}{V}\sum_{I=1}^{n}\left(u_{Ix}m_y^I + u_{Iy}m_x^I\right) && \bar{\gamma}_{yz} = \frac{1}{V}\sum_{I=1}^{n}\left(u_{Iy}m_z^I + u_{Iz}m_y^I\right)
\end{aligned} \tag{23}$$

3.2. Stiffness Matrix

As the linear elastic material, the strain energy expression of the element can be obtained as:

$$W_D = \frac{VE}{2(1+v)}\left[\frac{v}{1-2v}(\bar{\varepsilon} : \boldsymbol{U})^2 + \bar{\varepsilon} : \bar{\varepsilon}\right] \tag{24}$$

In this formula, V denotes the hexahedron volume, E expresses the elastic modulus, v expresses the Poisson's ratio, and U denotes the unit tensor.

Then, by substituting Equation (19) into Equation (24), the strain energy can be obtained as:

$$W_D = \frac{E}{4V(1+v)}\left[\frac{2v}{1-2v}\left(\boldsymbol{u}_I \times \boldsymbol{m}^I\right)^2 + \left(\boldsymbol{u}_I \cdot \boldsymbol{u}_J\right)m^{IJ} + \left(\boldsymbol{u}_I \cdot \boldsymbol{m}^J\right)\left(\boldsymbol{u}_J \cdot \boldsymbol{m}^I\right)\right] \tag{25}$$

Based on Equation (25), the force applying to the node $I(A, B, \ldots G)$ on the element can be expressed as:

$$f^I = \frac{\partial W_D}{\partial u^I} = K^{IJ} \cdot u_J \tag{26}$$

Then, the stiffness matrix K of the can be obtained as:

$$K^{IJ} = \frac{E}{2V(1+v)}\left[\frac{2v}{1-2v} m^I \otimes m^J + m^{IJ} U + m^J \otimes m^I\right] \tag{27}$$

Here, $m^I = m^I_i e_i = m^I_x e_x + m^I_y e_y + m^I_z e_z$, and $m^{IJ} = m^I \cdot m^J$.

By transforming Equation (27) into a Descartes coordinate system (x, y, z), the stiffness matrix K can be described to be:

$$
\begin{aligned}
K^{IJ} = \frac{E}{2V(1+v)}\Big[& e_x \otimes e_x\left(\tfrac{2v}{1-2v} m^I_x m^J_x + m^I_x m^J_x + m^I_y m^J_y + m^I_z m^J_z + m^I_x m^J_x\right) \\
& + e_x \otimes e_y\left(\tfrac{2v}{1-2v} m^I_x m^J_y + m^I_y m^J_x\right) + e_y \otimes e_x\left(\tfrac{2v}{1-2v} m^I_y m^J_x + m^I_x m^J_y\right) \\
& + e_x \otimes e_z\left(\tfrac{2v}{1-2v} m^I_x m^J_z + m^I_z m^J_x\right) + e_z \otimes e_x\left(\tfrac{2v}{1-2v} m^I_z m^J_x + m^I_x m^J_z\right) \\
& + e_y \otimes e_z\left(\tfrac{2v}{1-2v} m^I_y m^J_z + m^I_z m^J_y\right) + e_z \otimes e_y\left(\tfrac{2v}{1-2v} m^I_z m^J_y + m^I_y m^J_z\right) \\
& + e_y \otimes e_y\left(\tfrac{2v}{1-2v} m^I_y m^J_y + m^I_x m^J_x + m^I_y m^J_y + m^I_z m^J_z + m^I_y m^J_y\right) \\
& + e_z \otimes e_z\left(\tfrac{2v}{1-2v} m^I_z m^J_z + m^I_x m^J_x + m^I_y m^J_y + m^I_z m^J_z + m^I_z m^J_z\right)\Big]
\end{aligned}
\tag{28}
$$

or

$$
K^{IJ} = \frac{E}{2V(1+v)}
\begin{bmatrix}
\frac{2-2v}{1-2v} m^I_x m^J_x + m^I_y m^J_y + m^I_z m^J_z & \frac{2v}{1-2v} m^I_x m^J_y + m^I_y m^J_x & \frac{2v}{1-2v} m^I_x m^J_z + m^I_z m^J_x \\
\frac{2v}{1-2v} m^I_y m^J_x + m^I_x m^J_y & \frac{2-2v}{1-2v} m^I_y m^J_y + m^I_x m^J_x + m^I_z m^J_z & \frac{2v}{1-2v} m^I_y m^J_z + m^I_z m^J_y \\
\frac{2v}{1-2v} m^I_z m^J_x + m^I_x m^J_z & \frac{2v}{1-2v} m^I_z m^J_y + m^I_y m^J_z & \frac{2-2v}{1-2v} m^I_z m^J_z + m^I_x m^J_x + m^I_y m^J_y
\end{bmatrix}
\tag{29}
$$

in which m^I_i can be calculated:

$$
\begin{aligned}
m^I_x &= \tfrac{1}{4} S\left(n_{\alpha x} + n_{\beta x} + n_{\gamma x}\right) \\
m^I_y &= \tfrac{1}{4} S\left(n_{\alpha y} + n_{\beta y} + n_{\gamma y}\right) \\
m^I_z &= \tfrac{1}{4} S\left(n_{\alpha z} + n_{\beta z} + n_{\gamma z}\right)
\end{aligned}
\tag{30}
$$

where $(I, J = 1, 2, 3 \ldots, 8)$ are the nodes of an element, and $n_{IJ}(I = \alpha, \beta, \gamma; J = x, y, z)$ is the normal vector component of the air I about coordinate axes J.

4. Random Model of RAC

4.1. Aggregate Number

When considering the actual specimen of RAC, the aggregates are randomly distributed inside the test piece. Therefore, to obtain a more realistic mesostructure of RAC, the Fuller grading curve was adopted to calculate the amount of aggregate in the specimen.

By assuming that the aggregate shape is spherical, the simple equation is as follows:

$$P = 100\left(\frac{d}{D_{\max}}\right)^n \tag{31}$$

where P denotes the cumulative distribution of the aggregate that was filtered through the diameter of sieve pore, d denotes the diameter of sieve pore, $D_{\max}$ denotes the maximum size of aggregate, n denotes the index, and, in this paper, $n = 0.5$.

The volume of all aggregates in the grading interval $[d_c, d_{c+1}]$ is defined as:

$$V_i = 0.5 \times V \times (P_{ci} - P_{ci+1}) \tag{32}$$

where V is the specimen volume.

Then, the numbers of random spherical coarse aggregate particles with different diameters can be calculated:

$$N_i = V_i / \left(\pi D_i^3 / 6 \right) \tag{33}$$

where D_i represents the sizes of the spherical coarse aggregate particles.

According to Equations (31)–(33), the amount of random spherical coarse aggregates with different radii can be calculated. The amount of recycled aggregate (RA) and natural aggregate (NA) with different replacement ratios are displayed in Table 1.

Table 1. The amount of the aggregate.

Replacement Ratio	Aggregate Radius (mm)					
	7.5		12.5		17.5	
	RA	NA	RA	NA	RA	NA
0%	0	468	0	77	0	23
15%	70	398	12	65	3	20
30%	140	328	23	54	7	16
50%	234	234	38	39	11	12
100%	468	0	77	0	23	0

4.2. Placing Algorithm

Following to Monte Carlo random method, three independent random numbers R_n, E_n, F_n between 0 and 1 were generated to calculate the x_n, y_n, z_n position of the aggregate particles.

It should be noted that the placing algorithm should satisfy the following conditions: the named boundary condition and overlapping condition:

(1) The aggregate particles must be completely located in the specimen.
(2) The aggregate particles must not overlap with each other.
(3) The distance between centers of any two adjacent aggregate particles must be larger than that of $1.20 \left(d_a + d_b \right)$, where d_a and d_b are the radii of the two adjacent aggregates.

The placement process used can be summarized as follows:
Step 1: Generate three random numbers to get the particles coordinate.
Step 2: Check the boundary condition and the overlapping condition; if they do not meet the requirements, go back to Step 1.
Step 3: Place the aggregate into the specimen.
Step 4: Repeat the above steps for each aggregate.
The obtained aggregate coordinates are depicted in Figure 5.

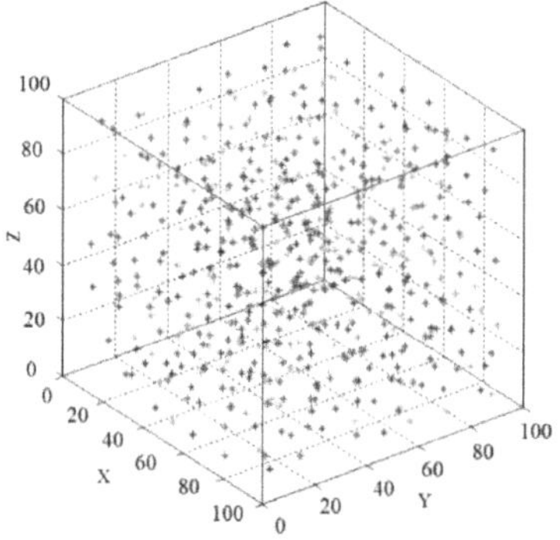

Figure 5. The aggregate coordinates scatter plot.

4.3. Numerical Model of RAC

The numerical models of RAC are displayed in Figure 6. Here, the dimension was $100 \times 100 \times 100$ mm, and the replacement ratio was 50%. It should be mentioned that the dark blue aggregates and purple aggregates represent the natural and recycled aggregates, respectively.

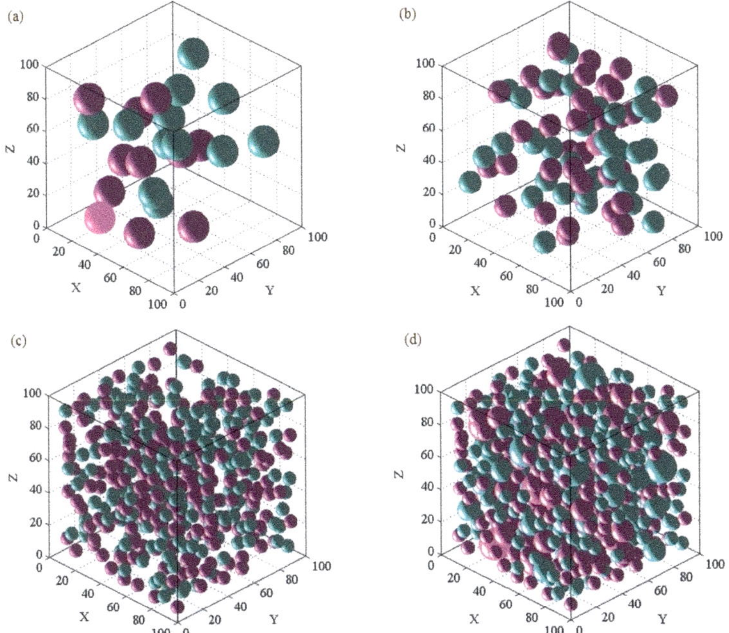

Figure 6. 3D Random aggregate model of recycled aggregate concrete (RAC). (**a**) The radius is 8.75 mm; (**b**) the radius is 6.25 mm; and (**c**) the radius is 3.75 mm; (**d**) All aggregates.

As can be observed in Figure 6, the aggregates had a good distribution and did not overlap with one another. In addition, four slices were extracted from the model to verify the accuracy of the placing algorithm. Here, the mesh size was 0.8×0.8 mm, as detailed in Figure 7a. Meanwhile, the five-phase system is also indicated in Figure 7b.

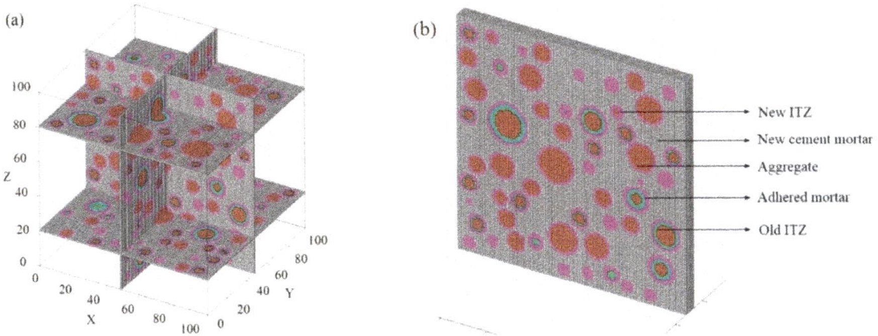

Figure 7. Slices and five-phase in the model. (**a**) Four slices in the model. (**b**) Five-phase system.

4.4. Mechanical Parameters

It is well accepted that the concrete is regarded as a quasi-brittle material, and lots of damage constitutive models have been presented [26–28]. In this work, the failure mechanical behavior of the five-phase system was described by a bilinear failure model [7].

According to the experiment results of [29–32] and numerical results of [7,23,33], it has been accepted that the mechanical properties of ITZs are weaker than those of the corresponding cement mortars. In addition, their elastic modulus is randomly distributed as 0.5–0.85 times that of cement mortar. Therefore, for this paper, the elastic modulus of the ITZs was selected as 0.55 times of the corresponding cement mortars, as noted in Table 2.

Table 2. Mechanical properties of the five-phase system.

Mechanical Properties	Five Phases				
	Natural Aggregate	Old ITZ	Adhered Mortar	New ITZ	New Cement Mortar
Elastic modulus/GPa	75	13.75	25	16.50	30
Poisson's ratio	0.16	0.20	0.22	0.20	0.22
Tensile strength/MPa	10.0	2.0	2.5	2.0	3.0

5. Simulation of Uniaxial Compressive Test

For this section, several RAC models were applied to the uniaxial compression loading. All nodes of the bottom elements and the nodes of the mid-bottom elements were restrained in the vertical direction and horizontal direction, respectively. In addition, the displacement loading was applied to the nodes of the top elements at 0.005 mm/step. These models were used to explore the influences of recycled aggregate substitution and the adhered mortar percent on the elastic modulus, the compressive strength, and the crack pattern. Therefore, (1) five different substitutions of recycled aggregate (0%, 15%, 30%, 50%, and 100%) were established for these models, and the adhered mortar was chosen as 40%; (2) another RAC model with one recycled aggregate was established, and six levels of adhered mortar percentage (0%, 5%, 10%, 30%, 40%, and 50%) of the recycled aggregate were designed; and (3) the RAC model with one aggregate was cut off to display the occurrence and development of cracks inside the specimen, with the percentage of the adhered mortar being 40%.

5.1. Effect of Aggregate Substitution

It is well known that the substitution of recycled aggregate is a major factor that affects mechanical properties. When increasing the substitution portion of natural aggregates by the recycled aggregate, both the elastic modulus and compressive strength decrease [28]. The obtained data are shown in Table 3.

Table 3. Effect of substitution on the mechanical properties of RAC.

Mechanical Properties	Replacement Ratio				
	0%	15%	30%	50%	100%
Elastic modulus/GPa	25.56	24.29	22.36	22.06	21.34
Compressive stress/MPa	28.09	26.08	25.23	25.19	25.06

As can be seen from Table 3, when the substitution was less than 30%, despite the reduction in the elastic modulus and compressive strength, only slight differences were obtained. However, when the substitution was increased from 0% to 100%, both the modulus of elasticity and the compressive strength showed reductions of up to 15.6% and 10.8%, respectively.

These phenomena can be attributed to the increasing substitution of recycled aggregate, which resulted in an increase in the mortar adhering to recycled aggregate and the old ITZ between them. These two increased phases were considered to be the weak phase in the specimen and had lower mechanical properties. Therefore, as the substitution increased, the elastic modulus of the specimen decreased. These observed data coincide with other results in the literature [9,34–37].

5.2. Effect of Adhered Mortar Percentage

Due to how waste concrete is dealt with, it is inevitable that some adhered mortar will remain around the surface of the aggregate. The physical property of the recycled aggregate depends on the percent and property of the adhering mortar. Previous studies have shown that the adhered mortar is a major factor that weakens the mechanical properties of RAC [36,38,39].

In this section, it should be mentioned that only one recycled aggregate was placed to test the effect of adhered mortar on the mechanical properties of RAC. This design avoided the influences of aggregate grading and aggregate distribution on its properties. Therefore, it was meaningless to use this model to investigate the values of the mechanical properties of RAC. Consequently, only the relative values of compressive strength and modulus of elasticity are given, as listed in Table 4.

Table 4. Effect of the percent of adhering mortar on the mechanical properties of RAC.

Mechanical Properties	Percentage of Adhered Mortar					
	0%	5%	10%	30%	40%	50%
Elastic modulus/GPa	1	0.96	0.93	0.86	0.84	0.82
Compressive stress/MPa	1	0.97	0.95	0.92	0.91	0.88

Note: 0% corresponds to natural aggregate concrete.

As listed in Table 4, with the increasing percentage of adhered mortar, both the modulus of elasticity and compressive strength decreased. This was the same effect as that of the substitution—when the percent of the adhering mortar was less than 10%, the value of the mechanical properties showed a slight reduction. As the percentage of the adhered mortar increased from 0% to 50%, the compressive strength and elastic modulus decreased by 12% and 18%, respectively. These results can be attributed to the lower mechanical properties of the adhered mortar than that of the new cement mortar and aggregate. Consequently, this weak phase weakened the compressive strength and elastic modulus of RAC.

5.3. Failure Mechanism

As follows experimental and numerical works [7,9,40–42], the first cracks always appeared in old and new ITZs and then propagated into the old and new cement mortar, resulting in several continuous cracks.

Due to the limitation of the technology, it was difficult to view the failure pattern during the macroscopic experiment loading, especially that of the damage process of internal materials. To explore the damage mechanism of RAC, the crack pattern of a 3D model with one aggregate was created, as illustrated in Figure 8.

From Figure 8, one can see that the first crack was formed in the old ITZ zones and then appeared in the new ITZ and adhered mortar. As the loading increased, several isolated cracks appeared in the cement mortar, and approximately two-to-three continuous cracks were formed in the specimen. As displayed in Figure 8g,h, the continuous cracks were inclined in the direction of 45°, and the materials in the middle of the specimen appeared to fall off. In addition, due to the restraining action on the bottom and top of the specimen, there was no failure crack in these zones. These phenomena can be attributed to the inferior properties of the ITZ and adhering mortar materials to other media. In a five-phase system, the mechanical properties of ITZs are the weakest, so cracks always appeared in these areas. Then, the stress of the adhered mortar adjacent to the damaged ITZs increased, and several isolated cracks were formed. When increasing the loading, a plurality of cracks was formed in the new cement mortar near the damaged ITZ and the adhered mortar. Finally, two-to-three continuous cracks were distributed in the specimen. These results showed a good agreement with other numerical and experimental data [13,22,40–43].

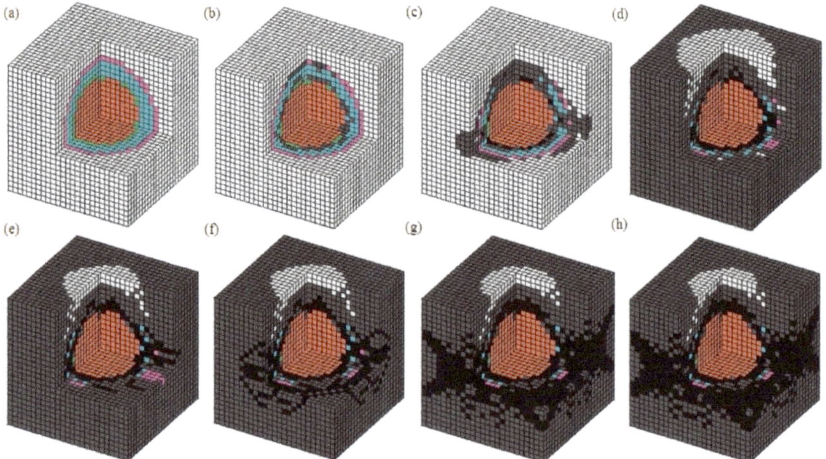

Figure 8. Failure pattern of RAC. (**a**) Initial model; (**b**) Elastic deformation; (**c**) First crack; (**d~f**) Crack propagation; (**g,h**) Continuous cracks.

6. Conclusions

This paper developed a new finite element method named the 3D base force element method (BFEM) that can be applied to analyze the damage behavior of materials. According to the 3D BFEM, the stiffness matrix and node displacement of a hexahedron element were derived. A 3D model of recycled aggregate concrete (RAC) with sphere aggregates was established. These models were applied to the uniaxial compression loading that was controlled by displacement loading. The effects of the replacement ratio of the recycled aggregate and the percentage of adhering mortar on the elastic modulus and compressive strength were explored. Additionally, the failure pattern was also displayed. According to the research and data described in this work, several conclusions can be reached as follows.

(1) The 3D BFEM can be used to explore the failure mechanism of heterogeneous materials. The stiffness matrix and the node displacement of a hexahedron element can be derived as an explicit expression and without the use of Gaussian integration.

(2) The 3D placing algorithm of the RAC numerical model is feasible to characterize the random structure of an aggregate in a specimen. The mesostructure and the mechanical behavior of RAC can be characterized by this numerical model.

(3) The replacement ratio of recycled aggregate has a major influence on mechanical properties. When increasing substitution, both the elastic modulus and compressive strength reduce. The substitution should be controlled below 30% in civil engineering.

(4) The percentage of the adhering mortar around the surface of the recycled aggregate has a negative influence on mechanical properties. The modulus of elastic and compressive strength decrease as the percentage increases. The waste concrete should be treated in a reasonable and inexpensive manner to advance the quality and performance of the recycled aggregate.

(5) The weak mechanical properties of an old ITZ and adhered mortar are major factors that cause the mechanical property degradation of RAC than that of NAC. In addition, these two phases have a significant influence on the failure mechanism of RAC.

Author Contributions: Y.W. carried out the numerical simulation and manuscript. H.Z. made the final approval of the manuscript. M.X. and C.W. collected and interpreted the data. J.F. and L.G. made the figures. M.M.A.K. modified the English grammar. All authors have read and agreed to the published version of the manuscript.

Funding: This research received no external funding.

Acknowledgments: The current work is supported by the Natural Science Foundation of the Jiangsu Higher Education Institutions of China (No. 19KJB560006), and a project supported by scientific research fund of Yancheng Polytechnic College, and Technology Innovation Team of Jiangsu Province, and National Natural Science Foundation of China (No. 10972015 and No. 11172015), and Beijing Natural Science Foundation (No. 8162008).

Conflicts of Interest: The authors declare no conflict of interest.

References

1. Paul, S.C.; van Zijl, G.P.A.G. Mechanical and durability properties of recycled concrete aggregate for normal strength structural concrete. *Int. J. Sustain. Constr. Eng. Technol.* **2013**, *4*, 89–103.
2. Mardani-Aghabaglou, A.; Yüksel, C.; Beglarigale, A.; Ramyar, K. Improving the mechanical and durability performance of recycled concrete aggregate-bearing mortar mixtures by using binary and ternary cementitious systems. *Constr. Build. Mater.* **2019**, *196*, 295–306. [CrossRef]
3. Meesala, C.R. Influence of different types of fiber on the properties of recycled aggregate concrete. *Struct. Concr.* **2019**, *20*, 1656–1669. [CrossRef]
4. Yu, Y.; Wu, B. Discrete Element Mesoscale Modeling of Recycled Lump Concrete under Axial Compression. *Materials* **2019**, *12*, 3140. [CrossRef]
5. Xie, J.; Zhao, J.; Wang, J.; Wang, C.; Huang, P.; Fang, C. Sulfate Resistance of Recycled Aggregate Concrete with GGBS and Fly Ash-Based Geopolymer. *Materials* **2019**, *12*, 1247. [CrossRef]
6. Paul, S.C.; Panda, B.; Garg, A. A novel approach in modelling of concrete made with recycled aggregates. *Measurement* **2018**, *115*, 64–72. [CrossRef]
7. Wang, Y.; Peng, Y.; Kamel, M.M.; Ying, L. 2D numerical investigation on damage mechanism of recycled aggregate concrete prism. *Constr. Build. Mater.* **2019**, *213*, 91–99. [CrossRef]
8. Clough, R. The finite element method in plane stress analysis. In Proceedings of the 2nd ASCE Conference on Electronic Computation, Pittsburgh, PA, USA, 8–9 September 1960.
9. Wang, Y.; Peng, Y.; Kamel, M.M.A.; Ying, L. Mesomechanical properties of concrete with different shapes and replacement ratios of recycled aggregate based on base force element method. *Struct. Concr.* **2019**, *20*, 1425–1437. [CrossRef]
10. Jin, L.; Yu, W.; Du, X.; Zhang, S.; Li, N.; Xiuli, D. Meso-scale modelling of the size effect on dynamic compressive failure of concrete under different strain rates. *Int. J. Impact Eng.* **2019**, *125*, 1–12. [CrossRef]
11. Xiao, J.; Li, W.; Corr, D.J.; Shah, S.P. Effects of interfacial transition zones on the stress–strain behavior of modeled recycled aggregate concrete. *Cem. Concr. Res.* **2013**, *52*, 82–99. [CrossRef]
12. Li, W.; Luo, Z.; Sun, Z.; Hu, Y.; Duan, W.H. Numerical modelling of plastic–damage response and crack propagation in RAC under uniaxial loading. *Mag. Concr. Res.* **2018**, *70*, 459–472. [CrossRef]
13. Peng, Y.; Wang, Q.; Ying, L.; Kamel, M.M.A.; Peng, H. Numerical Simulation of Dynamic Mechanical Properties of Concrete under Uniaxial Compression. *Materials* **2019**, *12*, 643. [CrossRef] [PubMed]
14. Wittmann, F.; Roelfstra, P.; Sadouki, H. Simulation and analysis of composite structures. *Mater. Sci. Eng.* **1985**, *68*, 239–248. [CrossRef]
15. Zaitsev, Y.B.; Wittmann, F.H. Simulation of crack propagation and failure of concrete. *Mater. Struct.* **1981**, *14*, 357–365.
16. Peng, Y.; Wang, Y.; Guo, Q.; Ni, J. Application of Base Force Element Method to Mesomechanics Analysis for Concrete. *Math. Probl. Eng.* **2014**, *2014*, 1–11. [CrossRef]
17. Walraven, J.; Reinhardt, H. Theory and experiments on the mechanical behavior of cracks in plain and reinforced concrete subjected to shear loading. *Heron* **1981**, *26*, 1–68.
18. Bazant, Z.P.; Tabbara, M.R.; Kazemi, M.T.; Pijaudier-Cabot, G. Random Particle Model for Fracture of Aggregate or Fiber Composites. *J. Eng. Mech.* **1990**, *116*, 1686–1705. [CrossRef]
19. Wriggers, P.; Moftah, S. Mesoscale models for concrete: Homogenisation and damage behaviour. *Finite Elements Anal. Des.* **2006**, *42*, 623–636. [CrossRef]

20. Chen, H.; Ma, H.; Tu, J.; Cheng, G.; Tang, J. Parallel computation of seismic analysis of high arch dam. *Earthq. Eng. Eng. Vib.* **2008**, *7*, 1–11. [CrossRef]

21. Sun, C.; Xiao, J.; Lange, D.A. Simulation study on the shear transfer behavior of recycled aggregate concrete. *Struct. Concr.* **2018**, *19*, 255–268. [CrossRef]

22. Chen, A.; Xia, X.; Zhang, Q.; Wu, M. The Meso-level Numerical Experiment Research of the Mechanics Properties of Recycled Concrete. *J. Softw.* **2012**, *7*, 1932–1940. [CrossRef]

23. Jayasuriya, A.; Adams, M.P.; Bandelt, M.J. Understanding variability in recycled aggregate concrete mechanical properties through numerical simulation and statistical evaluation. *Constr. Build. Mater.* **2018**, *178*, 301–312. [CrossRef]

24. Gao, Y.C. A new description of the stress state at a point with applications. *Arch. Appl. Mech.* **2003**, *73*, 171–183. [CrossRef]

25. Peng, Y.; Liu, Y. Base force element method of complementary energy principle for large rotation problems. *Acta Mech. Sin.* **2009**, *25*, 507–515. [CrossRef]

26. Ottosen, N.S. Constitutive model for short-time loading of concrete. *J. Eng. Mech. Div.* **1979**, *105*, 127–141.

27. Zhang, Z.; Chen, C.; Zhang, J. Yield Criterion in Plastic-Damage Models for Concrete. *Acta Mech. Solida Sin.* **2010**, *23*, 220–230. [CrossRef]

28. Darwin, D.; Pecknold, D.A. Nonlinear biaxial stress-strain law for concrete. *J. Eng. Mech. Div.* **1977**, *103*, 229–241.

29. Lutz, M.P.; Monteiro, P.J.; Zimmerman, R.W. Inhomogeneous Interfacial Transition Zone Model for the Bulk Modulus of Mortar. *Cem. Concr. Res.* **1997**, *27*, 1113–1122. [CrossRef]

30. Li, W.; Xiao, J.; Sun, Z.; Kawashima, S.; Shah, S.P. Interfacial transition zones in recycled aggregate concrete with different mixing approaches. *Constr. Build. Mater.* **2012**, *35*, 1045–1055. [CrossRef]

31. Bittnar, Z.; Bartos, P.J.; Nemecek, J.; Smilauer, V.; Zeman, J. (Eds.) *Nanotechnology in Construction 3: Proceedings of the NICOM3*; Springer: Berlin/Heidelberg, Germany, 2009.

32. Mondal, P.; Shah, S.P.; Marks, L.D. Nanomechanical Properties of Interfacial Transition Zone in Concrete. In *Nanotechnology in Construction 3*; Springer: Berlin/Heidelberg, Germany, 2009; pp. 315–320.

33. Xiao, J.; Li, W.; Sun, Z.; Lange, D.A.; Shah, S.P. Properties of interfacial transition zones in recycled aggregate concrete tested by nanoindentation. *Cem. Concr. Compos.* **2013**, *37*, 276–292. [CrossRef]

34. Xiao, J.; Li, J.; Zhang, C. Mechanical properties of recycled aggregate concrete under uniaxial loading. *Cem. Concr. Res.* **2005**, *35*, 1187–1194. [CrossRef]

35. Abid, S.R.; Nahhab, A.H.; Al-Aayedi, H.K.; Nuhair, A.M. Expansion and strength properties of concrete containing contaminated recycled concrete aggregate. *Case Stud. Constr. Mater.* **2018**, *9*, e00201. [CrossRef]

36. Verian, K.P.; Ashraf, W.; Cao, Y. Properties of recycled concrete aggregate and their influence in new concrete production. *Resour. Conserv. Recycl.* **2018**, *133*, 30–49. [CrossRef]

37. Folino, P.; Xargay, H. Recycled aggregate concrete—Mechanical behavior under uniaxial and triaxial compression. *Constr. Build. Mater.* **2014**, *56*, 21–31. [CrossRef]

38. de Sánchez, J.M.; Gutiérrez, P.A. Study on the influence of attached mortar content on the properties of recycled concrete aggregate. *Constr. Build. Mater.* **2009**, *23*, 872–877.

39. Etxeberria, M.; Vázquez, E.; Mari, A.; Barra, M. Influence of amount of recycled coarse aggregates and production process on properties of recycled aggregate concrete. *Cem. Concr. Res.* **2007**, *37*, 735–742. [CrossRef]

40. Li, W.; Xiao, J.; Sun, Z.; Shah, S.P. Failure processes of modeled recycled aggregate concrete under uniaxial compression. *Cem. Concr. Compos.* **2012**, *34*, 1149–1158. [CrossRef]

41. Xiao, J.; Liu, Q.; Wu, Y.C. Numerical and experimental studies on fracture process of recycled concrete. *Fatigue Fract. Eng. Mater. Struct.* **2012**, *35*, 801–808. [CrossRef]

42. Akçaoğlu, T.; Tokyay, M.; Celik, T. Assessing the ITZ microcracking via scanning electron microscope and its effect on the failure behavior of concrete. *Cem. Concr. Res.* **2005**, *35*, 358–363. [CrossRef]

43. Huang, Y.; He, X.; Sun, H.; Sun, Y.; Wang, Q. Effects of coral, recycled and natural coarse aggregates on the mechanical properties of concrete. *Constr. Build. Mater.* **2018**, *192*, 330–347. [CrossRef]

Article

Feasible Use of Cathode Ray Tube Glass (CRT) and Recycled Aggregates as Unbound and Cement-Treated Granular Materials for Road Sub-Bases

M. Cabrera, P. Pérez, J. Rosales and F. Agrela *

Area of Construction Engineering, University of Cordoba, 14007 Cordoba, Spain; manuel.cabrera@uco.es (M.C.); ir2pegop@uco.es (P.P.); jrosales@uco.es (J.R.)
* Correspondence: fagrela@uco.es

Received: 13 January 2020; Accepted: 31 January 2020; Published: 6 February 2020

Abstract: In the last 15 years, new types of display technologies have increasingly replaced cathode ray tube (CRT) screens, which has led to an increase in landfill of old discarded CRT televisions, which present a great environmental challenge throughout the world due to their high lead content. In addition, environmental awareness has led to greater use of recycled aggregates to reduce the exploitation of existing reserves. This document aims to study the feasibility of incorporating CRT glass waste with recycled aggregate (RA) in combinations for use in civil engineering, more specifically in road bases and sub-bases. For the mechanical and environmental assessment of all of the samples and materials, the following procedures have been performed: the compliance batch test of UNE-EN 12457-4:2004 for RA, CRT, and mixtures; the Percolation Test according CEN/TS 14405 for the mixtures, CRT, and RA; Modified Proctor and load capacity (the California Bearing Ratio, or CBR) in all mixtures without cement addition, and finally, compressive strength of the material treated with cement at different ages of curing. The analysis of the mechanical and environmental properties through different techniques of lixiviation was positive, showing the ability to use CRT for certain dosage percentages mixed with recycled aggregates.

Keywords: cathode ray tube glass; recycled aggregates; civil infrastructures; recycled aggregates; cement-treated materials

---

1. Introduction

The use of recycled aggregates in road construction has been studied in recent years, and this type of application should be a priority in the future [1]. It is possible to apply several types of recycled aggregates in low-intensity traffic roads (liR). A conventional liR is built applying a course layer of concrete or asphalt on the top of other layers, such as the base, sub-base, and subgrade.

Traditionally, natural materials such as crushed rock, selected gravel, and stabilised materials are used in the base and sub-base of the road. In the last decade, several studies have been carried out to investigate the possibility of using recycled concrete aggregates (RCA) [2–4] and mixed recycled aggregates (MRA) in bases and sub-bases to provide a viable option for the use of recycled materials from construction and demolition waste (CDW) [5].

The application of recycled aggregates in road construction can be done by means of bound or unbound layers [6,7]. Currently, mixed recycled aggregates (MRA), which include different constituents of particles (concrete, ceramic, asphalt, masonry, natural, etc.), is the most common RA produced in different countries such as Spain, Portugal, Italy, etc. [8]. Some authors have studied the possibility of using these materials in applications with higher added value, such as the construction of unbound granular structural layers [5,9,10], and in cement-treated layers [11,12]. The main problem of the MRA is that they can have high amounts of sulphates, which cause dimensional changes due to the formation of ettringite [13].

Different authors have studied the mechanical behaviour of the RCA and MRA, and they demonstrated that cement-treated recycled materials presented an acceptable mechanical resistance up to 7 and 28 days when applying a 100% replacement rate of conventional materials (soil or gravel) by MRA or RCA [7].

Real-scale trials in which RCA and MRA were applied in cement-treated applications in roads have been conducted [12,14] in Malaga, Spain. In these studies, it was demonstrated that the behaviour of sections where RCA and MRA treated with cement were applied presented similar properties as those where conventional materials were used. In the work of Agrela et al. [12], it was observed that by applying a 100% replacement of natural aggregate (NA) by MRA with 3% of cement, an appropriate mechanical and deformational behaviour was achieved in the long term.

Another industrial waste that could be used in the sub-base of road layers are the residues from cathode ray tube televisions and screens (CRT waste). Although these types of residues are decreasing every year, there are still thousands of tons of CRT equipment still to be recycled. It is estimated that around the world, only about 26% of CRT waste is recycled, and the remaining 59% is deposited in landfills without recovery [15].

The main composition of CRT waste media consists of barium, strontium, and lead silicate [16,17]. These waste are also rich in silica, which makes their use as a building material interesting; however, other experiences demonstrate that too high content of silicates can degrade the mechanical strength [18]. Different investigations have been carried out for the second life cycle of CRT waste in the field of construction. Hui and Sun [19] studied the properties of CRT waste as a replacement for the fine natural aggregate for the manufacture of mortars. Ling et al. [20] used CRT waste cement mortar for X-ray radiation-shielding applications. Romero et al. [21] studied the mechanical properties of concrete manufactured with different CRT waste protocols as a substitute for natural aggregates, and Abdallah and Fan [22] studied the characteristics of concrete with waste glass.

The main problem of CRT waste is its lead content, and its contamination potential has been studied by different authors. Ling and Poon [23] observed that the back of the CRT equipment had a higher lead content than the front, so its use in construction applications could cause problems with respect to the environment and can cause public health problems. However, recycled aggregate (RA), particularly MRA, the sulphate content of the recycled aggregates (mostly gypsum), is one of the important quality properties for recycling and classification.

This aims of this study is to investigate the possibilities of applying MRA with CRT waste in a combined manner, including a reduced percentage of CRT to avoid the possibility of the leaching of contaminant elements. To this end, several studies will be carried out on mixtures of MRA and CRT waste in different proportions, studying both their mechanical behaviour and the potential contamination that they could present. It is applied to different mixtures using mainly the CRT waste coming from the front of the equipment with very small quantities from the back of the equipment, which could be more polluting.

2. Materials

Cement: The cement used in this study was CEM II/B-L 32.5 N (referred to as CEM-II). The main properties of the cement are shown in Table 1. This cement is the most commonly used material in this type of application due to its medium hydration heat and high resistance to chemical attack.

Table 1. Properties of the cement.

Cement	SiO$_2$	Al$_2$O$_3$	Fe$_2$O$_3$	CaO	MgO	SO$_3$	K$_2$O	Loss on Ignition (975 °C)
CEM II (%)	26.24	8.7	3.36	54.06	1.34	3.32	1.44	1.3

Mixed Recycled Aggregates (MRA): These came from the CDW treatment of the Aristerra S.L. plant, located in Malaga (Spain). The composition of the MRA, determined according to the standards UNE-EN 13242 [24] and UNE-EN 933-11 [25], must exceed 70% by weight in concrete, concrete products, mortars, pieces for the manufacture of concrete masonry, aggregates, and natural stones, as well as materials treated with hydraulic binders. It cannot exceed 2% by weight of glass. The rest consisted of ceramic masonry materials made of clay (bricks and tiles) or calcium silicate.

Recycled asphalt pavement (RAP): This material comes from the treatment and crushing of asphalt pavements, and it is usually used in the construction of agglomerate layers of new roads. Sometimes, they are waste that arrives at the RCD treatment plants, and they are mixed to obtain MRA. This material also comes from the plant Aristerra S.L.

Cathode ray tube glass (CRT): The material coming from the recycling of electronic devices was received in the construction engineering laboratory of the University of Córdoba from landfill. The blocks of this waste were large, so it was decided to crush it prior to testing. The CRT is divided in two types. One side is coming from the front part where the image is displayed, called p-CRT$_F$ (processed front CRT), and on the other side is obtained from the rear part composed by the hood or cathode ray tube, and it is called p-CRT$_R$ (processed rear CRT). Figure 1 shows both parts from which come these CRT types.

Figure 1. Left, front part cathode ray tube (CRT$_F$); **right**, rear part (CRT$_R$).

MRAa: A mixture of MRA and processed RAP was prepared in the laboratory by mixing 75% MRA and 25% RAP in dry weight. The mixture was called MRAa.

Figure 2 shows the particle size distribution of the materials. It was observed that the RAP contains a particle size distribution with a lower percentage of fine particles compared to the rest of the materials analysed. This is due to its crushing process. The rest of the materials presented a continuous particle size distribution.

In Table 2, the data of the physical and chemical properties of the materials tested in the laboratory are shown. The MRA presented a lower density of saturated surface and a greater absorption of water due to the content of ceramic particles [12,26]. The coarse fraction of the p-CRT presents a near-zero absorption. The sulphate content in all materials does not exceed the Spanish regulations. CRTs are made with two different glass formulations, one for the front, p-CRT$_F$ (panel) and one for the rear, p-CRT$_R$ (funnel). p-CRT$_R$ contains lead silicate glass with a composition of SiO_2 51.20 wt%, PbO 23.14 wt%, and other oxides 25.66 wt% p-CRT$_F$ contains barium–strontium silicate glass with SiO_2 58.13 wt%, BaO 10.50 wt%, SrO 9.46 wt%, and other oxides 21.91 wt%.

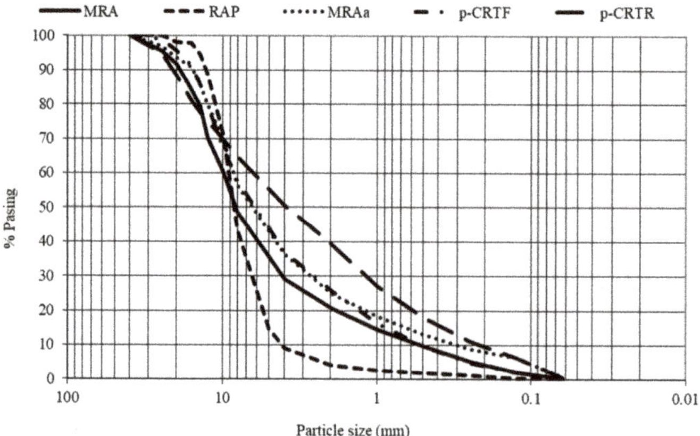

Figure 2. Particle size distribution.

Table 2. Physical and chemical properties of recycled aggregates and glass waste electrical. MRA: mixed recycled aggregates, RAP: recycled asphalt pavement, MRAa: mixture of 75% MRA and 25% processed RAP, p-CRT$_R$: cathode ray tube with lead silicate glass and a composition of SiO$_2$ 51.20 wt%, PbO 23.14 wt%, and other oxides 25.66 wt%. p-CRT$_F$: cathode ray tube with barium–strontium silicate glass and a composition of SiO$_2$ 58.13 wt%, BaO 10.50 wt%, SrO 9.46 wt%, and other oxides 21.91 wt%.

Properties	MRA	RAP	MRAa	p-CRT$_R$	p-CRT$_F$	Test Method
Acid-soluble sulphate (%SO$_3$)	0.7	0.3	0.39	0.05	0.03	UNE-EN 1744-1 [27]
Organic material (%)	1.37	1.13	1.21	-	-	UNE 103204 [28]
Oxide content (%)	-	-	-	-	-	-
SiO$_2$	53.12	-	50.75	51.2	58.13	UNE 196-2 [29]
Al$_2$O$_3$	13.15	-	10.6	4.15	2.71	-
TiO$_2$	1.74	-	0.98	0.1	0.5	-
CaO	10.12	-	16.56	3.56	2.4	-
MgO	5.15	-	2.28	2.45	0.80	-
Na$_2$O	2.8	-	1.97	7.3	8.10	-
K$_2$O	1.7	-	2.31	8.1	7.30	-
Fe$_2$O$_3$	8.4	-	3.55	-	0.10	-
BaO	-	-	-	-	10.50	-
SrO	-	-	-	-	9.46	-
PbO	-	-	-	23.14	-	-
Other	<4%	-	<11%			-
Density (kg/m^3)	-	-	-	-	-	UNE-EN 1097-6 [30]
0–4 mm	2.01	2.42	2.24	2.2	2.25	-
4–31.5 mm	2.08	2.26	2.12	2.49	2.51	-
Water absorption (%)	-	-	-	-	-	UNE-EN 1097-6 [30]
0–4 mm	10.27	5.1	9.41	6.27	5.76	-
4–31.5 mm	8.31	2.36	7.75	0.25	0.21	-
Plasticity	Non-plastic	Non-plastic	Non-plastic	Non-plastic	Non-plastic	UNE-EN-ISO 17892-12 [31]
Los Angeles	36	32	35	-	-	UNE-EN 1097-2 [32]
Friability ratio	27	21	24	31	33	UNE 146404 [33]

A classification test for the constituents of coarse recycled aggregates was carried out in accordance with the UNE-EN 933-11: 2009 [25] standard. Manual separation of the recycled aggregate components was carried out on particles of over 4 mm in size to obtain the results shown in Table 3.

Table 3. Composition.

	Concrete (Rc)	Natural Aggregates (Rc)	Ceramic (Rb)	Bituminous (Ra)	Glass (Rg)	Other (X)
MRA (%)	27.68	37.28	21.16	13.67	0.04	0.17
RAP (%)	-	-	-	100	-	-
MRAa (%)	22.61	31.31	19.73	26.72	0.03	0.11
p-CRT (%)	-	-	-	-	100	-

X-ray diffractometry was used to study both types of p-CRT. Figure 3 shows the XRD pattern of the samples analysed. The pattern obtained in both materials shows a mostly crystalline material. The p-CRT$_F$ shows mainly high contents of Cd, Mg, Si, and Ni; however, the p-CRT$_R$ is mainly formed by Pb and Si, which are elements present in both preliminary raw materials.

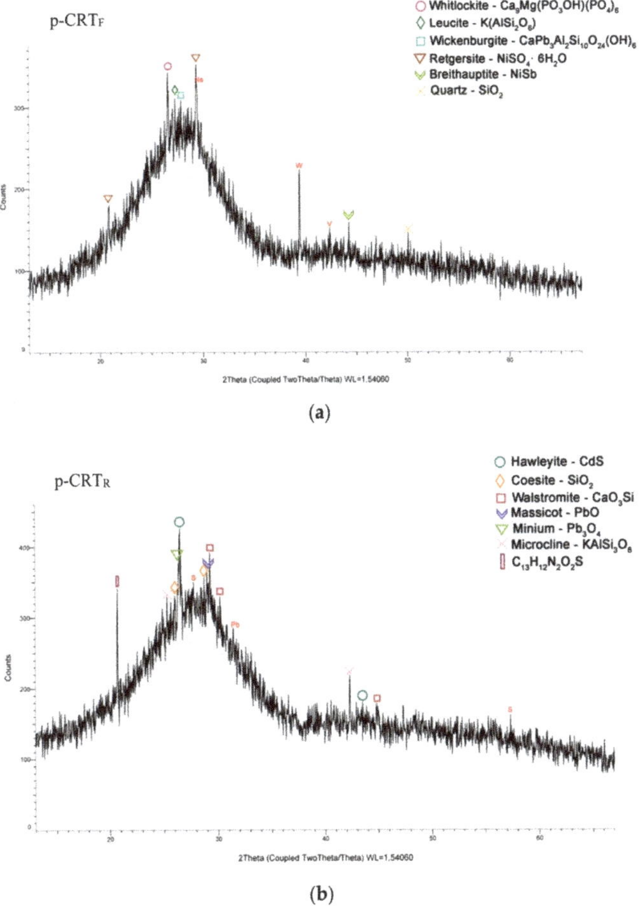

(a)

(b)

Figure 3. (a) X-ray diffractogram of p-CRT$_F$; (b) X-ray diffractogram of p-CRT$_R$.

Mixtures of Recycled Materials and p-CRT Studied

Different mixtures were made to study the possibility of use as a granular material and as cement-treated granular material (CTGM).

The p-CRT used for the study was a combination of the front and back. The proportion used was 2/3 p-CRT$_F$ and 1/3 p-CRT$_R$. The percentage of p-CRT that was added to the mixtures was 10% dry mass. Table 4 shows the mixtures and proportions used in the study.

Table 4. Dosages of the mixtures.

Mixtures	Materials (kg)		
	MRA	RAP	p-CRT
MRA	1000	-	-
MRAa	750	250	-
MRA + p-CRT	900	-	100
MRAa + p-CRT	675	225	100

3. Classification of Materials as a Function of their Pollutant Potential

3.1. Compliance Test UNE EN 12457-4:2004 [34]

Compliance testing was conducted to check whether the five materials satisfied European regulations. To classify these materials according to the EU Landfill Directive, not only heavy metals, but also inorganic anions were measured (sulphate, chloride, and fluoride). The UNE-EN 12457-4 [34] procedure consists of a two-step batch leaching test and is performed on materials with particle-size dimensions below 10 mm. The solid/liquid ratio is 1:10. The batch reactor is continuously stirred (10–12 rpm) for 24 h at a controlled temperature value equal to 20 ± 5 °C. At the end of the 24 h, the samples are left to decant and the pH, conductivity, and temperature are measured. The solution is filtered using a membrane filter (0.45 µm), and a sub-sample of the leachate was taken for each material.

Directive 2033/33/CE establishes the limits to be admitted to landfill in three categories (inert, non-hazardous, and hazardous waste) based on the concentration of heavy metals obtained in the UNE 12457-4 [34] compliance test. The measured concentrations in the leachate (mg/kg) are shown in Table 5 (data in bold indicate that the value exceeds the limit for inert waste).

Table 5. Leachate concentrations (mg/kg) obtained by the compliance test UNE EN 12457-4.

Element	MRA (mg/kg)	MRA + p-CRT (mg/kg)	p-CRT$_F$ (mg/kg)	p-CRT$_R$ (mg/kg)	p-CRT (mg/kg)	MRAa + p-CRT (mg/kg)	Limit Values (mg/kg) Inert	Non Hazardous
Cr	0.52496	0.43146	0.03208	0.00524	0.02291	0.470579	<0.5	0.5–10
Ni	0.00817	0.010147	0.00635	0.02166	0.01133	0.016748	<0.4	0.4–10
Cu	0.06137	0.036868	0.01116	0.09846	0.04015	0.036969	<2	2–50
Zn	0.04186	0.053005	0.16044	0.40719	0.24241	0.031649	<4	4–50
As	0.01705	0.019116	0.00822	0.00518	0.00074	0.020214	<0.5	0.5–2
Se	0.00143	0.01233	0.00275	0.00491	0.00394	0.010007	<0.1	0.1–0.5
Mo	0.07182	0.032939	0.02103	0.00604	0.01519	0.036853	<0.5	0.5–10
Cd	0.00016	0.00001	0.00000	0.00000	0.00000	0.00000	<0.04	0.04–1
Sb	0.04046	0.041975	0.28803	0.24131	0.26973	0.042393	<0.06	0.06–0.7
Ba	0.16046	0.288945	16.4451	8.26438	13.6741	0.281764	<20	20–100
Hg	0.00018	0.000921	0.00001	0.00000	0.00000	0.001562	<0.01	0.01–0.2
Pb	0.00048	0.00374	0.48885	**2.47351**	**1.5493**	0.00001	<0.5	0.5–10
Cl$^-$	21.5	50	300	40	197	18.5	800	15000
F$^-$	<2	<2	<2	<2	<2	<2	10	150
SO$_4^-$	1150	1280	127	138	129	1240	1500	20000

Comparing the measured concentrations of species released with the European environmental criteria at liquid-to-solid ratio of 10 L/kg (see Table 5), the calculated admission values for the MRA and p-CRT showed some critical parameters for their use: Cr and Pb.

The MRA was classified as non-hazardous waste due to the fact that the limit of inert waste was exceeded for Cr, according to other studies [35,36]. The presence of Cr was due to the content of ceramic particles.

The p-CRT contains 20–25% of PbO and is classified as hazardous waste [37]. Its content of Pb in leaching is a very important factor for its application as a construction material because it can represent a threat to the environment and human health. The tested sample of p-CRT$_R$ is classified as a non-hazardous waste. The mixture of MRA with p-CRT is classified as an inert material, its use being adapted from the environmental point of view.

3.2. Percolation Test CEN/TS 14405 [38]

The Column Leach Test describes a procedure for determining the leachability of inorganic components from solid, earthy, and stony materials and wastes as a function of the value of L/S. The method involves passing demineralised water upwards through a vertical column of particulate material (4 mm or smaller). Seven consecutive leachate fractions are collected, corresponding to a liquid-to-solid ratio range of 0.1 L/kg (v/m).

After the basic characterisation of the release of pollutants by the compliance test by which the most conflicting elements are identified, the percolation test was carried out according to the procedure CEN/TS 14405 [38] to evaluate the release of components with a liquid/solid ratio = 0.1 [39].

Table 6 shows the leachate concentrations obtained according to the percolation test. It is observed that mixtures with p-CRT behave as an inert material, corroborating the data obtained in the compliance test. Only p-CRT$_R$ exceeds the limits of an inert material.

Table 6. Release obtained according the percolation test CEN/TS 14405.

Element	RMA (mg/L)	RMA + p-CRT (mg/L)	p-CRT$_F$ (mg/L)	p-CRT$_R$ (mg/L)	p-CRT (mg/L)	RMAa + p-CRT (mg/L)	Limit Values Directive 2003/33/EC (mg/L) Inert	Non Hazardous
Cr	0.0052	0.0412	0.0087	0.0023	0.0065	0.0513	<0.1	0.1–2.5
Ni	0.0006	0.0052	0.0043	0.0020	0.0035	0.0057	<0.12	0.12–3
Cu	0.0010	0.0144	0.0219	0.0047	0.0160	0.0176	<0.6	0.6–30
Zn	0.0016	0.0079	0.0142	0.0160	0.0146	0.0097	<1.2	1.2–15
As	0.0006	0.0043	0.0030	0.0027	0.0028	0.0072	<0.06	0.06–0.3
Se	0.0016	0.0057	0.0011	0.0014	0.0012	0.0090	<0.04	0.04–0.2
Mo	0.0210	0.0811	0.0044	0.0038	0.0042	0.0820	<0.2	0.2–3.5
Cd	0.0001	0.0001	0.0000	0.0000	0.0000	0.0001	<0.02	0.02–0.3
Sb	0.0016	0.0077	0.0116	0.040	0.0208	0.0047	<0.1	0.1–0.15
Ba	0.0557	0.0615	0.2327	0.7725	0.4085	0.0972	<4	4–20
Hg	<0.0001	<0.0001	0.0001	0.0000	0.0001	0.0000	<0.002	0.002–0.03
Pb	0.0004	0.0018	0.0908	**0.1608**	0.1130	0.0028	<0.15	0.15–3

Theoretically, the Pb results of the percolation leaching test of MRA + p-CRT samples should be 0.0118 mg/L, but the results have a much lower value, which suggests that recycled aggregates absorb part of the contaminating power of p-CRT glass.

4. Experimental Methods and Results

4.1. Modified Proctor (UNE 103501:1994) [40]

The Modified Proctor test is a laboratory geotechnical testing method used to determine the soil compaction properties, specifically, to determine the optimal water content at which soil can reach its maximum dry density.

Modified Proctor has a real importance in the construction industry. The structures need a resistant base to lean on, and a poorly compacted soil could mean the collapse of a well-designed structure.

In some cases, such as on roads with little traffic or rural areas, the soil constitutes the rolling layer, so the importance of compaction becomes evident.

As shown in Figure 4, all materials have curves that are not very sensitive to changes in moisture content. Arulrajah et al. [41] reported that the materials with flat compaction curves can tolerate a greater amount of variation in moisture content without compromising much of the density obtained from compaction.

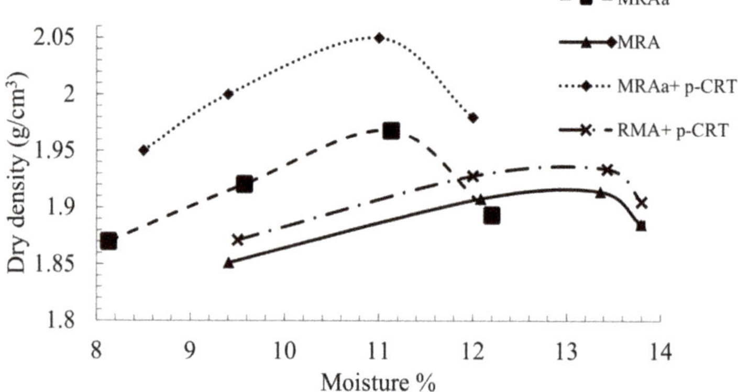

Figure 4. Moisture–density relationship.

The material with the highest humidity and the lowest density is the MRA (Table 7), due to its content of ceramic particles. This content of particles increases the porosity and therefore produces a lower density of the material and a greater absorption. The mixtures of MRA and MRAa with p-CRT increase the maximum dry density.

Table 7. Dry density and optimum moisture results.

	MRA	MRA + p-CRT	MRAa	MRAa + p-CRT
Dry density (g/cm³)	1.92	1.94	1.97	2.05
Moisture (%)	13.29	13.19	11.1	11.01

4.2. California Bearing Ratio (CBR)

The California Bearing Ratio (CBR) test is a strength test that compares the bearing capacity of a material with that of a well-graded crushed stone (thus, a high-quality crushed stone material should have a CBR of 100%). It is primarily intended for, but not limited to, evaluating the strength of cohesive materials having maximum particle sizes less than 20 mm according to UNE 103502-95 [42].

The CBR test involves comparing the application of one ratio of force per unit area required to penetrate a soil mass with a standard circular piston at the rate of 1.25 mm/min to that required for the corresponding penetration of a standard material.

Das [43] said that the bearing capacity of the soil is the capacity that the material has to support loads applied on it, without the failure that occurs by cutting or a large deformation.

The loads that a structural layer of a road transmits to the ground produce tensions and deformations on it. The deformation will depend on the tension and the physical–mechanical properties of the terrain. In pavements, the transmitted load is mobile, experiencing cycles of loading and unloading, where a part of the deformed land is recovered, and another part is not. It should be noted that the CBR test indicates the quality of support that a soil will have. Authors such as Crespo [44] classify the soil based only on the CBR index.

The CBR test was performed on the compacted samples at the optimum moisture content, for which the Modified Proctor compaction test was used. CBR tests were carried out in both unsoaked and 4-day soaked conditions, and the results are summarised in Figure 5.

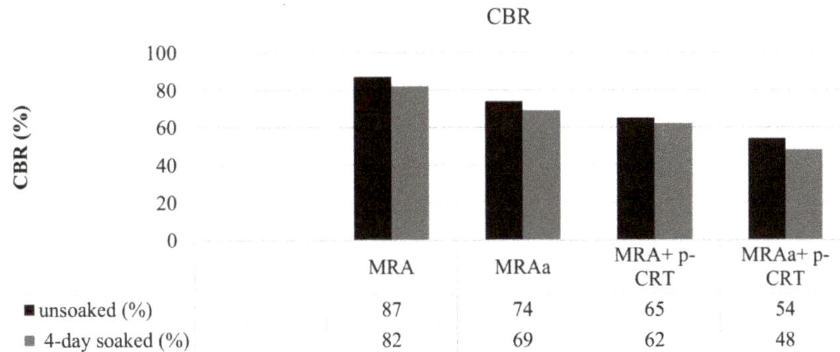

	MRA	MRAa	MRA+ p-CRT	MRAa+ p-CRT
■ unsoaked (%)	87	74	65	54
■ 4-day soaked (%)	82	69	62	48

Figure 5. California Bearing Ratio (CBR) values (unsoaked and 4-day soaked).

According to other studies [45], in an unsoaked condition, the materials showed the highest CBR value. The CBR value decreased with the p-CRT content in both conditions. One possible reason was the lower intrinsic resistance to the crushed p-CRT particles.

4.3. Vibrating Hammer Times

The materials were compacted in a CBR mould using a vibrating hammer in accordance with NLT-310/90 [46]. They were filled in three layers, each layer formed with a thickness of one-third of the length of the mould to produce the density/time plot. The compaction value required for each sample must be greater than 98%, according to the modified Proctor reference value. Three samples were manufactured using different vibration times of 5, 12, and 20 s.

The compaction value required for each specimen must be greater than 98%, according to the modified Proctor reference value. The compaction times of the vibrating hammer were different in all the mixtures. Materials not mixed with p-CRT showed high compaction times. Figure 6 shows the time required to obtain a vibration hammer of 98% Proctor density. Vibrating hammer compaction times range from 17 to 19 s.

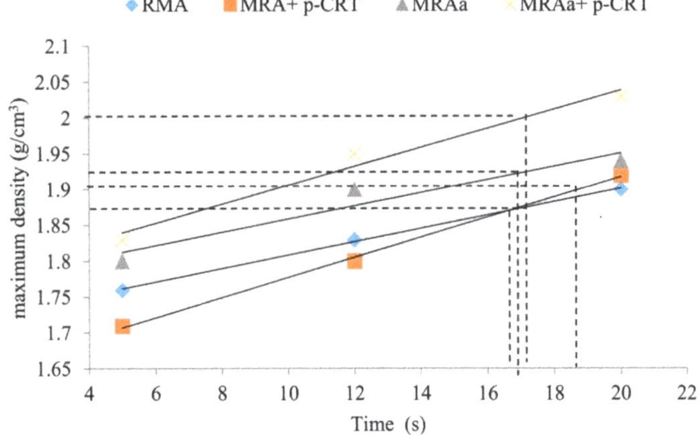

Figure 6. Vibrating hammer times.

4.4. Compressive Strength

The purpose of this test is the determination of the simple compressive strength of the materials treated with hydraulic binders. This test has been carried out according the methodology indicated by NLT-305/90 [46].

The simple compressive strength in mixtures treated with cement is an indicator of the degree of reaction of the soil with cement and water and the rate of hardening with respect to time. The values obtained depend on many factors: the content and type of cement, type of soil, the applied energy of compaction, the efficiency achieved in the mixing, amount of organic matter, size and shape of the test specimen, etc. Particularly, due to the morphology and low water absorption shown by the CRT, the interaction between the particles and the cement does not materialize as with a natural or recycled aggregate, obtaining a loss of resistance in the mixtures that incorporate CRT.

In the case of pavement structures, there are compressive strength values suggested according to the type of pavement and type of layer to be built. In Spain, the regulation requires a minimum resistance of 2.5 MPa to 7 days, with a cement percentage not lower than 3%.

Figure 7 shows the results obtained in all the CTGM mixtures produced at 7, 28, and 90 days. Dashed lines indicating the minimum compressive strength that must be obtained after 7 days (2.5 MPa), as well as the maximum resistance to 28 days (4.5 MPa) in accordance with the Spanish technical specifications, are also shown.

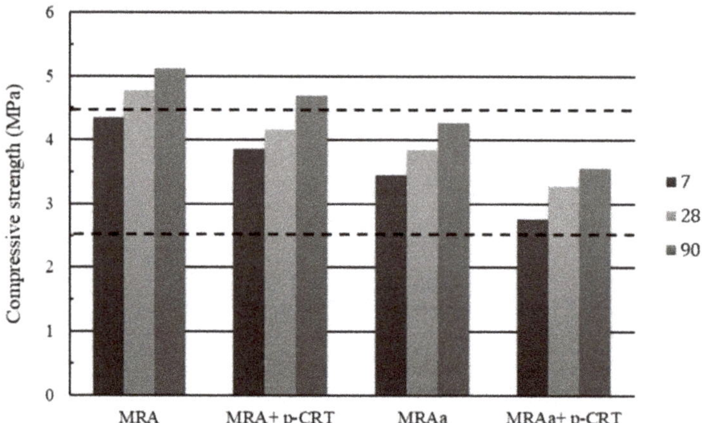

Figure 7. Compressive strength results with 3% cement.

It is observed that all the mixtures exceed the required limit at 7 days. The highest values of compression resistance were obtained in the MRA mixtures, with similar results obtained by other authors [13].

Figure 8 shows the evolution of the strength. At 28 days, compression strength values of approximately 90% were obtained in comparison with the resistance values obtained at 90 days. At 7 days, the results were approximately 80% compared to the results measured at 90 days. The addition of p-CRT decreases the resistance in both mixtures. If we compare the MRA with MRA + p-CRT and MRAa with MRAa+ p-CRT, there is a decrease in resistance at 28 days of 12.97% and 14.58%, respectively. The compressive strength of all the mixtures exceeds the minimum value of 2.5 MPa, corroborating the use of p-CRT in mixtures treated with cement.

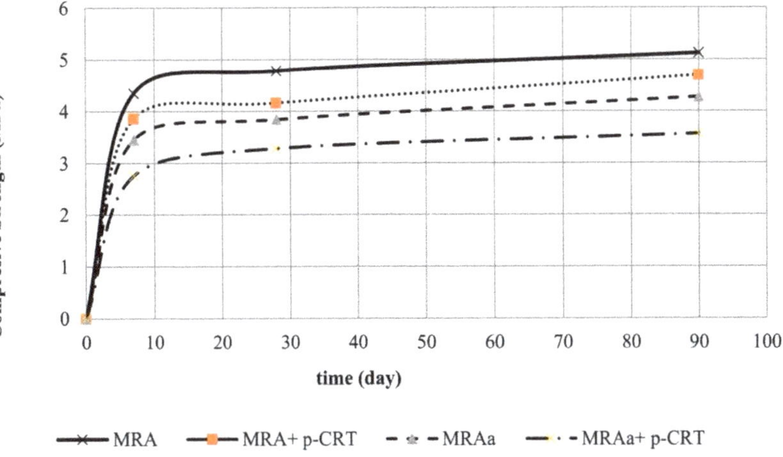

Figure 8. Compressive strength evolution.

5. Conclusions

This paper provides the results of research on cathode ray tube glass in combination with mixed recycled aggregates (p-CRT and MRA) with different compositions to be used as sub-bases and bases of roads pavements, obtaining the following conclusions:

- Regarding the physical and chemical properties of the materials, the mixed recycled aggregates presented a lower density and greater absorption of water due to its content of ceramic particles (21.16% of ceramic particles). Los Angeles abrasion testing indicates a good quality of all aggregates in accordance with current regulations (>40).
- Potential contamination by leaching was studied by means of two different methods, compliance testing (UNE-12457-4) [34] and percolation testing (CEN/TS-14405) [38]. The compliance test data revealed a potential contaminant in the funnel (p-CRT$_R$) of the analysed samples, due to the high content of Pb, confirming that they are an inappropriate material for the application as an isolated aggregate in any civil engineering application.

The present study indirectly demonstrates the neutralised capacity of mixed recycled aggregates from an environmental point of view, reducing the concentration of heavy metals when mixed with cathode ray tube glass (see Table 5).

The results show that the use of cathode ray tube glass as an aggregate, at 10%, not only had satisfactory levels of compressive strength and bearing capacity, but also does not exceed the limits established by the directive, all mixtures being classified as inert.

The viability of using cathode ray tube glass and mixed recycled aggregates as an aggregate for the production of structural road layers (base and sub-base) has been demonstrated.

Therefore, the present work proposes a solution that implies an environmental benefit for these agents in addition to verifying the samples (as unbound aggregates and as cement-treated granular materials mixtures) with potential to be applied as building materials during their second life cycle.

Author Contributions: Conceptualization, M.C. and F.A.; methodology, J.R. and M.C.; investigation, M.C. and P.P.; resources, F.A.; data curation, J.R. and M.C.; writing—original draft preparation, M.C. and F.A.; writing—review and editing, M.C. and J.R.; supervision, F.A. All authors have read and agreed to the published version of the manuscript.

Funding: The authors express their gratitude to London Weee and Arecosur (Companies for the treatment of waste in Málaga), for their financial support.

Conflicts of Interest: The authors declare no conflict of interest.

References

1. Agrela, F.; De Juan, M.S.; Ayuso, J.; Geraldes, V.; Jiménez, J.R. Limiting properties in the characterisation of mixed recycled aggregates for use in the manufacture of concrete. *Constr. Build. Mater.* **2011**, *25*, 3950–3955. [CrossRef]
2. Dhir, R.K.; Limbachiya, M.C.; Leelawat, T.; BS 5328; BS 882. Suitability of Recycled Concrete Aggregate for Use in Bs 5328 Designated Mixes. *Proc. Inst. Civ. Eng.* **1999**, *134*, 257–274. [CrossRef]
3. Poon, C.; Kou, S.; Lam, L. Use of recycled aggregates in molded concrete bricks and blocks. *Constr. Build. Mater.* **2002**, *16*, 281–289. [CrossRef]
4. Contrafatto, L.; Cosenza, R.; Barbagallo, R.; Ognibene, S. Use of recycled aggregates in road sub-base construction and concrete manufacturing. *Ann. Geophys.* **2018**, *61*, 223. [CrossRef]
5. Jiménez, J.R.; Agrela, F.; Ayuso, J.; López, M. A comparative study of recycled aggregates from concrete and mixed debris as material for unbound road sub-base. *Mater. Constr.* **2011**, *61*, 289–302. [CrossRef]
6. Molenaar, A.A.A.; Van Niekerk, A.A. Effects of Gradation, Composition, and Degree of Compaction on the Mechanical Characteristics of Recycled Unbound Materials. *Transp. Res. Rec.* **2002**, *1787*, 73–82. [CrossRef]
7. Xuan, D.; Houben, L.J.M.; Molenaar, A.A.A.; Shui, Z. Cement treated recycled demolition waste as a road base material. *J. Wuhan Univ. Technol. Sci. Ed.* **2010**, *25*, 696–699. [CrossRef]
8. De Brito, J.; Silva, R.V.; Agrela, F. *New Trends in Eco-Efficient and Recycled Concrete*; Woodhead Publishing: Duxford, UK, 2018.
9. Vegas, I.; Ibanez, J.; José, J.S.; Urzelai, A. Construction demolition wastes, Waelz slag and MSWI bottom ash: A comparative technical analysis as material for road construction. *Waste Manag.* **2008**, *28*, 565–574. [CrossRef]
10. Herrador, R.; Pérez, P.; Garach, L.; Ordóñez, J. Use of recycled construction and demolition waste aggregate for road course surfacing. *J. Transp. Eng.* **2011**, *138*, 182–190. [CrossRef]
11. Xuan, D.X.; Houben, L.J.M.; Molenaar, A.A.A.; Shui, Z.H. Mixture optimization of cement treated demolition waste with recycled masonry and concrete. *Mater. Struct.* **2012**, *45*, 143–151. [CrossRef]
12. Agrela, F.; Barbudo, A.; Ramírez, A.; Ayuso, J.; Carvajal, M.D.; Jiménez, J.R. Construction of road sections using mixed recycled aggregates treated with cement in Malaga, Spain. *Resour. Conserv. Recycl.* **2011**, *58*, 98–106. [CrossRef]
13. Agrela, F.; Cabrera, M.; Galvín, A.; Barbudo, A.; Ramirez, A. Influence of the sulphate content of recycled aggregates on the properties of cement-treated granular materials using Sulphate-Resistant Portland Cement. *Constr. Build. Mater.* **2014**, *68*, 127–134. [CrossRef]
14. Pérez, P.; Agrela, F.; Herrador, R.; Ordoñez, J. Application of cement-treated recycled materials in the construction of a section of road in Malaga, Spain. *Constr. Build. Mater.* **2013**, *44*, 593–599. [CrossRef]
15. Rashad, A.M. Recycled waste glass as fine aggregate replacement in cementitious materials based on Portland cement. *Constr. Build. Mater.* **2014**, *72*, 340–357. [CrossRef]
16. Yuan, W.; Li, J.; Zhang, Q.; Saito, F. Innovated Application of Mechanical Activation to Separate Lead from Scrap Cathode Ray Tube Funnel Glass. *Environ. Sci. Technol.* **2012**, *46*, 4109–4114. [CrossRef]
17. Tian, X.; Wu, Y. Recent development of recycling lead from scrap CRTs: A technological review. *Waste Manag.* **2016**, *57*, 176–186.
18. Contrafatto, L. Recycled Etna volcanic ash for cement, mortar and concrete manufacturing. *Constr. Build. Mater.* **2017**, *151*, 704–713. [CrossRef]
19. Hui, Z.; Sun, W. Study of properties of mortar containing cathode ray tubes (CRT) glass as replacement for river sand fine aggregate. *Constr. Build. Mater.* **2011**, *25*, 4059–4064. [CrossRef]
20. Ling, T.-C.; Poon, C.-S.; Lam, W.-S.; Chan, T.-P.; Fung, K.K.-L. Utilization of recycled cathode ray tubes glass in cement mortar for X-ray radiation-shielding applications. *J. Hazard. Mater.* **2012**, *199*, 321–327. [CrossRef]
21. Romero, D.; James, J.; Mora, R.; Hays, C.D. Study on the mechanical and environmental properties of concrete containing cathode ray tube glass aggregate. *Waste Manag.* **2013**, *33*, 1659–1666. [CrossRef]
22. Abdallah, S.; Fan, M. Characteristics of concrete with waste glass as fine aggregate replacement. *Int. J. of Eng. Tech. Res.* **2014**, *2*, 11–17.
23. Ling, T.-C.; Poon, C.-S. Utilization of recycled glass derived from cathode ray tube glass as fine aggregate in cement mortar. *J. Hazard. Mater.* **2011**, *192*, 451–456. [CrossRef] [PubMed]

24. *UNE-EN-13242:2003+A1:2008: Aggregates for Unbound and Hydraulically Bound Materials for Use in Civil Engineering Work and Road Construction*; Asociacion Española de Normalizacion (AENOR): Madrid, Spain, 2003.

25. *UNE-EN 933-11:2009: Tests for Geometrical Properties of Aggregates-Part 11: Classification Test for the Constituents of Coarse Recycled Aggregate*; Asociacion Española de Normalizacion (AENOR): Madrid, Spain, 2009.

26. Barbudo, A.; Agrela, F.; Ayuso, J.; Jiménez, J.R.; Poon, C. Statistical analysis of recycled aggregates derived from different sources for sub-base applications. *Constr. Build. Mater.* **2012**, *28*, 129–138. [CrossRef]

27. *UNE-EN 1744-1:2010+A1:2013: Tests for Chemical Properties of Aggregates-Part 1: Chemical Analysis*; Asociacion Española de Normalizacion (AENOR): Madrid, Spain, 2003.

28. *UNE 103204:2019: Organic Matter Dontent of a Soil by the Potassium Permanganate Method*; Asociacion Española de Normalizacion (AENOR): Madrid, Spain, 2019.

29. *UNE-EN 196-2:2014: Method of Testing Cement-Part 2: Chemical Analysis of Cement*; Asociacion Española de Normalizacion (AENOR): Madrid, Spain, 2014.

30. *UNE-EN 1097-6:2014: Tests for Mechanical and Physical Properties of Aggregates-Part 6: Determination of Particle Density and Water Absorption*; Asociacion Española de Normalizacion (AENOR): Madrid, Spain, 2014.

31. *UNE-EN ISO 17892-12:2019: Geotechnical Investigation and Testing-Laboratory Testing of Soil-Part 12: Determination of Liquid and Plastic Limits*; Asociacion Española de Normalizacion (AENOR): Madrid, Spain, 2019.

32. *UNE-EN 1097-2:2010: Tests for Mechanical and Physical Properties of Aggregates-Part 2: Methods for the Determination of Resistance to Fragmentation*; Asociacion Española de Normalizacion (AENOR): Madrid, Spain, 2010.

33. *UNE 146404:2018: Determination of the Coefficient of Friability of the Sands*; Asociacion Española de Normalizacion (AENOR): Madrid, Spain, 2018.

34. *UNE-EN 12457-4:2004: Characterisation of Waste-Leaching-Compliance Test for Leaching of Granular Waste Materials and Sludges-Part 4: One Stage Batch Test at a Liquid to Solid Ratio of 10 l/kg for Materials with Particle Size below 10 mm (Without or with Size Reduction)*; Asociacion Española de Normalizacion (AENOR): Madrid, Spain, 2004.

35. Galvín, A.P.; Ayuso, J.; Barbudo, A.; Cabrera, M.; López-Uceda, A.; Rosales, J. Upscaling the pollutant emission from mixed recycled aggregates under compaction for civil applications. *Environ. Sci. Pollut. Res.* **2018**, *25*, 36014–36023. [CrossRef] [PubMed]

36. Del Rey, I.; Ayuso, J.; Galvín, A.; Jiménez, J.R.; López, M.; García-Garrido, M. Analysis of chromium and sulphate origins in construction recycled materials based on leaching test results. *Waste Manag.* **2015**, *46*, 278–286. [CrossRef]

37. Lee, C.-H.; Chang, C.-T.; Fan, K.-S.; Chang, T.-C. An overview of recycling and treatment of scrap computers. *J. Hazard. Mater.* **2004**, *114*, 93–100. [CrossRef]

38. *CEN-EN 14405: Characterization of Waste-Leaching Behaviour Test-Upflow Percolation Test (Under Specified Conditions)*; European Committee for Standardization (CEN): Brussels, Belgium, 2017.

39. Roussat, N.; Méhu, J.; Abdelghafour, M.; Brula, P. Leaching behaviour of hazardous demolition waste. *Waste Manag.* **2008**, *28*, 2032–2040. [CrossRef]

40. *UNE 103501:1994: Geotechnic Compactation Test. Modified Proctor*; Asociacion Española de Normalizacion (AENOR): Madrid, Spain, 2004.

41. Arulrajah, A.; Rahman, M.A.; Piratheepan, J.; Bo, M.W.; Imteaz, M.A. Evaluation of interface shear strength properties of geogrid-reinforced construction and demolition materials using a modified large-scale direct shear testing apparatus. *J. Mater. Civ. Eng.* **2013**, *26*, 974–982. [CrossRef]

42. *UNE 103502:1995: Test Laboratory Nethod for Determining in a Soil the C.B.R. Index*; Asociacion Española de Normalizacion (AENOR): Madrid, Spain, 1995.

43. Das, B.M.; González, S.R.C. *Fundamentos de Ingeniería Geotécnica*; Thomson Learning: Madrid, Spain, 2001; pp. 445–494.

44. Crespo, C. *Mecánica de Suelos y Cimentaciones*; Limusa: Mexico City, Mexico.

45. Poon, C.S.; Chan, D. Feasible use of recycled concrete aggregates and crushed clay brick as unbound road sub-base. *Constr. Build. Mater.* **2006**, *20*, 578–585. [CrossRef]

46. *NLT 310/90: Vibrating Hammer Compaction of Granular Materials Treated*; CEDEX: Madrid, Spain, 1990.

Article

Recycling Aggregates for Self-Compacting Concrete Production: A Feasible Option

Rebeca Martínez-García [1], M. Ignacio Guerra-Romero [2], Julia M. Morán-del Pozo [2], Jorge de Brito [3] and Andrés Juan-Valdés [2,*]

1   Department of Mining Technology, Topography and Structures, University of León, Campus de Vegazana s/n, 24071 León, Spain; rmartg@unileon.es
2   Department of Agricultural Engineering and Sciences, University of León, Avenida de Portugal 41, 24071 Léon, Spain; ignacio.guerra@unileon.es (M.I.G.-R.); julia.moran@unileon.es (J.M.M.-d.P.)
3   Department of Civil Engineering, Architecture and Georresources, IST-Universidade de Lisboa, Av. Rovisco Pais, 1049-001 Lisbon, Portugal; jb@civil.ist.utl.pt
*   Correspondence: andres.juan@unileon.es; Tel.: +34-987-291-000 (ext. 5139); Fax: +34-987-291-810

Received: 28 December 2019; Accepted: 9 February 2020; Published: 14 February 2020

Abstract: The use of construction and demolition wastes (C&DW) is a trending future option for the sustainability of construction. In this context, a number of works deal with the use of recycled concrete aggregates to produce concrete for structural and non-structural purposes. Nowadays, an important number of C&DW management plants in the European Union (EU) and other countries have developed robust protocols to obtain high-quality coarse recycled aggregates that comply with different European standards in order to be used to produce new concrete. The development of self-compacting concrete (SCC) is another way to boost the sustainability of construction, due to the important reduction of energy employed. Using recycled aggregates is a relatively recent scientific area, however, studies on this material in the manufacture of self-compacting concrete have proven the feasibility thereof for conventional structural elements as well as high-performance and complex structural elements, densely reinforced structures, difficult-to-access formwork and difficult-to-vibrate elements. This paper presents an original study on the use of coarse recycled concrete aggregate (CRA) to obtain self-compacting concrete. Concrete with substitution ratios of 20%, 50% and 100% are compared with a control concrete. The purpose of this comparison is to check the influence of CRA on fresh SCC as well as its physical and mechanical properties. The parameters studied are material characterization, self-compactability, compressive strength, and tensile and flexural strength of the resulting concrete. The results conclude that it is feasible to use CRA for SCC production with minimal losses in the characteristics.

Keywords: self-compacting concrete; coarse recycled aggregate; sustainable concrete; construction and demolition waste management plant

1. Introduction

The construction sector has seen an exponential growth in recent decades throughout the European Union (EU). This growth has led to an increase in the production of construction and demolition waste (C&DW). The developed countries are a major consumer and generator of waste. According to the European Statistics Office, Eurostat, every EU citizen produces an average of 2000 kg of waste per year (not including waste from mining, if mining waste were included, this figure would be 5000 kg/person/year) [1].

More than a third of all waste generated in the EU comes from the construction sector considering that, after water, aggregates are the raw material most heavily-consumed by man and it is mainly used in concrete, which means the valorisation of this type of waste needs to be promoted. This would reduce the extraction of natural aggregates and eliminate difficult-to-manage waste [2,3].

Government agencies, the scientific community and the population, in general, are becoming increasingly aware of the depletion of natural resources and advocating sustainable development, which leads to the need of recycling waste so that resources can be part of a circular economy and can be sustained from generation to generation. Everyone is aware of the need for combining economic development with sustainability and environmental protection.

It is in this context that new applications and developments for C&DW have been studied. The production of concrete with recycled aggregates has been one of the most often studied applications. It is true that some reliable studies currently support the use of C&DW in conventional concrete [4–10]. In addition, some studies have been published in recent years on the use of this waste in self-compacting concrete (SCC) [11–27].

In these studies, on the one hand, a reference (or control SCC) concrete mix with natural aggregates is designed and, on the other hand, different mixes of SCC with various incorporation ratios of coarse recycled concrete aggregate (CRA) are tested to compare with the control SCC. Most of the studies only replace the coarse fraction of the aggregate and only some of them replace the sand fine fraction. We have selected the studies in which the recycled aggregates come from old concrete structures as in our case, discarding other normally using ceramic materials.

Grdic et al. [28] designed three types of mixes with the incorporation of 0%, 50% and 100% of CRA. The type of cement used was CEM/II/B-M 42.5N maintaining a constant quantity of 409.6 kg/m^3. The water absorption of the recycled aggregate was 5.08%, and the SP ratio was kept constant (0.7% relative to the weight of cement) and the water content was adjusted, increasing with the substitution ratio.

Safiuddin et al. [22] designed five types of mixes with substitutions of 0%, 30%, 50%, 70% and 100% of CNA with CRA. They employed CEM Type I, with a weight per m^3 of 342 kg/m^3. The W/C ratio and the content of SP are constant in all the mixes, 0.60 W/C and 1.50% respectively. The water absorption of the recycled aggregate used was 1.32%.

Tuyan et al. [29] design four types of mixes with substitutions of 0%, 20%, 40% and 60% of CNA with CRA. The cement was class C with a content of 315 kg/m^3. Three W/C ratios are used, 0.43, 0.48 and 0.53. The SP content was adjusted in relation to the percentage of replacement from 0.95% to 1.97% of the cement weight. The water absorption of the recycled aggregate was 4.80%.

Modani et al. [30] studied six types of mixes with substitutions of 0%, 20%, 40%, 60%, 80% and 100% employing a Grade 53 cement with 348 kg/m^3. The W/C ratio was 0.53. The SP was adjusted in relation to the replacement ratio from 1.3% to 1.58% of the cement weight. The water absorption of the recycled aggregate used was 5.64%.

Pereira de Oliveira et al. [21] designed four types of mixes with substitutions of 0%, 20%, 40% and 100% of CNA with CRA employing two types of recycled aggregates, using CEM I 42.5-R at 284.9 kg/m^3. The W/C ratio was 0.56 and 0.57. The SP was adjusted in relation to the replacement ratio from 1.19% to 2.10% of the cement weight. The water absorption of the two used recycled aggregates was 4.10% and 4.05%.

As a general rule, a number of studies specify that the addition of recycled aggregates in small proportions ($\leq$20%) in substitution of natural aggregates in concrete does not lower the performance of the resulting concrete. These studies have made it possible for these types of recycled materials to be included in standards and/or recommendations throughout the world. At a national level, Annex 15 "Recommendations for using recycled concrete" of the Spanish Structural Concrete Instruction EHE-08 [31] sets forth a few guidelines for its use: a) by limiting the content of coarse recycled aggregate (CRA) to 20% of the total weight of coarse aggregates, the final properties of recycled aggregates concrete are hardly affected and it allows experimental studies for substitution at higher ratios, b) not allowing the use of fine recycled aggregates irrespective of their nature, c) excluding the use of mixed recycled aggregates.

SCC is a high-performance concrete, the main characteristic of which is its fluid and viscous consistency allowing it to flow through densely reinforced structural elements without the addition of outside energy for compaction. Its composition is characterized by a high cement and fine particle content, a lower proportion of coarse aggregate and the use of next-generation additives [32].

The materials used to produce SCC are the same as for conventional concrete yet with some type of admixture, such as superplasticizer (SP) or viscosity modifiers, which are essential to avoid segregation and exudation of the mix. Special attention must be paid to the choice of materials in order to guarantee uniformity and consistency. This is complicated when using recycled aggregates for substitution purposes considering their great heterogeneity [32].

Some authors and studies recommend the use of Portland cement Type II (with a clinker content between 65% and 94%) for use in concrete with recycled aggregates to prevent the generation of alkali-aggregate reactions [33]. Other authors also recommend cement with a strength of no less than 42.5 MPa so that this characteristic is not a limiting factor. Moreover, the use of a sulphate-resistant cement seems interesting due to the possibility, according to some authors, of a reaction between the hydrated tricalcium aluminate of the hardened concrete and external sulphates, which leads to the formation of sulfoaluminate along with an increase in volume. Using sulphate-resistant cement would minimize this effect [10]. A number of studies ratify the higher absorption capacity of recycled aggregates at around 10% extra water and this is a premise to be considered. This higher demand for water is compensated by the use of superplasticizers [22,33–37].

Annex 17 "Recommendations for the use of SCC" of the Spanish Structural Concrete Instruction EHE-08 [31] provides a few guidelines: a) the maximum size of the aggregate shall be limited to 25 mm, b) the essential use of superplasticizers and/or viscosity modifiers in this type of concrete to guarantee self-compactability, c) it must comply with the project specifications, i.e., structural, operational and environmental requirements, d) the total fine particle content must be 450–600 kg/m^3, the cement content 250–500 kg/m^3, and the paste volume above 350 litres/m^3.

The use of recycled sand is not allowed by the Spanish Instruction for Structural Concrete [31] even though some studies show that substitutions of up to 10% do not produce any significant variations in the characteristics of concrete [15,17,19]. Currently, Instruction EHE-08 allows the partial addition (<20% weight) of coarse recycled concrete aggregates as replacement of natural aggregates.

The aim of this research study is to characterize recycled aggregates and design the optimal and feasible content to produce SCC by substituting coarse aggregate fractions with recycled aggregates at 20%, 50% and 100%. The purpose of this comparison is to examine the influence of CRA on SCC concerning the fresh and physical and mechanical properties. Four types of concrete mixes were produced for this study (CC or control concrete, Mix 20, Mix 50, Mix 100). The characterization of the materials in the mix was studied, compactability tests were performed on the fresh state as well as tests in the hardened state: compressive, tensile and flexural strength.

The use of recycled aggregates in the production of self-compacting concrete has been gradually increasing due to the economic and environmental advantages derived from the fusion of the two techniques. On the one hand, there are the advantages of SCC: optimal compaction without vibration, fluidity, complex or specially reinforced works, increase in the quality of the finish in exposed concrete, and greater adherence. On the other, there are the advantages of recycled aggregates: less demand for extraction of natural aggregates from quarries, reduction of the pollution produced by C&DW, savings in transport costs, low energy consumption, less investment as a good quality material is produced at a low cost, and resources are optimised. However, information on the quality of SCC with RCA is still scarce. This study (as well as other similar ones) provides very useful information for the practical use in concrete production. One of the main novelties of our study is the use of real coarse recycled aggregates, obtained from a C&DW management plant, with standard treatment, unlike some other studies that use a specific recycled aggregate obtained from only one specific demolition, construction or maintaining work.

2. Materials and Methods

2.1. Materials

This research focused on the study of the mechanical and rheological properties of SCC produced with coarse recycled aggregates from crushed concrete elements designed as per the Okamura recommendations [38] as well as the minimum requirements of Instruction EHE-08 [31].

The characteristics of the materials used in its production were carefully studied in the design of SCC as this has an immense impact on the behaviour of the mix. Figure 1 shows the study sequence.

Figure 1. Research sequence.

2.1.1. Paste

The paste for the concrete mixes was made with Portland CEM Type III/A 42.5 N/SR cement, which complies with the physical, chemical and mechanical requirements established in European standard EN 197-1:2013 [39]. The cement used, CEM type III, is cement with blast furnace slag addition. The content of clinker is 54% (EN-197 indicates that it must be 35–65%) and that of slag is 41% (the standard establishes 36–65% by weight), and it may contain up to a maximum of 5% of minor components, usually with gypsum as a setting regulator. Table 1 provides its chemical composition.

Table 1. Portland cement characteristics.

Chemical Composition	Value (wt.%)	Limit (wt.%) [39]
Clinker (SiO_2, Fe_2O_3, Al_2O_3, CaO, MgO and SO_3)	54	35–64
Blast-furnace slag	41	36–65
Minor components	5	≤5
LOI (Loss on ignition)	1.5	≤5
Physical Characteristics		
Le Chatelier (mm)	0.1	≤5
Setting time initial (min)	210	≥60
Setting time final (min)	260	≥60
Mechanical Characteristics		
Compressive strength (MPa) 2 days	20.1	≥13.5
Compressive strength (MPa) 28 days	56.6	≤42.5 and ≤62.5

Natural lime filler from a commercial plant was used as an addition. The chemical composition is shown in Table 2 and the particle size in Figure 2.

Table 2. Chemical composition: natural lime filler, coarse natural and fine natural aggregates

Chemical Composition	Filler Value (wt.%)	CNA [1] Value (wt.%)	FNA [2] Value (wt.%)
SiO_2	6%	98.09%	95.31%
Al_2O_3	1.2%	1.17%	2.24%
Fe_2O_3	0.69%	0.27%	1.06%
CaO	52.5%	0.05%	0.16%
MgO	1.4%	-	-
K_2O	-	0.14%	0.38%
TiO_2	-	-	0.17%
MnO	-	-	0.05%
CuO	-	-	0.03%
ZrO_2	-	-	0.03%
LOI (Loss on ignition)	38.21%	0.28%	0.57%

[1] CNA = coarse natural aggregates, [2] FNA = fine natural aggregates.

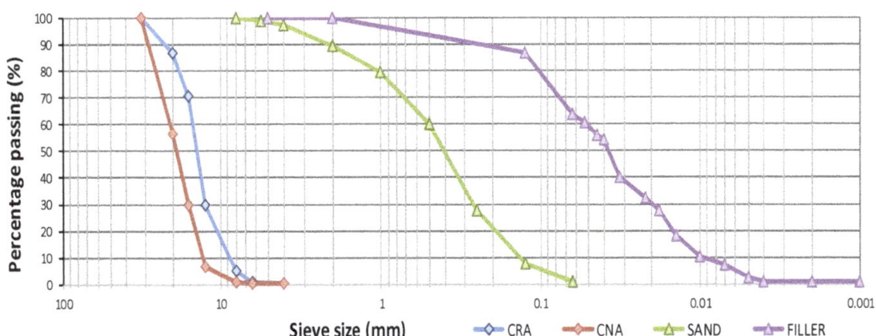

Figure 2. Particle size distribution of natural lime filler, natural sand, coarse natural aggregate (CNA) and coarse recycled aggregate (CRA).

2.1.2. Aggregates

Three aggregates were used to produce concrete: 0–4 mm fine siliceous aggregate, 4–12.5 mm natural siliceous river gravel and 4–12.5 mm recycled aggregate from concrete elements. The characteristics of the aggregates used are specified below. Figure 3 shows the appearance of natural and recycled coarse aggregates. The fineness module is 5.35 for CNA, 4.7 for CRA, and for natural sand. The mean diameters are 5.91 mm for natural sand, 11.20 mm for CRA and 12.27 for CNA.

(**a**)　　　　　　　　　　　　　　　　(**b**)

Figure 3. (**a**) Coarse natural aggregates, (**b**) Coarse recycled aggregates.

Composition and Characterization of the Recycled Aggregates

The recycled aggregates were supplied by C&DW recycling plant. The physical and chemical characteristics established by EHE-08 were used to check the degree of compliance of the results with the acceptable limits (Table 3). Figure 4 shows the macroscopic composition of the aggregates, which are mostly made of concrete, stone, low content of ceramic material and a residual quantity of gypsum EN 933-11:2009 [40]. As concerns the morphology, it is irregular in shape with sharp edges and surface roughness.

Table 3. Composition and physical and mechanical characterization of the recycled aggregates.

Parameter	Standard	Value	Limit EHE-08 [31]
Composition (%)			
Fl (floating particles) (%)		0	≤1
X (gypsum and impurities) (%)		0.04	≤1
Rc (concrete) (%)	EN 933-1:2012 [41]	70.7	-
Ru (natural stone) (%)		27	-
Rb (bricks and tiles) (%)		2.3	≤5
Ra (bituminous mat.) (%)		0	≤1
Rg (glass) (%)		0	≤1
Flakiness index (%)	EN 933-3:2012 [42]	5.7	≤35
Density and absorption			
P_a (apparent density) (Mg/m^3)		2.52	-
P_{od} (oven-dry density) (Mg/m^3)	EN 1097-6:2014 [43]	1.94	-
P_{ssd} (saturated surface dry density) (Mg/m^3)		2.17	-
WA^{24} (%)		4.8	≤5
Los Angeles coefficient (%)	EN 1097-2:2010 [44]	36	≤40

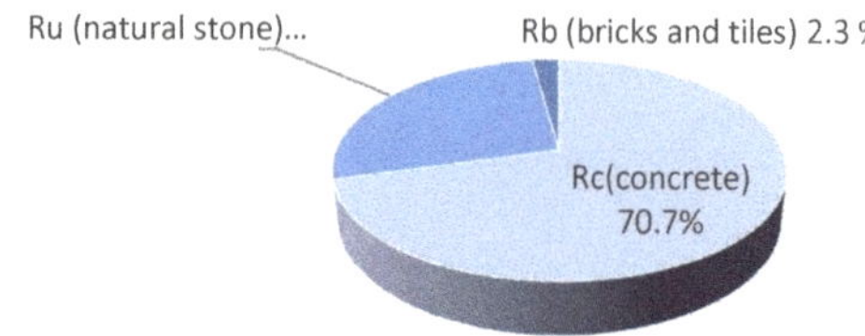

Figure 4. Coarse recycled aggregates composition according to EN 933-11:2009 [40].

The absorption of aggregate is slightly lower than the EHE-08 limit. This type of aggregate often shows high absorption values, but this specific aggregate is not affected by a higher absorption capacity due to the small quantity of ceramic material and attached mortar. Therefore, in principle, it does not need to be offset with a larger quantity of mixing water. Figure 2 shows the size distribution of the coarse recycled aggregates as per EN 933-3:2012 [42].

Composition and Characterization of the Natural Coarse and Fine Aggregates

The natural siliceous river coarse aggregate was supplied by a local company. It has a grain size of 4-12.5 mm. Table 4 shows its characteristics.

Table 4. Physical and mechanical characterization of the coarse natural aggregates.

Parameter	Standard	Value	Limit EHE-08 [31]
Flakiness index (%)	EN 933-3:2012 [42]	3.6	≤ 35
Density and absorption P_a (apparent density) (Mg/m^3) P_{rd} (oven-dry density) (Mg/m^3) P_{ssd} (saturated surface dry density) (Mg/m^3) WA^{24} (%)	EN 1097-6:2014 [43]	3.01 2.66 2.78 4.3	- - - ≤ 5
Los Angeles coefficient (%)	EN 1097-2:2010 [44]	34	≤ 40

The quantitative analysis of the chemical composition of the natural coarse aggregates was carried out by means of X-ray fluorescence (XRF). Table 2 shows the results of the analysis, confirming the siliceous nature of the aggregates. Concerning the morphology, a visual inspection indicated it is more regular in shape with rounded edges and a smoother, more polished and impermeable surface than the recycled aggregate. Figure 2 shows the coarse natural aggregates size distribution.

The fine natural aggregates also have a siliceous nature with a grain size of 0-4 mm. The grain size curve is described in Figure 2. The quantitative analysis of the chemical composition of the fine natural aggregates is also carried out by X-ray fluorescence (XRF). Table 2 shows the results of the siliceous nature of this aggregate.

2.1.3. Superplasticizer Additive

A third generation commercially available superplasticizer (SP), an aqueous-based modified polycarboxylate was used.

2.1.4. Water

Finally, the mixing water used came from the city of León drinking water supply. It meets the standards established in EHE-08.

2.2. Compositions

SCC was designed considering the instructions found in EHE-08 [31] and following Okamura method's procedures [32], modified as proposed by other authors. A cement content of 400 kg/m^3 and a W/C ratio of 0.47 were used as indicated by the standard.

Concrete was mixed in a vertical axis planetary concrete mixer. Once mixing ended, the mix was poured into different moulds, according to the proposed test and then submerged in a curing chamber for 28 days.

A reference concrete or control mix was designed as well as mixes substituting 20%, 50% and 100% of coarse natural aggregates with coarse recycled aggregates (Table 5).

Table 5. Proportion of natural and recycled aggregates in the mixes.

Mixes	Coarse Natural Aggregate (%)	Coarse Recycled Aggregate (%)
CC [1]	100%	0%
Mix 20 [2]	80%	20%
Mix 50 [3]	50%	50%
Mix 100 [4]	0%	100%

[1] CC = control concrete, [2] Mix 20 = mix 20% coarse recycled aggregates, [3] Mix 50 = mix 50% coarse recycled aggregates, [4] Mix 100 = mix 100% coarse recycled aggregates.

The amounts of materials used in the mix are shown in Table 6.

Table 6. Mix proportion (kg/m^3).

Content (m^3)	CC [1]	Mix 20 [2]	Mix 50 [3]	Mix 100 [4]
Cement (kg)	400	400	400	400
Lime filler (kg)	58	58	58	58
Water (kg)	190	190	190	190
Natural sand (kg)	904	904	904	904
Coarse natural aggregate (kg)	700	560	350	0
Coarse recycled aggregate (kg)	0	140	350	700
Superplasticizer (% weight of cement)	0.8	1	1.2	1.35
W/C ratio	0.47	0.47	0.47	0.47

[1] CC = control concrete, [2] Mix 20 = mix 20% coarse recycled aggregates, [3] Mix 50 = mix 50% coarse recycled aggregates, [4] Mix 100 = mix 100% coarse recycled aggregates.

The base mix was modified by increasing the superplasticizer volume (SP:CC = 0.8%, Mix20 = 1%, Mix 50 = 1.2% and Mix 100 = 1.35%) to achieve the self-compactability parameters, finally determining an optimal quantity of admixture of 0.8% with respect to the weight of cement for the control concrete. The W/C ratio was maintained in the mixes, for comparability purposes.

Different batches have been done to test the rheological and mechanical characteristics of the mixes. Four different batches per mix (CC, Mix20, Mix50 and Mix100) were produced (1) for J-ring test, (2) for compressive strength, (3) for flexural strength, and (4) for tensile strength tests. The number of samples per batch was at least three, to ensure enough representativeness of the results.

2.3. Mixing Protocol

The concrete mixing procedure used for all samples was based on standard EN 12390-2:2009 [45], following the steps shown in Figure 5.

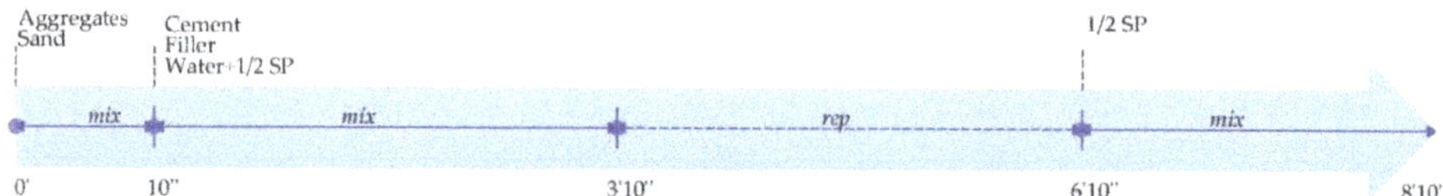

Figure 5. Mixing protocol.

Firstly, the mixer was moistened with water to prevent accumulation. The coarse natural and/or recycled aggregates were added from largest to smallest and later sand was added and then mixed for 10 s. The cement and filler were added and mixed for three minutes, slowly pouring in the mixing water with half of the superplasticizer, with time beginning upon completely pouring in the mixing water. The mix was left to stand for three minutes and was then mixed for another two minutes. The rest of the superplasticizer was poured homogeneously during no more than the first 10 s. The mixing was completed and testing began with fresh concrete.

To determine the self-compacting parameters prescribed, a flow test was done as per EN 12350-8:2011 [46], as well as a flow test using the J-ring test as per EN 12350-12:2011 [47] Both tests were performed immediately after mixing was finished.

Compressive strength was determined in the hardened state as per EN 12390-3:2009 [48] at 7, 14 and 28 days with cylindrical samples (300 × 150 mm).

Tensile and flexural strength were determined in the hardened state as per EN 12390-6:2010 [49] and EN 12390-5:2009 [50] respectively at 28 days old samples, 300 × 150 mm cylindrical samples were used for tensile strength and 100 × 100 × 400 mm prismatic samples for flexural strength.

3. Results and Discussion

3.1. Assessment of Coarse Recycled Aggregate

It is a concrete-sourced material that comes from C&DW of concrete elements with little rock, ceramic and gypsum impurities. As can be observed, the grain size of the recycled aggregates as well as of the natural aggregates is 4.5–12 mm, which exceeds the minimum size requirements for aggregates established by the Spanish code EHE-08. Although the minimum size of 4 mm is limiting, the grain size is believed to favour the self-compactability and workability of concrete adequate for these applications. Other studies often use grain sizes closer to 8–16 mm [11].

3.2. Composition

The amount of cement and W/C ratio limit values established by EHE-08 were complied with. According to the conclusions of other experimental studies [15,19], the adequate composition is somewhat enhanced and different from the one obtained by the strict use of the Okamura method [51]. Many of the tests consulted use cement contents ranging from 285 kg/m^3 to 440 kg/m^3 and compensating this range of values with more or less fine particles [11,19,26]. Other authors use 52.5 cement since its greater fineness helps with the workability and flow.

In the mixes proposed herein, the W/C ratio was maintained at 0.47 and no extra mixing water was added even though other authors correct this ratio as the proportion of recycled aggregates is increased, with up to 10% extra mixing water [21]. Other authors also use other valid techniques such as pre-saturating the aggregates [52]. Although the recycled aggregates used show a water absorption below the limits established by the standard, as shown in Table 3, it continues to be higher than the absorption of natural aggregate and, as has been proved, it was necessary to increase the percentage of superplasticizer based on the ratio of recycled aggregates. Other tests by other authors also consider this extra superplasticizer proportional to the increase in recycled aggregates [11,18].

3.3. Self-Compactability

The mixes were repeated twice, and the slump flow was measured with the J-Ring test (Figure 6). The Japanese ring flow test, J-Ring, assesses the blocking resistance of SCC through reinforcement bars under free-flow conditions. Moreover, any segregation, exudation and a higher concentration of the coarse aggregates in the central area can be observed. The standards in effect for this test are EN 12350-8:2011 [46] and EN 12350-12:2011 [47]. All mixes comply with the reference values shown in Table 7, self-compactability criteria.

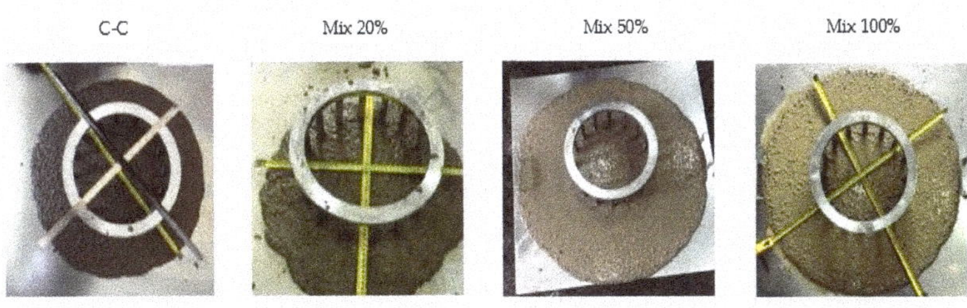

Figure 6. Japanese Ring test.

Table 7. Self-compactability of the samples designed.

| Mix | T_{500} (s) [1] | Limit EHE-08 [31] | SF (mm) [2] | Flow | J-Ring |
				Limit EHE-08 [31]	PJ (mm) [3]
CC	4	≤8	600	550 ≤ SF ≤ 850	14
	4	≤8	600	550 ≤ SF ≤ 850	14
Mix 20	5	≤8	590	550 ≤ SF ≤ 850	15
	5	≤8	580	550 ≤ SF ≤ 850	10.25
Mix 50	5	≤8	770	550 ≤ SF ≤ 850	11
	5	≤8	780	550 ≤ SF ≤ 850	5.3
Mix 100	4	≤8	730	550 ≤ SF ≤ 850	5.6
	4	≤8	770	550 ≤ SF ≤ 850	5.5

[1] T_{500} = Time until the concrete reaches the 500 mm circle. [2] **SF** = flow, [3] **PJ** = passage capacity calculated using the blocking scale.

The slump flow time T_{500} results of the concrete designed with different CRA contents are shown in Table 7. The T_{500} slump flow time varies between 4 and 5 s (Figure 7). Figure 8 shows a comparison between slump flow and substitution of coarse aggregates. As per EHE-08 [31], EFNARC [53] and other standards and guides, it should be within ranges of 2 to 5 s. Therefore, the times achieved are within the acceptable ranges and recycled aggregates SCC meets the requirements of the various standards (Figure 9).

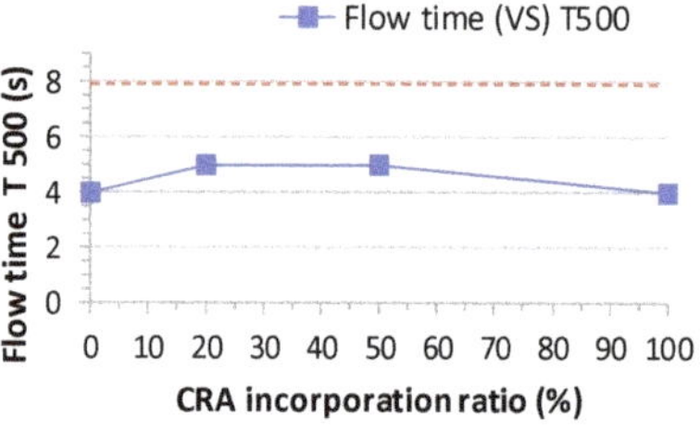

Figure 7. Flow time T_{500} test results.

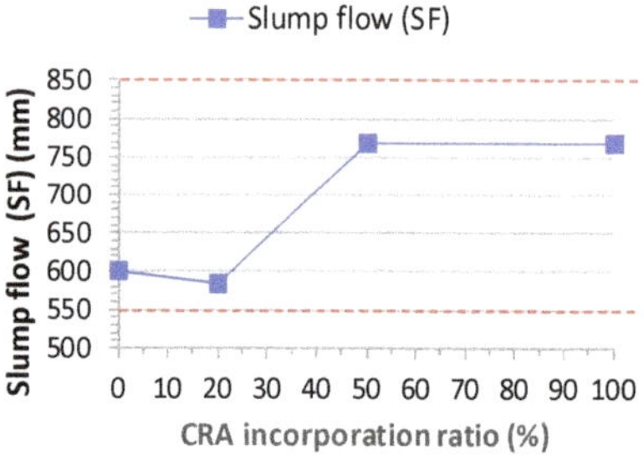

Figure 8. Slump flow test results.

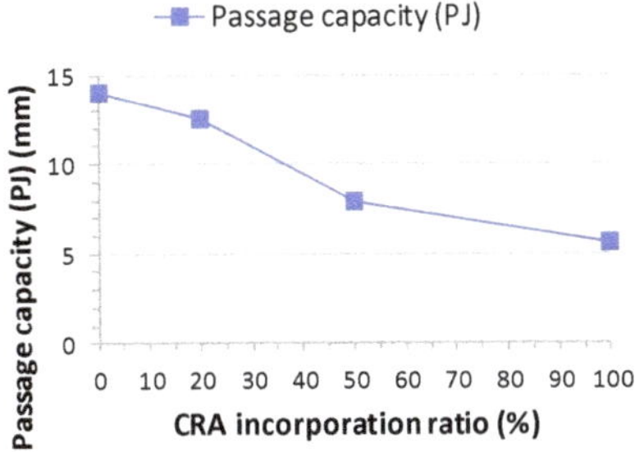

Figure 9. Passage capacity test results.

In terms of slump flow, a higher ratio of CRA leads to a larger final diameter and smaller passage capacity. The effect of CRA on the T_{500} slump flow time of the SCC should have been obvious for the 50% and 100% mixes, but they did not become more viscous with the increase in the percentage of superplasticizer. Visually, Mix 20 and Mix 50 reflected a good distribution of coarse aggregate and an absence of segregation and bleeding (Figure 6).

Generally, slump flow decreases with the CRA incorporation ratio since more water is absorbed as the ratio increases. Some authors, such as Safiuddin et al. [22], Tuyan et al. [29] and Modani and Mohitkar [30], have found that the slump flow increases for relatively small CRA incorporation ratios (20–40%) and the slump flow decreases for higher ratios (70–100%) because the fine aggregates content tend to increase due to the fracture during mixing and the resulting higher water absorption. However, according to Pereira-de-Oliveira el al. [21], slump flow increases with CRA's incorporation ratio given the gradual increase of SP content. This theory matches our results (Figures 10 and 11).

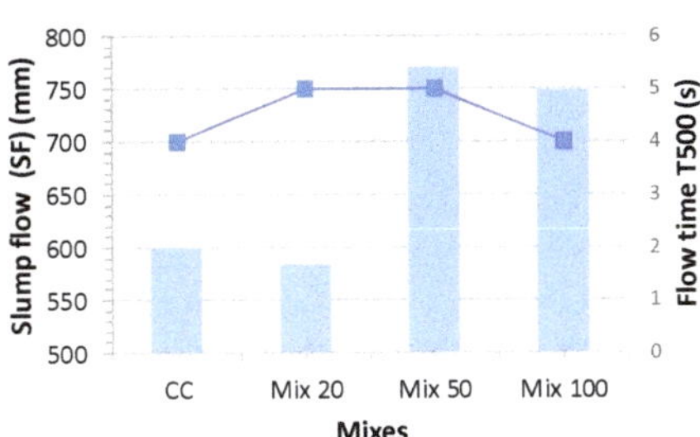

Figure 10. Slump flow time test results.

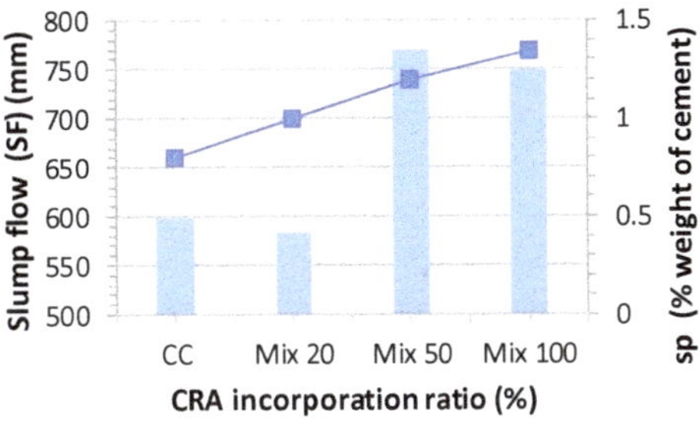

Figure 11. Slump flow and superplasticizer content.

Minimal variation was observed in the content of SP in the mixes at just 0.1% (relative to the weight of cement), which greatly influences the self-compactability of the mix. Readjusting the content of SP in some of the mixes could be considered: e.g., adjusting Mix 20 to 1.05–1.10% of SP and mix 100 to 1.325% of SP would be appropriate.

Figure 12 shows the flow time (T500) and the passage capacity (mm) of the different mixes in order to classify the self-compactability of concrete.

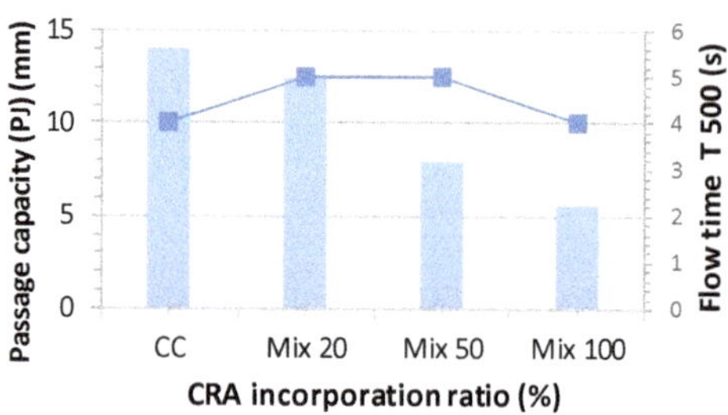

Figure 12. PJ and T_{500} results.

The classification obtained for concrete as per the criteria found in EHE-08, Annex 17 [31] and standard EN 12350-8:2011 [46] would be as follows:

- Based on the flow class: Class AC-E1,
- Based on the viscosity class: AC-V1,
- Based on the blocking resistance class: AC-RB2.

In general, class AC-E1 is the most adequate for most structural elements normally built, although it would be better to use a more fluid concrete such as type AC-E3 with 750 mm $\leq$ SF $\leq$ 850 mm for densely reinforced structures and formwork that is difficult to access or demands a long horizontal pouring displacement [31].

3.4. Compressive Strength

The compressive strength test was determined as per standard EN-12390-3 [48]. The results are shown in Figure 13. In terms of compressive strength, the recycled aggregates studied herein led to a noticeable difference. The compressive strength of Mix 20 and Mix 50 is higher than that of CC. For Mix 20 it is 55.58 MPa, which is 20% higher than that of CC (46.36 MPa). The average value for Mix 50 is 18% higher than for CC and Mix 100's is 5% lower than CC. This nonconformity may be due to the fact that the recycled aggregates in the composition, as shown in Table 3, have a high content of concrete that may positively influence the mechanical behaviour. A well-known effect is that the compressive strength decreases as the W/C ratio grows. In this study, the W/C ratio remains constant for all mixes, but the content of SP increases with the incorporation of CRA. This may explain why the compressive strength drops for Mix 100.

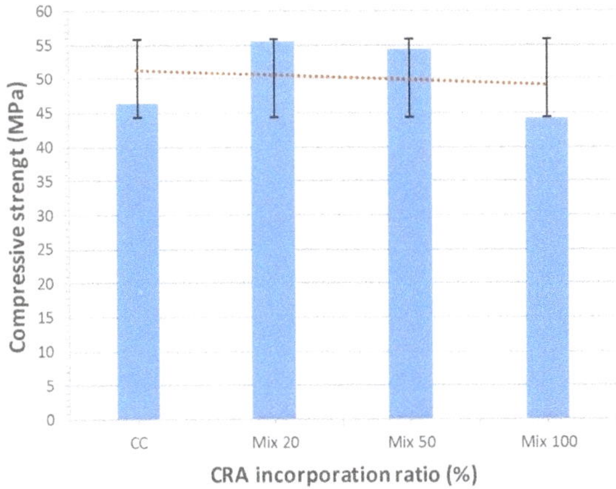

Figure 13. 28-day compressive strength results.

Grdic et al. [28] concluded that compressive strength decreases slightly as replacement increases (3.88% for 50% CRA and 8.55% for 100% CRA). The explanation for this reduction in strength can be found in the microstructure, i.e., the irregular composition of the CRA (formed by natural aggregate and cement paste) which provokes this slightly reduction.

Tuyan et al. [29] found that compressive strength increases as the W/C ratio decreases. Compressive strength increases slightly in mixes containing up to 40% CRA, about 5% with respect to the control mix. For higher replacement ratios (60%), compressive strength decreases. This is due to the microstructure, pores and cracks formed in the interfacial transition zones (ITZ) between the CRA and the paste slightly weakening the overall structure of the SCC.

The study of Modani et al. [30] concluded that, as the percentage of CRA increases, compressive strength decreases. This reduction is justified by the mortar and impurities adhering to CRA, creating areas of weakness in the SCC. For the 40% CRA mix, the loss was 5.80% compared to the control concrete. The mix containing the highest amount of CRA did not maintain the gain in strength over time due to the presence of unhydrated cement in CRA.

Pereira de Oliveira et al. [21] established that the replacement of CRA does not significantly influence the mechanical behaviour of the SCC, and finds a 5% difference between the control SCC and the 100% replacement mix.

The comparison with previous studies showed that, for a 20% substitution ratio, our results share the same trend and range of Tuyan et al. (a) [29] and Pereira de Oliveira et al. [21], with values between 54 MPa and 56 MPa. The trend line is very similar in all studies, i.e., up to 40–50% substitution, the compressive strength slightly increases relative to the control concrete and, for higher values of substitution, the compressive strength decreases. Figure 14 shows the comparison of our study with the results obtained by other authors. For total substitution (100%), our results match the figures obtained by Modani et al. [30]. Most of the authors studied [22,28–30] generally conclude that compressive strength slightly decreases with the CRA incorporation ratio. Some authors [21] conclude that compressive strength does not vary significantly with the incorporation of CRA because of the high volume of paste involved.

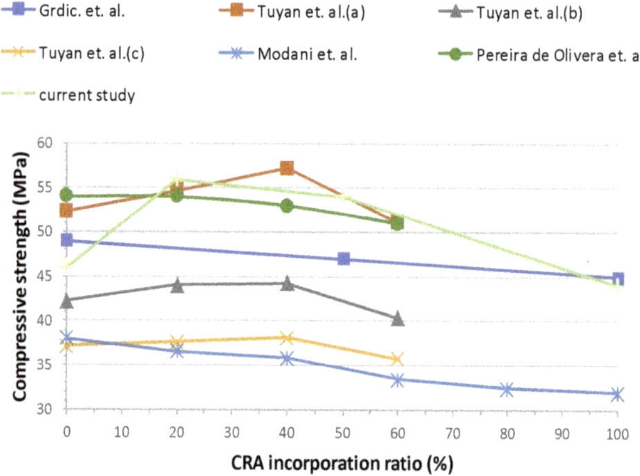

Figure 14. 28-day compressive strength results overview.

With these compressive strength values, it is perfectly possible to reach f_{ck} of 40 MPa, coherent with the W/C ratio and the amount of cement used.

3.5. Tensile Strength

In terms of splitting tensile strength, the results of all mixes are shown in Figure 15. Tensile strength was determined in the hardened state as per EN 12390-6:2010 [49] at 28 days, 300 × 150 mm cylindrical samples were used.

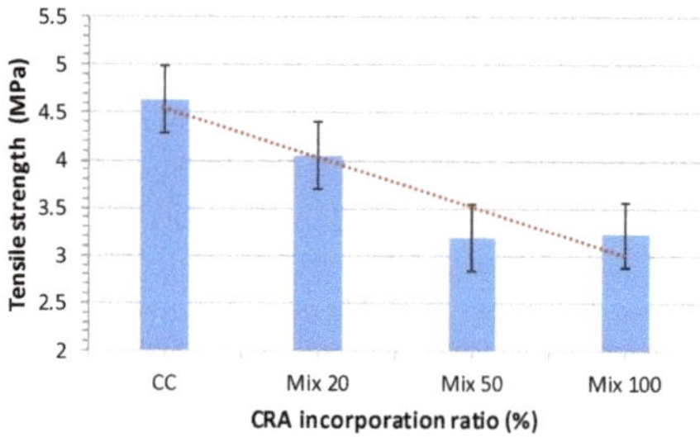

Figure 15. 28-day tensile strength test results.

Tensile strength decreases with the replacement of CNA with CRA, as expected. For an increase in RCA content, the tensile strength decreased by 12.71%, 31.25% and 30.60% in the mixes with CRA ratios of 20%, 50% and 100% respectively.

Tuyan et al. [29] concluded that the tensile strength of CRA mixes is slightly lower than that of the control mix. This loss is attributed to the ITZ between the CRA and the old bonded mortar paste and to the characteristics of the recycled aggregates. The strength loss goes from 8.8% to 16%.

Modani et al. [30] concluded that, as the percentage of CRA increases, the tensile strength decreases. The fracture surface occurs in the ITZ between the CRA and the CNA.

Concerning this parameter, our work matches previous results of other authors for substitution ratios between 20% and 50%, keeping the same trend. There is hardly a difference of ±1 MPa between them. Equally, for a 100% substitution, our results are in line with results obtained by Modani et al. [30]. Figure 16 shows the comparative results.

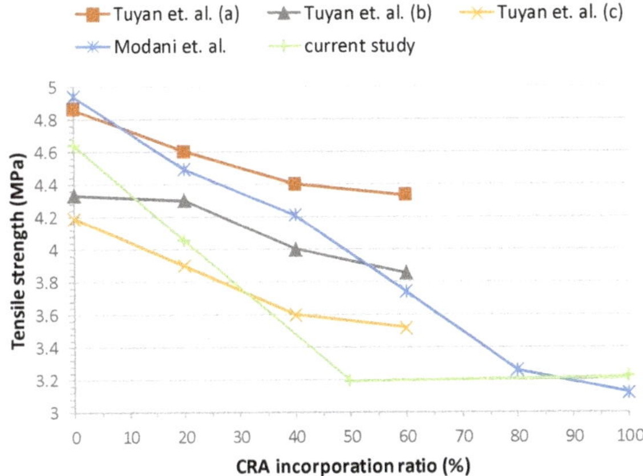

Figure 16. 28-day splitting tensile strength test results overview.

Other authors had similar test results and concluded that tensile strength decreases as the CRA incorporation ratio increases, which can be justified by the lower strength and greater porosity of CRA compared with CNA [21,27,29,30] (Figure 16). In our study, it is observed than the increase of CRA up to 20% led to a slight decrease of 13%, and the incorporation of 50% and 100% of CRA led to a similar effect with a decrease of 30%.

3.6. Flexural Strength

The results obtained for flexural strength for all mixes are shown graphically in Figure 17. Flexural strength was determined in the hardened state as per EN 12390-5:2009 [50] at 28 days, $100 \times 100 \times 400$ mm prismatic samples were used.

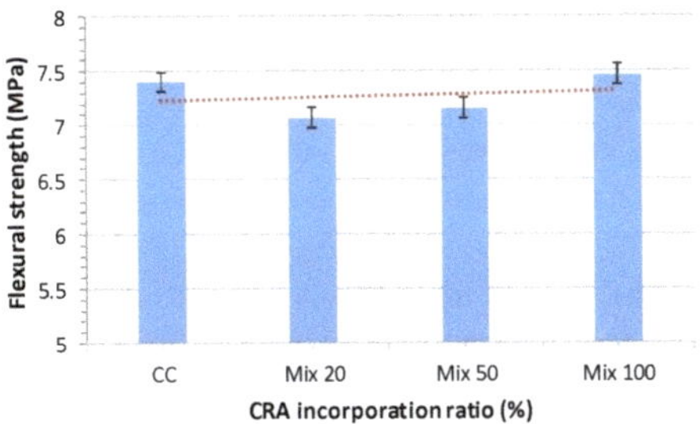

Figure 17. 28-day flexural strength test results.

In terms of flexural strength, the recycled aggregates studied herein did not lead to a notable difference relative to CC. The values are almost constant, and hardly suffer a variation of 4%.

Grdic et al. [28] concluded that flexural strength decreases with the increase of CRA. This reduction is due to the microstructural changes in the SCC. The reduction is 2.49% at SCC 50% CRA and 13.95% at SCC 100% CRA. The rest of the studies do not perform flexural tests.

Generally, authors found that flexural strength decreases as the replacement ratio increases, which is explained by the old mortar adhered to CRA [28] (Figure 18). In this case, the CRA used in the plant maintains its characteristics almost constant, has been washed and prepared and does not contain attached particles of mortar or gypsum, which explains this minimal variation in flexural strength.

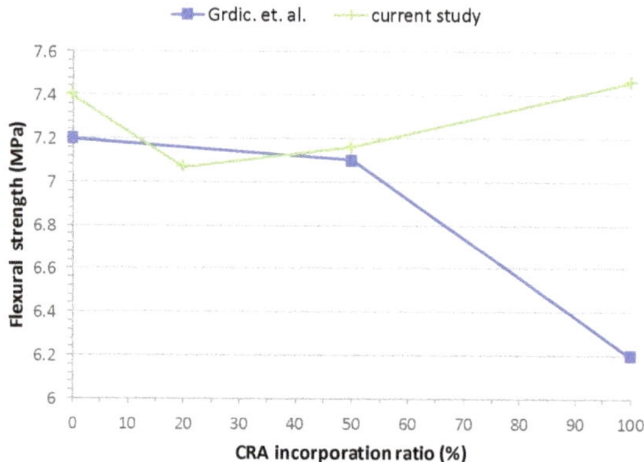

Figure 18. 28-day flexural strength test overview.

Figure 18 shows a comparison between the study of Grdic et al. [28] and our work. For substitution ratios up to 50%, the results obtained are very similar, but for higher ratios, the results are very far apart. While our tests remain practically the same (even slightly higher than CC for a 100% substitution), the study of Grdic et al. [28] shows a strong decrease. As discussed above, this is probably due to the attached mortar in the recycled aggregates.

4. Conclusions

The following conclusions have been drawn from the development and results of this research study:

- It is feasible to achieve an SCC with recycled aggregates. SCC was produced with this C&DW as aggregate in the proportions recommended by EHE-08 of 20% substitution and above 50% and 100% with minimal loss in the properties. The SP content needs to be increased as RCA content increases, so CC concrete contains 0.8% of the weight cement, Mix 20 contains 1%, Mix 50 contains 1.2%, and Mix 100 contains 1.35%,

- The aggregates studied showed characteristics within the limits established by EHE-08 for recycled aggregates, a large quantity of concrete (71%) and natural stone (27%) in composition and a low proportion of ceramic, which means absorption is not penalized and it is within the limits allowed by the standard,

- The consistency of recycled aggregate mixes is not significantly affected by adding coarse recycled aggregates and it shows adequate workability. The slump flow increases with the CRA incorporation ratio due to the gradual increase in SP content,

- The compressive strength is excellent and coherent with the composition used when producing concrete and is completely feasible for use in structural elements. The compressive strength increases with the CRA incorporation ratio. Mix 20 has 20% more strength than CC and Mix 50 about 18%. Only the total replacement mix, Mix 100, suffered a small drop in compressive strength, 5% less than the CC,

- Tensile strength decreases with the replacement of CNA with CRA by 12.71%, 31.25% and 30.60% in the mixes with CRA ratios 20%, 50% and 100% respectively. Regarding flexural strength, recycled aggregates concrete mixes did not show a notable difference with respect to CC. The values are almost constant, and hardly suffer a variation of 4%,

- The good self-compactability, mechanical resistance and durability results obtained should be an incentive to reconsider the use of this type of waste as recycled aggregates in concrete, which would also help make the construction process more sustainable.

The following could be considered as future lines of research for this study:

- Readjusting the percentage of SP in some of the mixes, especially in Mix 20,
- Partial or total replacement of FRA (fine recycled aggregates) in the mixes,
- Microscopy and durability studies of the resulting concrete.

The results of this research show that coarse recycled aggregates may be considered feasible and workable for use in SCC as appropriate mechanical properties can be achieved with contents that are completely satisfactory for structural purposes. Their use would be a significant contribution to the sustainability of the construction sector.

Author Contributions: Conceptualization, A.J.-V., R.M.-G.; Investigation, R.M.-G., A.J.-V., M.I.G.-R.; Writing—Original Draft Preparation, R.M.-G., A.J.-V., J.M.M.-d.P.; Writing—Review & Editing, R.M.-G., A.J.-V. and J.d.B.; Supervision, A.J.-V. and J.d.B.; Project Administration, J.M.M.-d.P.; Funding Acquisition, A.J.-V. All authors have read and agreed to the published version of the manuscript.

Funding: This work has been financially supported by the Spanish Ministry of Economy and Competitiveness through the research project grant BIA2017-83526-R.

Acknowledgments: Reutiliza S.L. for offering the recycled and natural aggregate used in this study free of cost. CERIS, from Instituto Superior Técnico (Lisbon), and FCT for financial support.

Conflicts of Interest: The authors state they have no conflicts of interest.

References

1. Eurostat. Recycling Rate of Waste Excluding Major Mineral Wastes. Available online: https://ec.europa.eu/eurostat/tgm/refreshTableAction.do?tab=table&plugin=1&pcode=ten00106&language=en (accessed on 24 April 2019).
2. Eusrostat. Generation of Waste by Economic Activity. Available online: https://ec.europa.eu/eurostat/tgm/refreshTableAction.do?tab=table&plugin=1&pcode=ten00106&language=en (accessed on 3 December 2019).
3. Anefa. *Informe de Situación Económica Sectorial*; Eurostat: Madrid, Spain, 2018.
4. Evangelista, L.; de Brito, J. Mechanical Behaviour of Concrete Made with Fine Recycled Concrete Aggregates. *Cem. Concr. Compos.* **2007**, *29*, 397–401. [CrossRef]
5. Tabsh, S.W.; Abdelfatah, A.S. Influence of Recycled Concrete Aggregates on Strength Properties of Concrete. *Constr. Build. Mater.* **2009**, *23*, 1163–1167. [CrossRef]
6. Juan-Valdés, A.; Rodríguez-Robles, D.; García-González, J.; Guerra-Romero, M.I.; Morán-del Pozo, J.M. Mechanical and Microstructural Characterization of Non-Structural Precast Concrete Made with Recycled Mixed Ceramic Aggregates from Construction and Demolition Wastes. *J. Clean. Prod.* **2018**, *180*, 482–493. [CrossRef]
7. Pacheco, J.; de Brito, J.; Chastre, C.; Evangelista, L. Experimental Investigation on the Variability of the Main Mechanical Properties of Concrete Produced with Coarse Recycled Concrete Aggregates. *Constr. Build. Mater.* **2019**, *201*, 110–120. [CrossRef]
8. Mas, B.; Cladera, A.; Del Olmo, T.; Pitarch, F. Influence of the Amount of Mixed Recycled Aggregates on the Properties of Concrete for Non-Structural Use. *Constr. Build. Mater.* **2012**, *27*, 612–622. [CrossRef]
9. Etxeberria, M.; Vázquez, E.; Marí, A.; Barra, M. Influence of Amount of Recycled Coarse Aggregates and Production Process on Properties of Recycled Aggregate Concrete. *Cem. Concr. Res.* **2007**, *37*, 735–742. [CrossRef]
10. Rodríguez-Robles, D.; García-González, J.; Juan-Valdés, A.; Morán-del Pozo, J.M.; Guerra-Romero, M.I. Effect of Mixed Recycled Aggregates on Mechanical Properties of Recycled Concrete. *Mag. Concr. Res.* **2015**, *67*, 247–256. [CrossRef]
11. Güneyisi, E.; Gesoglu, M.; Algın, Z.; Yazıcı, H. Effect of Surface Treatment Methods on the Properties of Self-Compacting Concrete with Recycled Aggregates. *Constr. Build. Mater.* **2014**, *64*, 172–183. [CrossRef]
12. Herbudiman, B.; Saptaji, A.M. Self-Compacting Concrete with Recycled Traditional Roof Tile Powder. *Procedia Eng.* **2013**, *54*, 805–816. [CrossRef]
13. González-Taboada, I.; González-Fonteboa, B.; Eiras-López, J.; Rojo-López, G. Tools for the Study of Self-Compacting Recycled Concrete Fresh Behaviour: Workability and Rheology. *J. Clean. Prod.* **2017**, *156*, 1–18. [CrossRef]

14. Esquinas, A.R.; Ramos, C.; Jiménez, J.R.; Fernández, J.M.; de Brito, J. Mechanical Behaviour of Self-Compacting Concrete Made with Recovery Filler from Hot-Mix Asphalt Plants. *Constr. Build. Mater.* **2017**, *131*, 114–128. [CrossRef]

15. Carro-López, D.; González-Fonteboa, B.; Martínez-Abella, F.; González-Taboada, I.; De Brito, J.; Varela-Puga, F. Proportioning, Microstructure and Fresh Properties of Self-Compacting Concrete with Recycled Sand. *Procedia Eng.* **2017**, *171*, 645–657. [CrossRef]

16. Silva, P.; de Brito, J. Experimental Study of the Mechanical Properties and Shrinkage of Self-Compacting Concrete with Binary and Ternary Mixes of Fly Ash and Limestone Filler. *Eur. J. Environ. Civ. Eng.* **2017**, *21*, 430–453. [CrossRef]

17. Santos, S.A.; da Silva, P.R.; de Brito, J. Mechanical Performance Evaluation of Self-Compacting Concrete with Fine and Coarse Recycled Aggregates from the Precast Industry. *Materials* **2017**, *10*, 904. [CrossRef]

18. Esquinas, A.R.; Álvarez, J.I.; Jiménez, J.R.; Fernández, J.M.; de Brito, J. Durability of Self-Compacting Concrete Made with Recovery Filler from Hot-Mix Asphalt Plants. *Constr. Build. Mater.* **2018**, *161*, 407–419. [CrossRef]

19. Carro-López, D.; González-Fonteboa, B.; Martínez-Abella, F.; González-Taboada, I.; de Brito, J.; Varela-Puga, F. Proportioning, Fresh-State Properties and Rheology of Self-Compacting Concrete with Fine Recycled Aggregates. *Hormigón y Acero* **2018**, *69*, 213–221. [CrossRef]

20. Ghalehnovi, M.; Roshan, N.; Hakak, E.; Shamsabadi, E.A.; de Brito, J. Effect of Red Mud (Bauxite Residue) as Cement Replacement on the Properties of Self-Compacting Concrete Incorporating Various Fillers. *J. Clean. Prod.* **2019**, *240*, 118213. [CrossRef]

21. Pereira-de-Oliveira, L.A.; Nepomuceno, M.C.S.; Castro-Gomes, J.P.; Vila, M.F.C. Permeability Properties of Self-Compacting Concrete with Coarse Recycled Aggregates. *Constr. Build. Mater.* **2014**, *51*, 113–120. [CrossRef]

22. Safiuddin, M.; Salam, M.A.; Jumaat, M.Z. Effects of Recycled Concrete Aggregate on the Fresh Properties of Self-Consolidating Concrete. *Arch. Civ. Mech. Eng.* **2011**, *11*, 1023–1041. [CrossRef]

23. Kebaïli, O.; Mouret, M.; Arabi, N.; Cassagnabere, F. Adverse Effect of the Mass Substitution of Natural Aggregates by Air-Dried Recycled Concrete Aggregates on the Self-Compacting Ability of Concrete: Evidence and Analysis through an Example. *J. Clean. Prod.* **2015**, *87*, 752–761. [CrossRef]

24. Khaleel, O.R.; Abdul Razak, H. Mix Design Method for Self Compacting Metakaolin Concrete with Different Properties of Coarse Aggregate. *Mater. Des.* **2014**, *53*, 691–700. [CrossRef]

25. Nieto, D. *Estudio de Hormigón Autocompactante Con Árido Reciclado*; Universidad Politécnica de Madrid: Madrid, Spain, 2015.

26. Iglesias Oria, R. *Hormigón Autocompactante Mejorado Con Aditivos Químicos y Uso de Áridos Reciclados*; Universidad de Cantabria: Cantabria, Spain, 2012.

27. Santos, S.; da Silva, P.R.; de Brito, J. Self-Compacting Concrete with Recycled Aggregates–A Literature Review. *J. Build. Eng.* **2019**, *22*, 349–371. [CrossRef]

28. Grdic, Z.J.; Toplicic-Curcic, G.A.; Despotovic, I.M.; Ristic, N.S. Properties of Self-Compacting Concrete Prepared with Coarse Recycled Concrete Aggregate. *Constr. Build. Mater.* **2010**, *24*, 1129–1133. [CrossRef]

29. Tuyan, M.; Mardani-aghabaglou, A.; Ramyar, K. Freeze–Thaw Resistance, Mechanical and Transport Properties of Self-Consolidating Concrete Incorporating Coarse Recycled Concrete Aggregate. *Mater. Des.* **2014**, *53*, 983–991. [CrossRef]

30. Modani, P.O.; Mohitkar, V.M. Self-Compacting Concrete with Recycled Aggregate: A Solution for Sustainable Development. *Int. J. Civ. Struct. Eng.* **2014**, *4*, 430–440.

31. EHE-08. Instrucción de Hormigón Estructural. Ministerio de Fomento: España. Available online: https://www.fomento.gob.es/organos-colegiados/mas-organos-colegiados/comision-permanente-del-hormigon/cph/instrucciones/ehe-08-version-en-castellano (accessed on 8 May 2019).

32. Ouchi, M.; Hibino, M.; Okamura, H. Effect of Superplasticizer on Self-Compactability of Fresh Concrete. *Transp. Res. Rec.* **1997**, *1574*, 37–40. [CrossRef]

33. Pereira-de Oliveira, L.A.; Nepomuceno, M.; Rangel, M. An Eco-Friendly Self-Compacting Concrete with Recycled Coarse Aggregates. *Inf. Construcción* **2013**, *65*, 31–41. [CrossRef]

34. Available online: http://www.hormigonespecial.com/~{}pdfs/MONOGRAFIA_RECICLADO.pdf (accessed on 13 February 2020).

35. Señas, L.; Priano, C.; Marfil, S. Influence of Recycled Aggregates on Properties of Self-Consolidating Concretes. *Constr. Build. Mater.* **2016**, *113*, 498–505. [CrossRef]

36. González Taboada, I.; González Fonteboa, B.; Martínez Abella, F.; Rojo López, G. Influencia de Las Variaciones En Los Materiales Sobre La Reología de Hormigones Autocompactantes Reciclados. In *V Congreso Iberoamericano de Hormigón Autocompactante y Hormigones Especiales*; València de, E.U.P., Ed.; Polytechnic University of Valencia: Valencia, Spain, 2018. [CrossRef]

37. Guerrero Vilches, I.M.; Rodriguez Jeronimo, G.; Rodriguez Montero, J. Valorización Como Árido Reciclado Mixto de Un Residuo de Construcción y Demolición En La Confección de Hormigones Autocompactantes Durables En Terrenos Con Yesos. In *HAC2018-V Congreso Iberoamericano de Hormigón Autocompactante y Hormigones Especiales*; Llano-Torre, A., Martí-Vargas, J.R., Ros, P.S., Eds.; Polytechnic University of Valencia: Valencia, Spain, 2018. [CrossRef]

38. Okamura, H.; Ouchi, M. Self-Compacting Concrete. *J. Adv. Concr. Technol.* **2003**, *1*, 5–15. [CrossRef]

39. *UNE-EN 197-1:2013. Cemento. Parte 1: Composición, Especificaciones y Criterios de Confromidad de Los Cementos Comunes*; Aenor: Madrid, España, 2013.

40. *UNE-EN 933-11:2009. Ensayos Para Determinar Las Propiedades Geométricas de Los Áridos. Parte 11: Ensayo de Clasificación de Los Componentes de Los Áridos Gruesos Reciclados*; Aenor: Madrid, España, 2009.

41. *UNE-EN 933-1:2012. Ensayos Para Determinar Las Propiedades Geométricas de Los Áridos. Parte 3: Determinación de La Forma de Las Partículas. Índice de Lajas*; Aenor: Madrid, España, 2012.

42. *UNE-EN 933-3:2012. Ensayos Para Determinar Las Propiedades Geométricas de Los Áridos. Parte 1: Determinación de La Granulometría de Las Partículas. Método Del Tamizado*; Aenor: Madrid, España, 2012.

43. *UNE-EN 1097-6:2014. Ensayos Para Determinar Las Propiedades Mecánicas y Físicas de Los Áridos. Parte 6: Determinación de La Densidad de Partículas y La Absorción de Agua*; Aenor: Madrid, España, 2014.

44. *UNE-EN 1097-2:2010. Ensayos Para Determinar Las Propiedades Mecánicas y Físicas de Los Áridos. Parte 2: Métodos Para La Determinación de La Resistencia a La Fragmentación*; Aenor: Madrid, España, 2010.

45. *UNE-EN 12390-2:2009 Ensayos de Hormigón Endurecido. Parte 2: Fabricación y curado de probetas para ensayos de resistencia*; Aenor: Madrid, España, 2009.

46. *UNE-EN 12350-8:2011. Ensayos de Hormigón Fresco. Parte 8: Hormigón Autocompactante*; Aenor: Madrid, España, 2011.

47. *UNE-EN 12350-12:2011. Ensayos de Hormigón Fresco. Parte 12: Hormigón Autocompactante. Ensayo Con El Anillo Japonés*; Aenor: Madrid, España, 2011.

48. *UNE-EN 12390-3:2009 Ensayos de Hormigón Endurecido. Parte 3: Determinación de La Resistencia a Compresión de Probetas*; Aenor: Madrid, España, 2009.

49. *UNE-EN 12390-6:2010. Ensayos de Hormigón Endurecido. Parte 6: Resistencia a Tracción Indirecta de Probetas*; Aenor: Madrid, España, 2010.

50. *UNE-EN 12390-5:2009. Ensayos de Hormigón Endurecido. Parte 5: Resistencia a Flexión de Probetas*; Aenor: Madrid, España, 2009.

51. Okamura, H.; Ouchit, M. Self-Compacting High Performance Concrete Development of Self-Compacting Concrete. *Prog. Struct. Mater. Eng.* **1998**, *1*, 378–383. [CrossRef]

52. García González, J. Concrete with Ceramic Mixed Construction and Demolition Waste: Optimization of Physical, Mechanical and Durability Properties by Aggregate Pre-Saturation, Use of Superplasticizers and Microbially Induced Carbonate Precipitation. Ph.D. Thesis, University of Ghent, Ghent, Belgium, 2016.

53. EFNARC. Specification and Guidelines for Self-Compacting Concrete. *Rep. EFNARC* **2002**, *44*, 32.

Article

Study on the Microstructure of the New Paste of Recycled Aggregate Self-Compacting Concrete

Kheira Zitouni [1], Assia Djerbi [2,*] and Abdelkader Mebrouki [1]

[1] Civil Engineering and Architecture Department-LCTPE, Faculty of Sciences and Technology, Abdelhamid Ibn-Badis University, 27000 Mostaganem, Algeria; kheira.zitouni@hotmail.com (K.Z.); mebroukiaek@yahoo.fr (A.M.)

[2] MAST Department-FM2D, University Gustave Eiffel, IFSTTAR, F-77447 Marne la Vallée, France

* Correspondence: assia.djerbi@univ-eiffel.fr; Tel.: +33-(0)1-81-66-83-76

Received: 3 March 2020; Accepted: 21 April 2020; Published: 2 May 2020

Abstract: Previous literature indicates a decrease in the mechanical properties of various concrete types that contain recycled aggregates (RA), due to their porosity and to their interface of transition zone (ITZ). However, other components of the RA concrete microstructure have not yet been explored, such as the modification of the new paste (NP) with respect to a reference concrete. This paper deals with the microstructure of the new paste of self-compacting concrete (SCC) for different levels of RA. The water to binder ratio (w/b) was kept constant for all concrete mixtures, and equal to 0.5. The SCC mixtures were prepared with percentages of coarse RA of 0%, 30%, 50% and 100%. Mercury intrusion porosimetry test (MIP) and scanning electron microscope (SEM) observations were conducted on the new paste of each concrete. The results indicated that the porosity of the new paste presents a significant variation for replacement percentages of 50% and 100% with respect to NP0 and NP30. However, RA contributed to the refinement of the pore structure of the new paste. The amount of macrospores the diameter of which is in the 50–10,000 nm range was reduced to 20% for NP50 and NP100, while it was about 30% for NP0 and NP30, attributed to the water released by RA. Compressive strength loss for SCC50 and SCC100 concretes are both influenced by porosity of RA, and by the NP porosity. The latter is similar for these two concretes with the 26% increase compared to a reference concrete.

Keywords: recycled aggregate; self-compacting concrete; mercury intrusion porosimetry; SEM observation; new paste; compressive strength

---

1. Introduction

In the context of sustainable development, recycling of concrete waste is carried out with the main purpose of protecting the environment through the reduction of generated greenhouse gases and the preservation of natural resources. The reuse of concrete from demolition as aggregates in new concrete compositions reduces overheads and eases costs associated to waste management. Thus, this approach promotes the protection of natural resources, which are becoming increasingly difficult to obtain. Recycling of demolition waste has then witnessed a rather significant development. The European Directive on waste (2008/98/EC) [1] has fixed thresholds of waste from deconstruction/demolition to be implemented in new concretes at 70% by 2020.

Using recycled aggregates (RA) in self-compacting concrete (SCC) improves the ecological impact of concrete, since their microstructure properties are greater than normal vibrated concrete [2–4]. The use of the fine content in SCC refines the microstructure and hence the pore network of the material, which improves SCC mixtures' durability compared to that of ordinary vibrated concrete [2]. In fact, the replacement of natural aggregates with RA can affect the other properties of SCCs. Some research has been carried out for a constant volume of binder in order to evaluate the effect of the replacement rate of RA and to quantify their effects on different properties [5–9]. Sasanipour et al. [5] showed that the replacement of 25% of coarse aggregates had no significant effect on the durability properties of self-compacting concrete including electrical resistivity and chloride ion resistance. However, the use of silica fume improved the durability performance of SCC-containing recycled aggregates [6]. Pereira-de-Oliveira et al. [7] concluded it is viable to replace natural coarse aggregates by recycled coarse aggregates in SCCs. Their results showed that the recycled coarse aggregate incorporation did not significantly affect the water permeability, while water capillarity coefficient is slightly decreased when 100% of coarse RA is used. In addition, the water penetration depth is reduced when increasing the amount of RA in SCC. Kou et al. [8] found that the properties of the SCCs made from river sand and crushed recycled aggregates showed only slight differences with respect to a reference concrete. However, Omrane et al. [9] showed the RA content should be limited to 50% to achieve good performance and to fulfil all the recommended conditions of SCCs.

Few investigations have been conducted on the microstructure of SCCs made with RA. Unlike studies on conventional concrete, which showed the performance of recycled aggregate concrete is mainly due to the porosity caused by the old paste in RA [10–12], this porosity leads to a weak interfacial zone between the new paste and the old one [13]. Hence, this low bond is attributed to both porosity and to the high absorption capacity of the RA [14]. It is clear, therefore, that the microstructure of the recycled aggregate concrete (RAC) is different from conventional one, since RA contain a proportion of old cement paste. In order to understand the behavior of RAC, it is necessary to study its porosity and its pore distribution, not only in the interface of transition zone (ITZ) between an RA and the new paste, but also in the bulk new paste. The most widely used method to determine the pore structure is mercury intrusion porosimetry (MIP) [15–19]. However, MIP has limitations in measuring actual pore size distributions and detecting pore diameters below 3 nm and above 375 µm, because pores are too small or too isolated to be filled by mercury. Nonetheless, MIP still has a great capacity to estimate total porosity and a characteristic pore size [19]. Some authors have investigated the microstructure of RAC with this technique [20–25]. Results show that the substitution of natural aggregates (NA) with RA increased porosity. Shicong et al. [20] showed that RA modify tailings pore size distribution with higher intrusion volumes in pores which diameters larger than 0.01 µm in RAC compared to natural aggregate concrete (NAC). The conclusion was that the porous old paste adhered to RA. Similar findings were observed by Uchikawa et al. [21]. Previous literature [22–25] also indicated that the replacement of RA correlates with the total volume and pore size. In addition, it demonstrated that the influence of RA was more important at lower ages, and that its effect was reduced while increasing the curing duration. Major quantitative changes were observed with the increase in pore volume for those with a radius below 30 nm. Gonzalez-Corominas et al. [24] used MIP to investigate the influence of steam curing on the pore structure of concrete containing different qualities of RA. Their results showed the effectiveness of steam curing in the refinement of pore structure. Furthermore, although the lowest quality RA had a coarser pore size distribution, the highest reductions—especially with respect to the capillary pores (10 µm–0.01 µm)—were observed in this material. This observation is explained by the internal curing effect of porous aggregates that enlarges binder hydration. The use of silica fume [25] caused a 32% reduction in the cement paste porosity with respect to reference concrete. However, the representational value of the sample for the MIP test in the case of (RAC) was not taken into account [20–25].

The different components of an RAC (RA, new paste and NA) each generally occupy a significant volume, in the range of tens of cubic centimeters. On the other hand, MIP does need a small volume of the sample for testing (a few cm^3), raising the question of RAC sample representativity of a RAC sample with respect to the real mix. Therefore, the current porosity will be fully identified using other techniques. Hence, it is necessary to propose a new approach to study the microstructure of RAC by MIP test. A suitable sampling method, especially considering the lower volume of coarse aggregate in SCC compared to conventional concrete, might ease the sampling of the new paste. Accordingly, this work proposes a sampling method to study the microstructure of the new paste (NP) of self-compacting concrete (SCC) for different levels of recycled aggregates (RA). The binder content was kept constant for all concrete mixtures, with a water to binder ratio (w/b) of 0.5. SCC mixtures were prepared with 0%, 30%, 50% and 100% of coarse RA. The pore distribution of the new paste was evaluated by MIP test, and its anhydrous content was quantified by SEM images analysis. Additionally, the contribution of the porosity of NP and RA to the decrease in compressive strength was investigated and demonstrated.

2. Materials and Methods

2.1. Materials

The type of cement used in all mixtures was CEM I/42.5 N with a real density of 3.16 g/cm^3 and a blaine specific surface of 3150 g/cm^2. A limestone filler (LF) was used with a real density of 2.70 g/cm^3 and a blaine specific surface of 4200 g/cm^2. The chemical composition of both materials is provided in Table 1. A superplasticizer (SIKA VISCOCRETE TEMPO 12) enhanced the workability of SCC.

Table 1. Chemical composition of cement and limestone filler (LF).

Compound (%)	SiO$_2$	CaO	MgO	Fe$_2$O$_3$	Al$_2$O$_3$	SO$_3$	K$_2$O	Na$_2$O	LOI
Cement	21.8	64.7	1.62	6.37	3.88	0.42	0.33	0.22	0.66
LF	5.19	49.35	1.29	1.29	1.50	0.06	0.27	0.12	40.93

To obtain an SCC, the volume of the aggregates and their maximum size has been reduced to mitigate friction and to avoid blockages in confined areas. The maximum diameter of the coarse aggregates was fixed in our study at 15 mm. In this study, siliceous sand and crushed limestone sand were used as the fine natural aggregates. Natural coarse aggregates were made of crushed limestone aggregates and recycled coarse aggregates were obtained from construction and demolition waste (CDW). The grading curves of aggregates are presented in Figure 1.

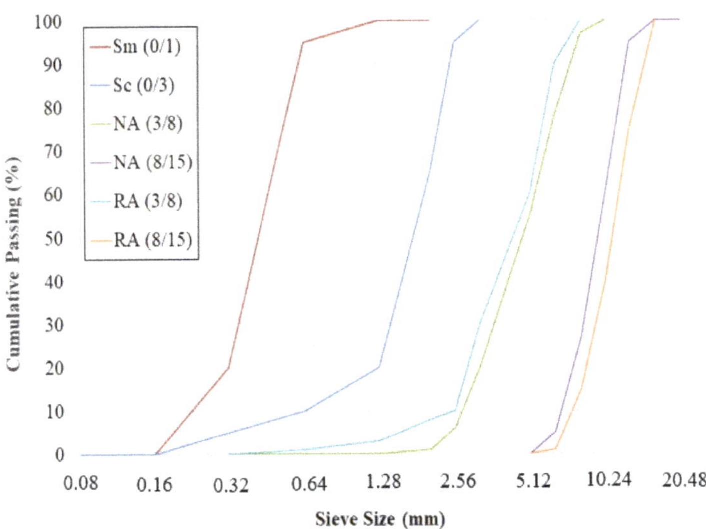

Figure 1. Grading curves of different types of aggregates.

The main properties of the aggregates are displayed in Table 2, where Dr is the relative density of the particles, A is the water absorption, and LA is the Los Angeles fragmentation coefficient. These properties were measured according to the NF-EN-1097-6 standard [26]. The porosity P was determined according to the NF P 18-459 standard [27,28].

Table 2. Properties of the aggregates.

Aggregate Size (mm)	Dr (g/cm^3)	A (%)	P (%)	LA (%)
Siliceous sand Sm (0/1)	2.60	1	-	-
Crushed limestone sand Sc (0/3)	2.68	1.8	-	-
Natural coarse aggregates NA (3/8)	2.66	1.4	-	-
Natural coarse aggregates NA (8/15)	2.64	1.3	-	21
Recycled coarse aggregates RA (3/8)	2.35	6.1	18.7	-
Recycled coarse aggregates RA (8/15)	2.42	4.8	16.4	33

2.2. RA Pre-Saturation Process

In order to achieve equal consistency, and due to the higher water rate absorption of RA, the mixing of water had to be adjusted. This adjustment was achieved by using RA under the saturated surface dry (SSD) condition. If the aggregates are fully saturated, the surface moisture would increase the effective w/c ratio, the aim being that RA would have the same water absorption coefficient as NA. Thus, for each fraction of RA (3/8) and (8/15), the amount of water pre-saturation (M W.pre-sat) is defined as the difference between the water absorbed by RA and NA, and is calculated as follows:

$$\% \text{ Wpre-sat} = \% \text{ abs (RA)} - \% \text{ abs (NA)} \tag{1}$$

$$M \text{ W.pre-sat} = \% \text{ Wpre-sat} \times M \tag{2}$$

With M the mass of RA during mixing

Despite the very rapid evolution of water absorption of the RA in the first five minutes, a pre-saturation time for 24 h was chosen. This is consistent with the relevant literature, in which it is indicated that absorption lasts for hours after its initial quick step [29,30]. This guaranteed that no water was absorbed after mixing, when pre-saturation is carried out for 5 min. Before each mix, the aggregates were placed in a large plastic bottle for 24 h and the pre-saturation water was added residually. Great care was taken when rolling the bottle to avoid any air entrance and to ensure all the aggregates were impregnated. This operation was conducted for a period of 30–45 min. The bottle method was also adopted in the case of recycled sand [31], and the aggregates were left in their dedicated flask.

2.3. Design of Concrete Mixes

The mixing method used to design the compositions of the SCC is based on a composition of one with limestone fillers. The amount of components required for making one cubic meter of concrete was constant. They consisted in a paste volume of 37.5% (375 L/m^3), a water to binder ratio w/b ratio of 0.5 and a gravels/sand (G/S) ratio close to one. SCC-RA mixes were designed by replacing a volume of NA with RA. Four mixes were prepared with 0%, 30%, 50% and 100% recycled aggregates respectively labelled to SCC0, SCC30, SCC50 and SCC100 respectively. The mix designs of the concrete are shown in Table 3.

Table 3. Mix composition of the concretes, per cubic meter.

Mixture(kg/m^3)	SCC0	SCC30	SCC50	SCC100
Cement	382	382	382	382
LF	65	65	65	65
Sm (0/1)	578	578	578	578
Sc (0/3)	253	253	253	253
NA (3/8)	335	235	168	0
NA (8/15)	495	347	248	0
RA (3/8)	0	89	147	296
RA (8/15)	0	136	227	454
Effective water	224	224	224	224
w/b (C + LF)	0.5	0.5	0.5	0.5
Superplasticizer (%)	2.67	2.67	2.67	2.67

All SCC mixes showed a slump flow in the 700 mm-range and a good resistance to segregation. For each composition, cylindrical specimens of 16 × 32 cm were casted in steel molds, and cured in water at 20 ± 2 °C. The compressive strength of these concrete cylinders was measured on three samples according to NF EN 12390-3 [32] with a hydraulic press capacity of 2400 kN.

2.4. MIP Tests

Mercury intrusion porosimetry (MIP) is one of the main methods for investigating the mesoporous structure (pore radius between 2 and 50 nm) and the macro-porous nature (apertures greater than 50 nm) of cementitious materials. Its effectiveness is based on the principle that to fill a non-wetting fluid into a pore of diameter d, a pressure P inversely proportional to this diameter must be applied. This pressure is given by the Washburn Equation (3):

$$P = \frac{-4\,\gamma cos\theta}{d} \tag{3}$$

Here, the surface tension γ of mercury is of 485 N/m, and the contact angle θ between mercury and the pore wall of 141.3°. MIP tests were conducted on an AutoPore IV 9500 V1.09, from Micrometrics Instrument Corporation, under a maximum pressure of 413 MPa to reach pores with a radius of 3.6 nm.

RA concretes (SCC30 and SCC50) are considered as multiphase composites in which the new paste (NP) is joined into coarse NA and coarse RA. The new paste consists of the binder and the sand, and represents 69% v/v of concrete. SCC0 consists in NP and coarse NA, while SCC100 in NP and coarse RA. As the common phase between the different concretes is the new paste, the study and the comparison between mixes focused on its microstructure. Each concrete type was sampled (approximate volume 2 cm^3), as shown in Figure 2, and extracted from 16 × 5 cm discs. The tested NP samples extracted from SCC0, SCC30, SCC50 and SCC100 were respectively labelled NP0, NP30, NP50 and NP100.

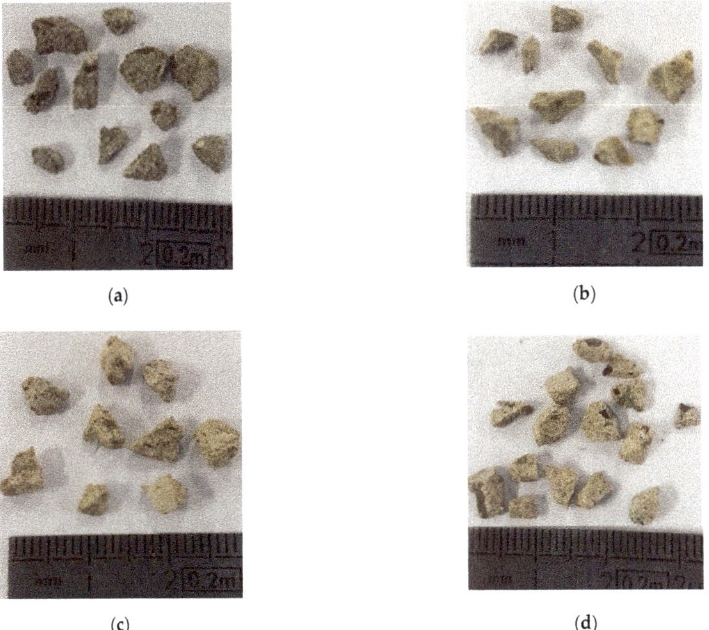

(a)   (b)

(c)   (d)

Figure 2. Samples of new pastes before MIP tests (**a**) NP0, (**b**) NP30, (**c**) NP50 and (**d**) NP100.

These samples were first introduced into the liquid nitrogen for 5 min in order to stop the hydration. They were then lyophilized in an alpha freeze-dryer 1–4 Ld for 48 h to extract the liquid water by sublimation. The samples were stored in a desiccator until testing. This technique has proven to be the most appropriate one to limit microstructural damage [33].

2.5. SEM Observations

The study focused on the anhydrous content for the new paste (NP0, NP30, NP50 and NP100) of each concrete type. The study of the concrete microstructure of all RA-SCC samples was performed using a scanning electron microscope (SEM), Quanta 400 from FEI Company. Their observation was carried out through electron backscattered imaging (BSE) to identify the different phases of the microstructure. Before initiating SEM observation, to remove the free water, small samples of size 35 × 35 × 10 mm were dried under vacuum with silica gel at 45 °C for 14 days. The dry samples were then impregnated with epoxy resin; then, the samples were polished in various steps to create a smooth plane surface for SEM imaging. Because our samples were not conductive, we had to add a very thin gold metallic coat. Obtained BSE images were processed with the Stream Motion software to distinguish the different phases of the samples using a toll dedicated to contrast the grey levels. Images of NP were taken at random spots, with an analysis over a 10,000 μm^{2-} square on images with a magnification of 800 times, as illustrated in Figure 3. As shown in a previous study [34], the porosity appears red, the anhydrous are blue, and the hydrated cement gel appears green (see Figure 3). It took approximately 30 images to cover new paste on three samples of each type of concrete.

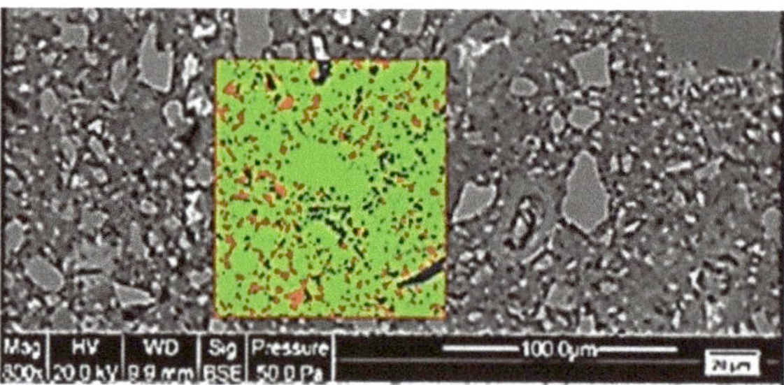

Figure 3. Example of SEM image analysis with a magnification of 800 on a NP0 sample.

3. Results

3.1. Compression Strengths

Figure 4 shows the compressive strengths of SCC as a function of their RA content. The results obtained have good repeatability with a standard deviation almost constant. A 90-day compressive strength decrease can be seen with the increase in the RA content.

This result is widely accepted and consistent with the literature [8,35]. Some authors attribute this compressive strength decrease to a high total porosity of RCA concrete [22], while others to the high porosity in the ITZ [36,37]. Strengths losses are then linked to the presence of old paste, without taking into account the modifications of the NP microstructure. The presence of the old paste leads to high porosity and, consequently, the compressive strength decreases. However, and as will be demonstrated later in this study, the microstructure of the NP is affected by the presence of RA once a replacement threshold is reached. This is why a specific study of the NP microstructure based on the MIP test was undertaken.

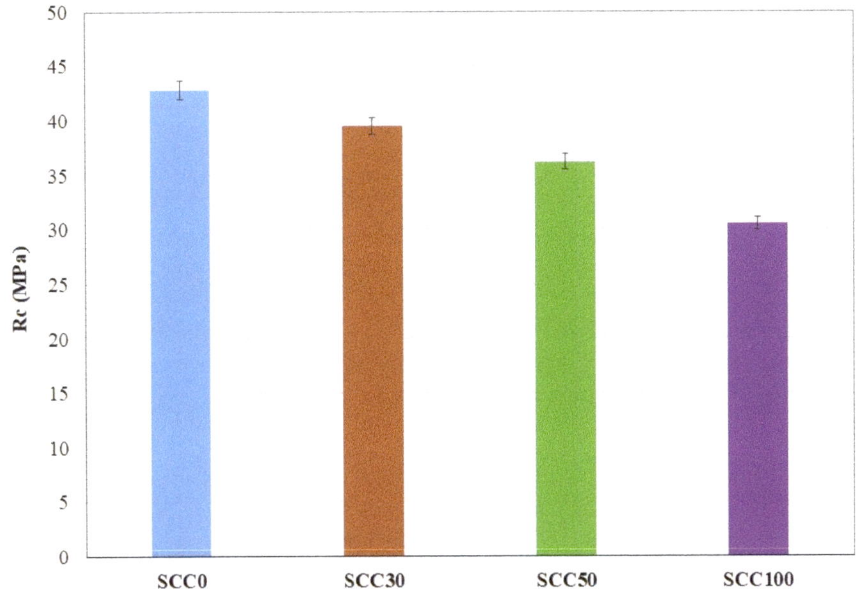

Figure 4. Compressive strengths results at 90 days.

3.2. Influence of Recycled Aggregates (RA) on the Porous Structure of the New Paste (NP)

Figure 5 shows the cumulative intrusion volume as a function of pore diameter for various the tested NP. The greater the RA percentage, the larger the total mercury intrusion. Two groups of pore size distributions can be distinguished. The first one for NP0 and NP30 evolves similarly with a total introduced pore volume of 0.069 mL/g, while a second one for NP50 and N100 pastes present a total introduced pore volume of 0.09 mL/g. In both groups, a slope change is observed for pore diameter of 100 nm.

Differential intrusion curves of the NP in Figure 6 provide extended information on pore structure. Generally, in cured cement paste, two different pore diameters are considered. The pore diameter corresponding to the first peak occurs when the mercury intrusion is through a porous network connected to the surface of the sample. It is defined as the critical pore diameter (Dc) of the capillary pores that varies from 10 nm to 10,000 nm [38]. The second peak is related to gel pores, and according to study [39], its value is less than 10 nm, while some other authors are providing a value in the 20–40 nm range [40].

In the present study, NP0 and NP30 have a pore diameter distribution spread all over the whole measured range. No clear visible peaks in the capillary pores were observed besides the ones at 5 and 4.35 nm. A less rounded peak at the capillary pores for NP0 and NP30 is attributed to a large pore size distribution for this material phase, as stated by [38]. It is possible that these peaks correspond to an intrusion in the material phase with a distinct network of smaller pores. A less pronounced peak in these pastes, therefore, corresponded to a more or less connected pore path.

For NP50 and NP100 pastes, a sharp peak is visible around 50–60 nm. It is associated with the minimum diameter of an interconnected capillary network, called the critical diameter. This diameter is determined from the inflection point of the curves in Figure 5, and summarized in Table 4. Other peaks could also be observed for the range of mesopores, at 11 nm for the NP100, and at 4.60 nm for both NP50 and NP100 related to gel pores.

Table 4. Porosity, total intruded mercury volume and critical pore size of NP and RA.

	Porosity (%)	Mercury Volume (mL/g)	Dc(nm)
NP0	14.99	0.0693	-
NP30	14.80	0.0691	-
NP50	18.75	0.093	40.24
NP100	18.92	0.094	33.80
RA	17.26	0.084	21.19

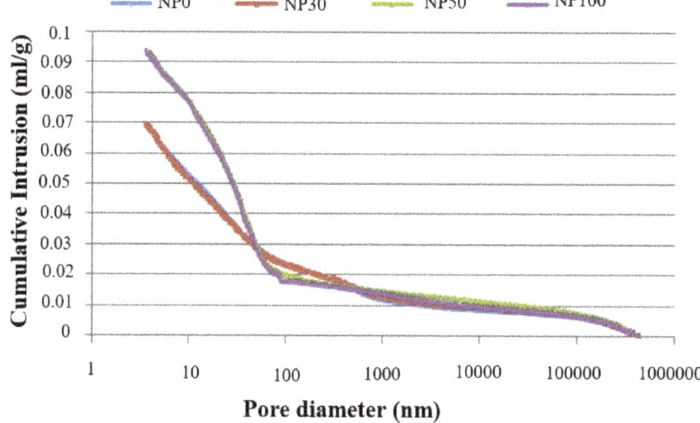

Figure 5. Cumulative intruded pore volume of new pastes (NP).

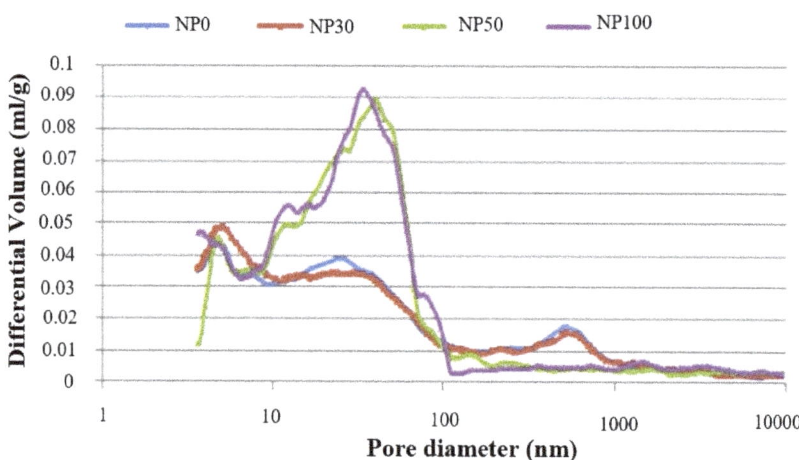

Figure 6. Pore size distribution for the different new pastes (NP).

Figure 7 shows the total mercury intrusion pore volume for several pore size ranges. NP50 and NP100 pastes, despite their high total porosity, have a finer pore distribution than NP0 and NP30 pastes in the pore aperture below 50 nm. By means of the wet RA, a water movement between the new paste and the RA is created, which disturbs the hydration of the new paste and initiates two hypotheses. In the first, moisture contained in the pores of the paste is gradually released to allow a continuous hydration [41,42], which leads to the refinement of the pore structure [24]. In the second, RA absorbs mixing water, reducing the w/c ratio in the NP. This observation made it necessary to implement further investigations. For this, SEM observations were carried out to examine the microstructure of the new paste.

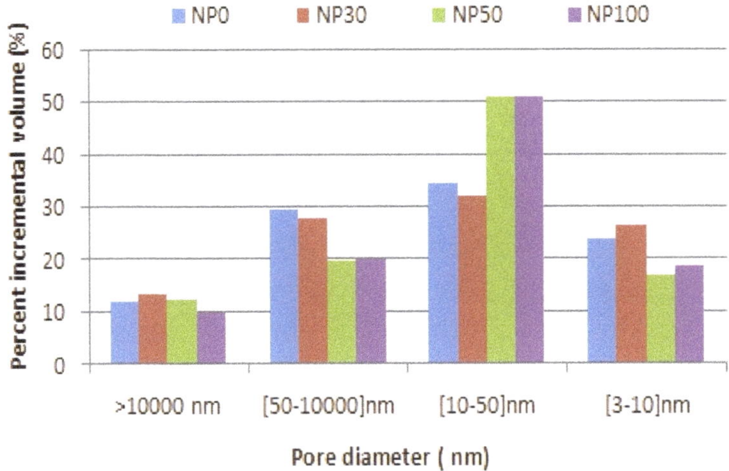

Figure 7. The pore percentage of new pastes (NP) for different pore diameter ranges.

3.3. Influence of Recycled Aggregates (RA) on the Anhydrous Content of the New Paste (NP)

The SEM observations conducted on the NP of each SCC enables us to evaluate anhydrous content. The content of the anhydrous grains (white elements in Figure 8) in the new paste change from one concrete to the next. Image analysis evaluations show that the anhydrous content in the new paste decreases with an increasing RA replacement percentage. The anhydrous content was 5.75% for NP0 and 5.04% for NP30, while it was 3.63 % for NP50 and 1.21% for NP100. This validates the first hypothesis, which is that the water absorbed by the RA during the pre-saturation process can migrate into the new paste, which generates less anhydrous and macrospores for high RA-substitution rate. Cement hydration continues due to the availability of additional water, as has already been observed by Djerbi [29].

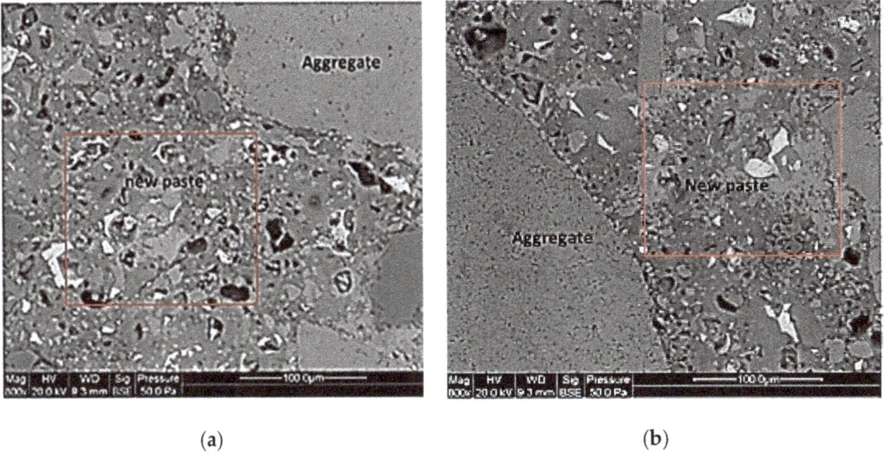

(a) (b)

Figure 8. *Cont.*

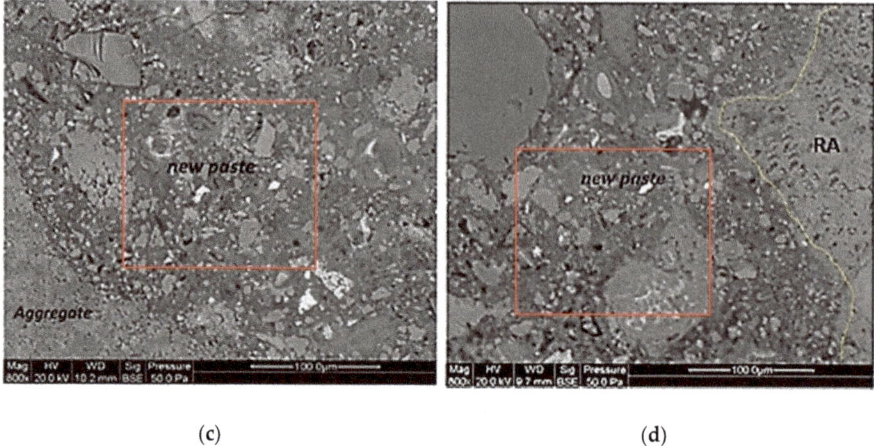

(c)   (d)

Figure 8. SEM images of new pastes (**a**) NP0, (**b**) NP30, (**c**) NP50 and (**d**) NP100.

Figure 7 shows that the percentage of macrospores of sizes between 50 and 10,000 nm is nearly 30% for both NP0 and NP30, respectively, and only 20% for NP50 and NP100. This phenomenon has made it possible to generate a finer pore distribution in the existing mesopore network. It can be seen that the percentage of pores in the 10–50 nm range is of the order of 51% for NP50 and NP100, compared to only 34% and 32%, respectively, for NP0 and NP30. Within the 3–10 nm pore radius range, the NP0 and NP30 develop higher gel pore rates than the other two pastes. This explains the convergence of the peaks towards the areas of the small pores and the disappearance of that at the level of the capillary pores. Therefore, as the hydration moves forward, the hydrates formed fill the pores and lead to the movement of the first peak towards a finer pore diameter [43]. The comparison between the porosity of RA and NP from Table 4 shows that RA has a total porosity 15% greater than that of NP0 and NP30. For high substitution rates, it shows a decrease of about 9% for NP50 and NP100.

3.4. Contribution of the Porosity of NP and RA to the Decrease in Compressive Strength

In order to study the effect of the porosity of each component of concrete, it is important to relate the porosity to the volume of each component present in the concrete. The volume of the coarse aggregates (>3 mm) is 31% in the total volume of concrete (see Table 3). An estimate of the porosity of the coarse RA evaluated by MIP test for each substitution rate is 30% for RA30-SCC30, 50% for RA50-SCC50, and 100% for RA100-SCC100. The volume of the NP is 69%, also taking into account the sand fraction (see Table 3). Figure 9 shows the contribution of the porosity of RA and that of NP for each concrete type in the decrease in compressive strength (Rc). The reduction in compressive strength between SCC0 and SCC30 can be attributed to RA porosity. While the reduction in the compressive strength between SCC30 and SCC50 is due to NP porosity increase and RA porosity of about 26% and 66%, respectively; this increase in porosity causes a decrease in Rc of about 8%. On the other hand, the decrease in Rc of 16% between SCC50 and SCC100 is only generated by the increase in RA porosity by a factor of 2.

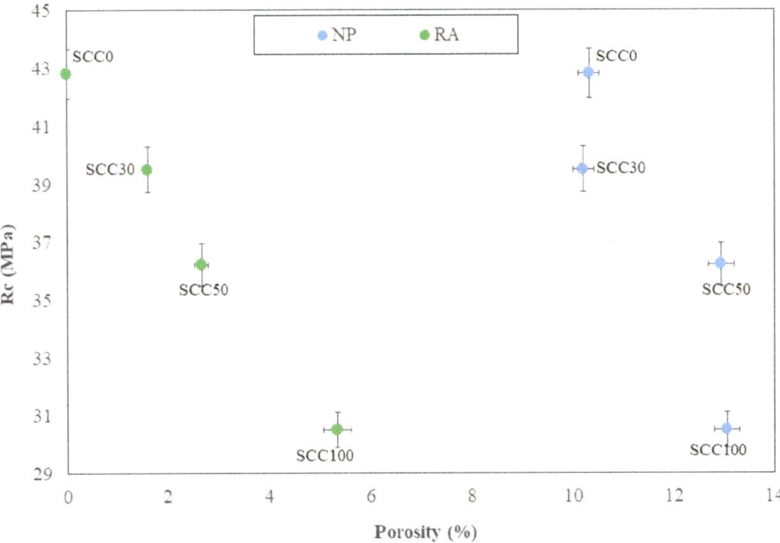

Figure 9. Relationship between porosity of RAC component and compressive strength.

4. Conclusions

The modification of the microstructure of the new paste (NP) of self-compacting concrete (SCC) for different levels of recycled aggregate (RA) at 90 days compared that to the reference concrete has been explored. The results obtained from this work can be summarized as follows:

- The microstructural characterization of self compacting concretes (SCC) containing recycled aggregates (RA) with the proposed sampling method enables us to evaluate the pore network of their different components as RA and NP.

- The replacement of coarse aggregates by RA at 30% does not affect the new paste of SCC. The pore size distribution of NP30 is similar to the reference concrete NP0 with a total introduced pore volume of 0.069 mL/g. While NP50 and N100 pastes present a total introduced pore volume of 0.09 mL/g.

- The porosity of the new paste presents a significant variation for 50% and 100% replacement in RA with respect to NP0 and NP30. NP50 and NP100 exhibit higher refinement of the porous structure. The number of macrospores of a diameter between 50 and 10,000 nm is reduced by 20% for NP50 and NP100 in comparison to NP0 and NP30, of which the reduction is about 30%. This is attributed to the release of water absorbed by the RA during the pre-saturation process that allows continuous hydration.

- The anhydrous rates in the NP decrease with increasing RA replacement percentage. The anhydrous content was 5.75% for NP0 and 5.04% for NP30, while it was 3.63% for NP50 and 1.21% for NP100. This validates the hypothesis that the water absorbed by RA during the pre-saturation process can migrate into the NP, generating less anhydrous and macrospores for high rate substitution of RA.

- The compressive strength loss for RA concretes SSC50 and SCC100 is influenced not only by the porous old paste of RA, but also by the porosity of the new paste. The paste is similar for these two concretes and shows an increase of 26% compared to the reference concrete.

- Further works will be conducted to study the microstructure evolution between 28 days and 90 days.

Author Contributions: Conceptualization, K.Z., A.D. and A.M.; methodology, K.Z., A.D. and A.M.; validation, A.D. and A.M.; formal analysis, K.Z. and A.D.; investigation, K.Z. and A.D.; resources, A.M and A.D.; data curation, K.Z. and A.D.; writing—original draft preparation, K.Z. and A.D.; writing—review and editing, K.Z., A.D. and A.M.; supervision, A.M. and A.D. All authors have read and agreed to the published version of the manuscript.

Materials **2020**, 13, 2114

Funding: This research received no external funding

Acknowledgments: The authors thank the PROFAS program of France and Algeria for PhD grant.

Conflicts of Interest: The authors declare no conflict of interest.

References

1. *Directive 2008/98/EC of the European Parliament and of the Council of 19 November 2008 on Waste and Repealing Certain Directives Text with EEA Relevance*; European Commission: Brussels, Belgium, 2008.
2. Kanellopoulos, A.; Petrou, M.F.; Ioannou, I. Durability performance of self-compacting concrete. *Constr. Build. Mater.* **2012**, *37*, 320–325. [CrossRef]
3. Perraton, D.; Aïtcin, P.C.; Carles-Gbergues, A. *Permeability as Seen by Researchers*; Malier, Y., Ed.; High Performance Concrete: From Material to Structure; E & FN Spon: London, UK, 1994; pp. 252–275.
4. Nazari, A.; Riahi, S. The effect of TiO2 nanoparticles on water permeability and thermal and mechanical properties of high strength self-compacting concrete. *Mater. Sci. Eng.* **2010**, *528*, 756–763. [CrossRef]
5. Sasanipour, H.; Aslani, F. Durability properties evaluation of self-compacting concrete prepared with waste fine and coarse recycled concrete aggregates. *Constr. Build. Mater.* **2020**, *236*, 117540. [CrossRef]
6. Sasanipour, H.; Aslani, F.; Taherinezhad, J. Effect of silica fume on durability of self-compacting concrete made with waste recycled concrete aggregates. *Constr. Build. Mater.* **2019**, *227*, 116598. [CrossRef]
7. Pereira-de-Oliveira, L.A.; Nepomuceno, M.C.S.; Castro-Gomes, J.P.; Vila, M.D.F.C. Permeability properties of self-compacting concrete with coarse recycled aggregates. *Constr. Build. Mater.* **2014**, *51*, 113–120. [CrossRef]
8. Kou, S.C.; Poon, C.S. Properties of self-compacting concrete prepared with coarse and fine recycled concrete aggregate. *Cem. Concr. Compos.* **2009**, *31*, 622–627. [CrossRef]
9. Omrane, M.; Kenai, S.; Kadri, E.-H.; Aït-Mokhtar, A. Performance and durability of self compacting concrete using recycled concrete aggregates and natural pozzolan. *J. Clean. Prod.* **2017**, *165*, 415–430. [CrossRef]
10. Abbas, A.; Fathifazl, G.; Isgor, O.B.; Razaqpur, A.G.; Fournier, B.; Foo, S. Durability of recycled aggregate concrete designed with equivalent mortar volume method. *Cem. Concr. Compos.* **2009**, *31*, 555–563. [CrossRef]
11. Butler, L.; West, J.S.; Tighe, S.L. The effect of recycled concrete aggregate properties on the bond strength between RCA concrete and steel reinforcement. *Cem. Concr. Res.* **2011**, *41*, 1037–1049. [CrossRef]
12. Gomez-Soberon, J.M.; Gomez-Soberon, J.C.; Gomez-Soberon, L.A. Shrinkage and creep of recycled concrete interpreted by the porosity of their aggregate. Challenges of Concrete Construction: Volume 5, Sustainable Concrete Construction. In Proceedings of the International Conference held at the University of Dundee, Scotland, UK, 9–11 September 2002.
13. Sucic, A.; Lotfy, A. Effect of new paste volume on performance of structural concrete using coarse and granular recycled concrete aggregate of controlled quality. *Constr. Build. Mater.* **2016**, *108*, 119–128. [CrossRef]
14. Poon, C.S.; Shui, Z.H.; Lam, L. Effect of microstructure of ITZ on compressive strength of concrete prepared with recycled aggregates. *Constr. Build. Mater.* **2004**, *18*, 461–468. [CrossRef]
15. Kumar, R.; Bhattacharjee, B. Study on some factors affecting the results in the use of MIP method in concrete research. *Cem. Concr. Res.* **2003**, *33*, 417–424. [CrossRef]
16. Zeng, Q.; Li, K.; Fen-Chong, T.; Dangla, P. Analysis of pore structure, contact angle and pore entrapment of blended cement pastes from mercury porosimetry data. *Cem. Concr. Compos.* **2012**, *34*, 1053–1060. [CrossRef]
17. Zhou, J.; Ye, G.; Van Breugel, K. Characterization of pore structure in cement-based materials using pressurization–depressurization cycling mercury intrusion porosimetry (PDC-MIP). *Cem. Concr. Res.* **2010**, *40*, 1120–1128. [CrossRef]
18. Choi, Y.C.; Kim, J.; Choi, S. Mercury intrusion porosimetry characterization of micropore structures of high-strength cement pastes incorporating high volume ground granulated blast-furnace slag. *Constr. Build. Mater.* **2017**, *137*, 96–103. [CrossRef]
19. Moro, F.; Böhni, H. Ink-bottle effect in mercury intrusion porosimetry of cement-based materials. *J. Colloid Interface Sci.* **2002**, *246*, 135–149. [CrossRef]
20. Shicong, K.; Poon, C.S. Compressive strength, pore size distribution and chloride-ion penetration of recycled aggregate concrete incorporating class-F fly ash. *J. Wuhan Univ. Technol. Sci. Ed.* **2006**, *21*, 130–136. [CrossRef]
21. Uchikawa, H.; Hanehara, S. *Recycled concrete waste. Concrete for Environment Enhancement and Protection*; E & FN Spon: London, UK, 1996; pp. 163–172.

22. Gómez-Soberón, J.M.V. Porosity of recycled concrete with substitution of recycled concrete aggregate. *Cem. Concr. Res.* **2002**, *32*, 1301–1311. [CrossRef]
23. Gómez-Soberón, J.M.V. Relationship between Gas Adsorption and the Shrinkage and Creep of Recycled Aggregate Concrete. *Cem. Concr. Aggreg.* **2003**, *25*, 1–7.
24. Gonzalez-Corominas, A.; Etxeberria, M.; Poon, C.S. Influence of steam curing on the pore structures and mechanical properties of fly-ash high performance concrete prepared with recycled aggregates. *Cem. Concr. Compos.* **2016**, *71*, 77–84. [CrossRef]
25. Chinchillas-Chinchillas, M.J.; Rosas-Casarez, C.A.; Arredondo-Rea, S.P.; Gómez-Soberón, J.M.V.; Corral-Higuera, R. SEM image analysis in permeable recycled concretes with silica fume. A quantitative comparison of porosity and the ITZ. *Materials* **2019**, *12*, 2201.
26. NF. EN. *1097-6, Tests for Mechanical and Physical Properties of Aggregates—Part 6: Determination of Particle Density and Water Absorption*; AFNOR—Association Française de Normalization: Paris, France, 2014.
27. N.F. *P18-459, Concrete-Testing Hardened*; AFNOR—Association Française de Normalization: Paris, France, 2010.
28. Omary, S.; Ghorbel, E.; Wardeh, G. Relationships between recycled concrete aggregates characteristics and recycled aggregates concretes properties. *Constr. Build. Mater.* **2016**, *108*, 163–174. [CrossRef]
29. Djerbi Tegguer, A. Determining the water absorption of recycled aggregates utilizing hydrostatic weighing approach. *Constr. Build. Mater.* **2012**, *27*, 112–116. [CrossRef]
30. Tam, V.W.Y.; Gao, X.F.; Tam, C.M.; Chan, C.H. New approach in measuring water absorption of recycled aggregates. *Constr. Build. Mater.* **2008**, *22*, 364–369. [CrossRef]
31. Yacoub, A.; Djerbi, A.; Fen-Chong, T. Water absorption in recycled sand: New experimental methods to estimate the water saturation degree and kinetic filling during mortar mixing. *Constr. Build. Mater.* **2018**, *158*, 464–471. [CrossRef]
32. NF. EN. *12390-3, Tests on Hardened Concrete—Part 3: Compressive Strength of Test Specimens*; AFNOR—Association Française de Normalization: Paris, France, 2012.
33. Gallé, C. Effect of drying on cement-based materials pore structure as identified by mercury intrusion porosimetry. *Cem. Concr. Res.* **2001**, *31*, 1467–1477. [CrossRef]
34. Djerbi, A. Effect of recycled coarse aggregate on the new interfacial transition zone concrete. *Constr. Build. Mater.* **2018**, *190*, 1023–1033. [CrossRef]
35. Chakradhara Rao, M.; Bhattacharyya, S.K.; Barai, S.V. Influence of field recycled coarse aggregate on properties of concrete. *Mater. Struct.* **2011**, *44*, 205–220. [CrossRef]
36. Otsuki, N.; Miyazato, S.; Yodsudjai, W. Influence of Recycled Aggregate on Interfacial Transition Zone, Strength, Chloride Penetration and Carbonation of Concrete. *J. Mater. Civ. Eng.* **2003**, *15*, 443–451. [CrossRef]
37. Ryu, J.S. Improvement on strength and impermeability of recycled concrete made from crushed concrete coarse aggregate. *J. Mater. Sci. Lett.* **2002**, *21*, 1565–1567. [CrossRef]
38. Cook, R.A.; Hover, K.C. Mercury porosimetry of hardened cement pastes. *Cem. Concr. Res* **1999**, *29*, 933–943. [CrossRef]
39. Powers, T.C.; Brownyard, T.L. Studies of the physical properties of hardened Portland cement paste. *J. Proc.* **1946**, *43*, 101–132.
40. Ye, G.; Hu, J.; van Breugel, K.; Stroeven, P. Characterization of the development of microstructure and porosity of cement-based materials by numerical simulation and ESEM image analysis. *Mater. Struct.* **2002**, *35*, 603–613. [CrossRef]
41. Gonzalez-Corominas, A.; Etxeberria, M. Effects of using recycled concrete aggregates on the shrinkage of high performance concrete. *Constr. Build. Mater.* **2016**, *115*, 32–41. [CrossRef]
42. Yildirim, S.T.; Meyer, C.; Herfellner, S. Effects of internal curing on the strength, drying shrinkage and freeze-thaw resistance of concrete containing recycled concrete aggregates. *Constr. Build. Mater.* **2015**, *91*, 288–296. [CrossRef]
43. Yu, Z.; Ye, G. The pore structure of cement paste blended with fly ash. *Constr. Build. Mater.* **2013**, *45*, 30–35. [CrossRef]

materials

Article

Mechanical Behavior of Masonry Mortars Made with Recycled Mortar Aggregate

René Sebastián Mora-Ortiz *, Emmanuel Munguía-Balvanera, Sergio Alberto Díaz, Francisco Magaña-Hernández, Ebelia Del Angel-Meraz and Álvaro Bolaina-Juárez

División Académica de Ingeniería y Arquitectura (DAIA-UJAT), Universidad Juárez Autónoma de Tabasco, Carretera Cunduacán-Jalpa de Méndez km. 1, Cunduacán, Tabasco CP 86690, Mexico; balvanerae@hotmail.com (E.M.-B.); alberto.diaz@ujat.mx (S.A.D.); francisco.magana@ujat.mx (F.M.-H.); ebelia.delangel@ujat.mx (E.D.A.-M.); aalvarado914130@gmail.com (Á.B.-J.)
* Correspondence: rene.mora@ujat.mx or renemora1121@hotmail.com

Received: 25 April 2020; Accepted: 19 May 2020; Published: 21 May 2020

Abstract: Recycling is an important habit to avoid waste. This paper evaluates the performance of masonry mortar, elaborated by replacing natural sand with recycled fine aggregate (RFA) obtained from mortar. Five families of mixtures were prepared with different replacement proportions: 20%, 40%, 60%, and 100%. A 1:4 volumetric cement-to-aggregate ratio was used for all mixtures by experimentally adjusting the amount of water to achieve the same consistency of 175 ± 5 mm. The effects of the following procedures were analyzed: (1) the use of a deconstruction technique to collect the RFA, (2) pre-wetting of the aggregates, and (3) the use of a commercial plasticizer. Experimental results show that it is possible to use this type of recycled fine aggregate as a substitute for natural sand by up to 60% in the manufacture of masonry mortar without significantly affecting its properties.

Keywords: construction and demolition waste; recycled fine aggregate; mortars; sustainable construction; recycled aggregates

---

1. Introduction

A guideline for developing new construction materials is to improve material performance, optimize supplies and reduce manufacturing costs. Therefore, it is necessary to develop new techniques and take advantage of materials which are considered waste in this and other industries. For decades, the amount of construction and demolition waste (CDW) has been increasing globally, becoming one of the main agents of environmental pollution.

Using the CDW produced during the demolition of concrete structures as a substitute for thick and fine aggregates in masonry mortar mixtures reduces the amount of pollutant waste released to the environment, compensates for the lack of stone aggregates, and represents an innovation in the development of construction materials [1–3]. Recycled aggregates (RA) which are used in replacement of natural gravel, are known as recycled thick aggregates (RTA), whereas RA used to replace sand are called recycled fine aggregates (RFA). RFA may be classified into two types: those obtained from concrete, and those obtained from other materials, such as mortar or ceramic.

The use of RA, especially RFAs obtained from mortar elements, has some drawbacks, most of them associated with the nature of these materials, such as their porosity, their high-water absorption potential, and the possibility of containing pollutants. For these reasons, CDW is not recycled as it should be, which causes its accumulation in landfills, generating pollution. This research presents two possible alternatives for the reuse of mortar RFA as a substitute for sand in new mixtures. The analysis was divided into two stages: first, the characteristics of RAs obtained by a "deconstruction" procedure were compared against those obtained through a conventional demolition. In the second stage, two possible scenarios were analyzed: (i) pre-wetting the RFA before mixing, and (ii) using a commercial plasticizer to reduce the amount of water in the mixture. The performance of mortars made with RFA was compared against a conventional mortar mixture (cement, natural sand, and water).

Currently, several researchers support the use of RTA obtained from the demolition of concrete elements as a partial substitute for natural gravel in the preparation of concrete mixtures [4–6]. However, they do not recommend the replacement of natural sand by RFA, because they consider this substitution detrimental to the mechanical properties and durability of concrete [4,7]. Notwithstanding, some researchers, such as Pereira et al. [8], Mefteh et al. [9], Evangelista and de Brito [10], and Cartuxo et al. [11] believe that natural sand can be replaced by RFA in a proportion of up to 20% without significantly affecting the mechanical properties of the concrete. Using RFA as a substitute for natural sand in masonry mortar mixes is the best option, because it has fewer structural requirements than concrete. RFA obtained from concrete elements have been studied by researchers such as Braga et al. [12], Neno et al. [13], Saiz-Martínez et al. [14] and Ng and Engelsen [15], who have proven their viability as a partial substitute for natural sand. However, research on the use of RFA obtained from mortar elements is scarce, due to the inferior physical, mechanical and chemical characteristics of this type of material, such as its high-water absorption potential, porosity, and susceptibility to contain contaminants [16,17]. This has resulted in the accumulation of debris from the demolition of prefabricated mortar elements in sanitary landfills and clandestine landfills, in the over overexploitation of river or quarry sandbanks, and in the increase of energy consumption and CO_2 emissions as a consequence of crushing rocks to produce fine aggregates [11]. Silva et al. [18] showed that incorporating 10% ceramic RFA improves most of the mortar´s properties. In the research conducted by Jiménez et al. [19], natural sand was replaced by RFA composed by 54% ceramic and 40% masonry mortar. Experimental results showed that replacing natural sand with this type of RFA in a proportion of 40% does not significantly affect mortar properties in fresh and hardened state. On the other hand, Silva et al. [20] demonstrated the technical feasibility of recycling RFA resulting from the demolition of bricks or red clay tiles. They concluded that mortars made with replacement ratios of at least 20% generally show a better performance than conventional mortar, giving emphasis to aspects such as flexural and compressive strengths.

Recycled mortar properties depend on RFA quality, the substitution rate of natural sand, cement content, and water-cement (W/C) ratio. Among these factors, RFA quality is perhaps the most important. It is known that RAs usually contain a certain number of sulfates, chlorides, and other contaminants [21]. These impurities are, for the most part, the product of a lack of demolition planning. Since the reuse of demolition rubble is not considered from the beginning, it is deposited outdoors, or in places where it may be contaminated, allowing its mixture with other materials. Researchers such as Rahal [22], Cachim [23], and Debieb et al. [24] (just to mention a few), have studied the effect of the presence of these impurities on the behavior of concrete which has been manufactured with replacement of natural aggregates by RA. The content of impurities and contaminants in RA can be greatly reduced by means of selective demolition techniques [25,26]. Kumbhar et al. [27], as well as Coelho and de Brito [28], describe deconstruction techniques for obtaining good quality CDW.

Water absorption potential is a determining factor in RFA quality. This is because, in the RA, mortar is attached to the natural aggregate [29,30]. The implication of this RA characteristic is the W/C reduction ratio in the cement paste, which results in poor workability, a greater number of pores, less compression resistance, as well as drying contractions [31,32]. Therefore, to guarantee the workability of the cement paste there are two possible solutions: the incorporation of plasticizers in the mixture [7,11,33,34] or pre-wetting RAs before mixing [9,35–37]. Pereira et al. [8] conducted one of the first studies on the effect of superplasticizers on the properties of fresh and hardened concrete made with concrete RFA. Together with Cartuxo et al. [11] and Barbudo et al. [38], they showed that in mixtures in which sand is replaced by RFA, the use of plasticizers improves the mechanical characteristics of concrete. Zega and Maio [39] concluded that using a water-reducing additive produces recycled concrete with adequate performance, which follows the specifications established by different international construction standards.

Studies such as those directed by González et al. [35], Mefteh et al. [9], and Cuenca-Moyano et al. [40], have reported the benefits of pre-wetting RAs before making concrete mixtures. In all of these cases, humidity levels below 100% of absorption capacity were recommended. Researchers like Cabral et al. [41] and Zhao et al. [42] point out that the best results are obtained with humidities lesser than or equal to 80% of absorption capacity.

The main objective of this research was to separately analyze the influence of pre-wetting RFAs and using commercial plasticizers during the creation of new masonry mortars with partial substitution of natural sand by RFA obtained from mortar elements. It is intended that the results of this research contribute to an increase in the reuse of this type of RAs.

2. Materials and Methods

2.1. Obtaining Recycled Aggregates

The RFAs used in this research were obtained from the renovation of the second floor of the "K" building of the Academic Division of Information Technology and Systems, at the Universidad Juárez Autónoma de Tabasco (UJAT, Cunduacán, Tabasco, Mexico). This two-story building is 18 years old. As part of the renovation, two separation walls made of prefabricated mortar elements (mortar blocks) were demolished. Since the first objective of this research was to determine the effect of using demolition strategies in obtaining the RAs, the demolition of one of the walls was planned and coordinated with the builder. Demolition of the second wall proceeded according to the original plan of the builder (conventional demolition). Therefore, two types of RA were obtained: the RFA*, obtained through a deconstruction process, and the RFA, produced through a conventional demolition.

The proposed deconstruction plan was simple and consisted of five steps: (1) estimation of the CDW volume to be obtained, (2) location of the rubble storage site, (3) removal of surface materials other than mortar (wood, metals, plastics, crystals, etc.) prior to demolition, (4) manual demolition followed by handling and separation of the remaining waste, and (5) crushing and storage. Because the ground floor of the building was empty due to the renovation work, it was used as a CDW warehouse. Before starting with the demolition of the walls, the window glass was removed, and the largest possible amount of paint was removed with a wire brush and spatulas. After the demolition process of the walls, the rubble was moved to the storage point (the ground floor of the building). Once all of the debris was deposited on the ground floor, personnel with safety equipment removed materials such as cables and some electrical devices that remained among the debris. The next step was to crush the rubble, thus obtaining the RFA*. For grinding, a Los Angeles abrasion machine was used. Subsequently, the debris was screened by means of an sieve number 4 (4.75 mm) and stored separately on the ground floor of the building, protecting it from the weather.

The demolition of the second separation wall produced the RFA. The conventional demolition process used by the builder was one that is commonly carried out in this type of remodeling, which consists of three stages: (1) removal of elements contemplated for reuse in the project (in this case none), (2) demolishing of the wall of prefabricated elements using hand tools, and (3) storage of all debris in an area away from the construction zone, to be later transferred to an authorized dumping area. For research purposes, debris from the separation wall was collected before being transferred to the municipal landfill. As in the previous case, before crushing and sieving with the number 4 sieve, materials other than mortar were removed as much as possible. The RFA obtained from this process was stored in a container in front of the building that was being renovated.

2.2. Characterization of the Materials

All the RAs used in this research were characterized according to the UNE-EN 13139 [43] standard on mortar aggregates. Natural sand (NS) obtained from a riverbank was used as a reference element. The properties that were analyzed, as well as their reference standard, are shown in Table 1. Table 2 and Figure 1 show the particle size distribution of the sand and recycled aggregates. It was observed that in general, the granulometry of the aggregates obtained from prefabricated mortar pieces (RFA* and RFA) are similar.

The cement that was used is of the PCC 30R type Cemex® brand (Monterrey, México), referring to a Portland cement compound of resistant class 30 with rapid resistance (3 days). This cement meets the international requirements of ASTM C150 / C150M-09 [44] and ASTM C595 / C595M-19 [45]. Table 3 shows its chemical characteristics.

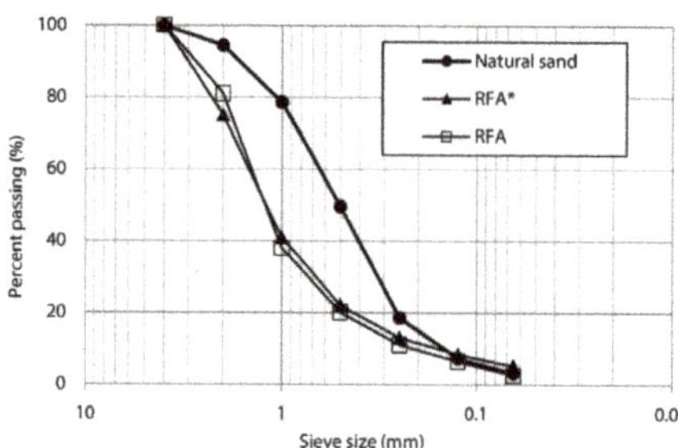

Figure 1. Particle size distribution of the sand and recycled aggregates.

Table 1. Standards used in the characterization of natural sand and recycled aggregates.

Fine Content (%)	Sand Equivalent (%)	Dry Sample Density (gr/cm³)	Water Absorption (%)	Acid Soluble Sulphates (% SO₃)	Water Soluble Chlorides (% Cl)	Total Sulphurs (% SO₃)
UNE-EN 933-1 [46]	UNE-EN 933-8 [47]	UNE-EN 1097-6 [48]	UNE-EN 1097-6 [48]	UNE-EN 1744-1 [49]	UNE-EN 1744-1 [49]	UNE-EN 1744-1 [49]

Table 2. Particle size distribution.

Sieve Size (mm)		4	2	1	0.5	0.25	0.125	0.063
	NS	100	94	79	50	19	7.3	3.2
Percent passing(%)	RFA*	100	75	41	22	13	8.4	5.2
	RFA	100	81	38	20	11	6.5	2.6

Table 3. Chemical composition of cement (%) given by the fabricant, Cemex®.

Composition	CaO	SiO₂	Al₂O₃	Fe₂O₃	MgO	K₂O	Na₂O	SO₃
%	63	22	6	2.5	2.6	0.6	0.3	2.0

2.3. Mixes

All mixtures were prepared with the same proportion and content of cement. A conventional mortar mixture (cement, natural sand, and water) was developed as a reference parameter. For the first phase of this investigation, two families of mixtures were prepared with the gradual replacement of NS with RA: RFA* was used in one and RFA in the other. For the second phase, two more types of mixtures were made. In all of them, the RFA* was taken as a substitute for sand. In one of them, the RFA* was subjected to pre-wetting before mixing (RFA* + h) and in the other, the RFA* was used with its natural moisture, but a commercial plasticizer was added to the mixture (RFA* + P) (Table 4). As a result of the above, five families of mortar were defined: NS, RFA*, RFA, RFA* + h, and RFA* + P.

Four replacement ratios were defined: 20%, 40%, 60%, and 100%. The replacement was carried out as a percentage of dry weight [19,50]. In total, 19 mixtures were made. Table 4 shows the nomenclature which was used.

Table 4. Nomenclature for mortar mixes in which RA was used as a substitute of NS.

Replacement Ratio (%)	First Phase		Second Phase	
	RFA*	RFA	RFA* + h	RFA* + P
0	—	—	RFA* + h_0	RFA* + P_0
20	RFA*$_{20}$	RFA$_{20}$	RFA* + h_{20}	RFA* + P_{20}
40	RFA*$_{40}$	RFA$_{40}$	RFA* + h_{40}	RFA* + P_{40}
60	RFA*$_{60}$	RFA$_{60}$	RFA* + h_{60}	RFA* + P_{60}
100	RFA*$_{100}$	RFA$_{100}$	RFA* + h_{100}	RFA* + P_{100}

The plasticizer which was used was Sikament 500, which is a medium-range water reducing liquid additive that does not contain chlorides. It complies with ASTM-C-494 Type D [51] and with ASTM-C-1017 Type II [52]. Its density is 1.20 ± 0.05 kg/L.

Mortar dosages were implemented according to the characteristics which were obtained for the materials. The following criteria were established:

- All the RAs that were used were smaller than 4 mm in particle diameter.
- The cement-aggregate ratio used in all mixtures was 1: 4.
- The amount of water was adjusted experimentally to achieve a consistency of 175 ± 5 mm in the mixtures.
- Pre-wetting of the aggregates was performed only in one type of mixture (RFA* + h). The rest of the aggregates were used with their natural humidity (2.3 ± 0.2). Pre-wetting was performed to reach 80% of the total absorption capacity of the RA, guaranteeing the presence of water in the aggregate and decreasing the migration of water from the mixture to the RA [5,41]. The procedure used to achieve the aforementioned wetting was based on that described by Fonseca et al. [53]: the aggregate was immersed in water for five minutes and then allowed to drain before its use.

The plasticizer was used in a proportion of 1% of cement weight. This value was recommended by the manufacturer. Table 5 shows mixture proportions used in this research. This table shows that according to the literature that was consulted [9,13,42], mortars containing RA need a larger amount of water to achieve project consistency.

The mixtures were made in a standard mixer, placing the cement and fine aggregate first, then mixing for a minute. Then, during the next 20 s, water was added while the mixer was still mixing the cement and aggregate. The mixing of these materials was continued for three minutes at a speed of 140 rpm. This procedure was followed for all mixtures to which no plasticizer was added. For the latter, we relied on the procedure described by Jiménez et al. [19]: water and additive were first placed in the mixer´s container, then mixed at low speed (140 rpm) for two minutes, after which cement and aggregate were slowly added. All these materials were mixed at low speed for three minutes.

Table 5. Mortar mixture proportions.

Mortar Type	NS/RA	NS (gr)	RA (gr)	CEM (gr)	Water (gr)	Consistency Index (mm)	W/C
AN	100/0	2307	0	332	285	176	0.86
RFA*$_{20}$	80/20	1845	462	332	304	172	0.92
RFA*$_{40}$	60/40	1384	923	332	328	171	0.99
RFA*$_{60}$	40/60	923	1384	332	352	177	1.06
RFA*$_{100}$	0/100	0	2307	332	371	174	1.12
RFA $_{20}$	80/20	1845	462	332	309	177	0.93

Table 5. *Cont.*

Mortar Type	NS/RA	NS (gr)	RA (gr)	CEM (gr)	Water (gr)	Consistency Index (mm)	W/C
RFA $_{40}$	60/40	1384	923	332	337	180	1.02
RFA $_{60}$	40/60	923	1384	332	358	173	1.08
RFA $_{100}$	0/100	0	2307	332	386	175	1.16
RFA* + h$_0$	100/0	2307	0	332	276	170	0.83
RFA* + h$_{20}$	80/20	1845	462	332	293	174	0.88
RFA* + h$_{40}$	60/40	1384	923	332	315	179	0.95
RFA* + h$_{60}$	40/60	923	1384	332	330	171	0.99
RFA* + h$_{100}$	0/100	0	2307	332	356	176	1.07
RFA* + P$_0$	100/0	2307	0	332	249	180	0.75
RFA* + P$_{20}$	80/20	1845	462	332	255	173	0.77
RFA* + P$_{40}$	60/40	1384	923	332	271	171	0.82
RFA* + P$_{60}$	40/60	923	1384	332	293	172	0.88
RFA* + P$_{100}$	0/100	0	2307	332	318	170	0.96

2.4. Rehearsal Program

Assessing the properties of the mortar in its fresh state is an important aspect, because its characteristics in this state have a great impact on the performance of the hardened mortar. To evaluate the properties of fresh mortar, bulk density and air content tests were used. Whereas to characterize hardened mortar, dry bulk density, compressive strength, adhesive strength, and water absorption coefficient due to capillary action were tested. Table 6 shows the standards used during the tests.

Table 6. Standards used in mortar characterization.

Test	Standard	Curing Time (Days)
Properties of fresh mortar		
Bulk density of the fresh mortar	UNE-EN 1015-6 [54]	—
Entrained air	UNE-EN 1015-7 [55]	—
Properties of hardened mortar		
Dry bulk density	UNE-EN 1015-10 [56]	28
Compressive strength	UNE-EN 1015-11 [57]	28
Adhesive strength	UNE-EN 1015-12 [58]	28
Water absorption coefficient due to capillary action	UNE-EN 1015-18 [59]	28

3. Results and Discussion

3.1. Deconstruction Process Effect

Table 7 shows the results of the characterization of natural sand and recycled aggregates. The percentage of fine content refers to particles smaller than 0.063 mm. The sand equivalence values for recycled aggregates are similar to those of natural sand. According to the background literature review, recycled aggregates have a lower dry density than natural sand (NS). As expected, the water absorption value in all recycled aggregates is high with respect to that of NS [29,30]. The absorption of RFA is slightly higher than that exhibited by RFA*, probably due to the impurities that the first of these aggregates possesses.

All recycled aggregates meet with the specifications of the UNE-EN 13139 [43] standard for mortar aggregates. The quantity of sulfates ($\leq$0.8), chlorides ($\leq$0.15), and total sulfur ($\leq$1) meet with the limits established by the UNE-EN 1744-1 standard [49]. Due to the characteristics of the processes through which the RAs were recovered, the presence of other contaminants that could change the properties of the mortars was not observed. However, it can be observed that although the total sulfate, chloride, and sulfur values satisfy the established quality standards, the contents of these substances in the

RAs obtained through a conventional demolition (RFA) are a bit higher than those obtained through a deconstruction process (RFA*). This shows the benefit of planning a demolition according to the recycling of CDW.

Table 7. Natural sand and recycled aggregate characterization.

Aggregate	Fine Content (%)	Sand Equivalent (%)	Dry Sample Density (gr/cm^3)	Water Absorption (%)	Acid Soluble Sulphates (% SO$_3$)	Water Soluble chlorides (% Cl)	Total Sulphurs (% SO$_3$)
NS	9.12	95	2.65	0.28	<0.010	<0.010	<0.010
RFA*	4.12	84	2.1	6.76	0.0027	0.041	0.0027
RFA	6.56	86	1.98	7.22	0.0039	0.054	0.0039

3.2. Fresh Mortar

3.2.1. Bulk Density of Fresh Mortar

Figure 2 shows the changes in bulk density of fresh mortar with respect to the percentages of substitution of natural sand (NS) with RA, as well as its corresponding W/C ratio.

It can be seen that relative to the control sample (NS), the density of mortars with RA decreases as the replacement ratio increases. This is a result of the low density of the RA and its high absorption [11]. The latter characteristic makes it necessary to increase the amount of water in the mixture to achieve project consistency (175 ± 5 mm). This increases the W/C ratio, thereby decreasing the density of the mixture. These results match those observed by other researchers [20,60]. The comparison of density values obtained for the families of RFA* and RFA mixtures, shows that the latter exhibit lower density for the same replacement ratios, as a consequence of their greater absorption.

On the other hand, the samples that used RFA* and underwent a pre-wetting process (RFA* + h), as well as those made with the commercial additive (RFA* + P), showed the highest densities for each of the replacement proportions. This increase in density is due to the fact that in these samples, the volume of free water was reduced in the mixture, allowing a larger quantity of solid particles in the mortar paste and therefore decreasing the W/C ratio.

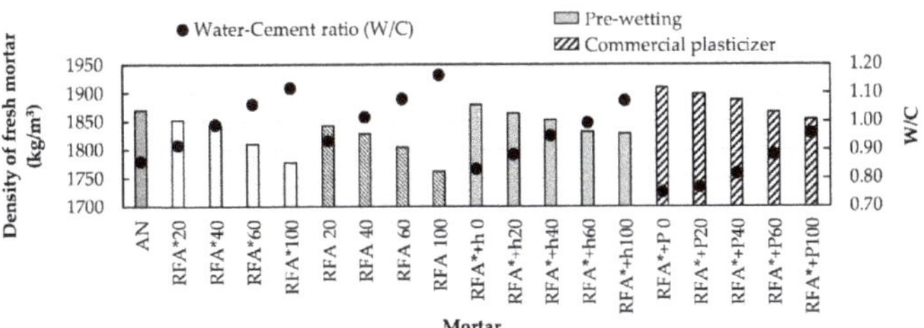

Figure 2. Bulk density of fresh mortar and W/C ratio.

3.2.2. Air Content in Fresh Mortar

Air content in mortar mixtures is an important parameter because it influences aspects such as durability and compressive strength. In spite of this, there are no specified limits for the amount of air in mortar mixtures. However, some researchers suggest that acceptable air contents are in the range of 5% to 20%, which was adopted as a reference in this work [40]. Figure 3 shows the variation of air content in fresh mortar with respect to the percentages of substitution of natural sand (NS) for RA, as well as its corresponding W/C ratio.

It is observed that the air content of the mortars manufactured with RFA* and RFA is greater than that of the control mixture, and also increases as the replacement ratio and the W/C ratio increase.

In general, both families of mixtures show similar behavior, however, mixtures of the RFA family exhibit a slightly higher air content, due to their higher W/C ratio. On the other hand, the mixtures of the RFA* + w and RFA* + P families proved to have the lowest air content values, because they reduce the amount of water in the cement paste. Mixtures of the RFA* + P family showed the lowest air content values, all of them within the recommended ranges.

Figures 2 and 3 clearly show the influence of the W/C ratio on the bulk density and on the air content of fresh mortar, respectively [19,40,61].

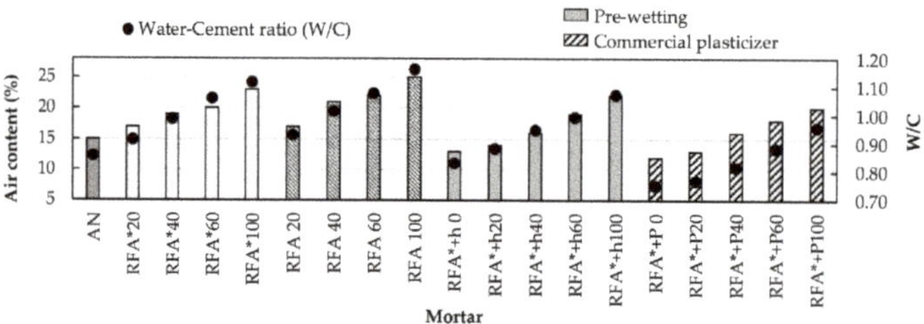

Figure 3. Air content of fresh mortar and W/C ratio.

3.3. Hardened Mortar

3.3.1. Dry Bulk Density

The bulk density of the hardened mortar at an age of 28 days and the W/C ratio for all mortar families are shown in Figure 4. As happens in the fresh state, it is observed that as the replacement percentage and the W/C ratio increase and density decreases, which is consistent with what was observed by other researchers [40]. It is noted that the control mortar (NS) develops a greater density than that of mortars of the RFA* and RFA families. An analysis among the latter makes it clear that mortars manufactured with aggregates obtained by a deconstruction process exhibit a higher dry bulk density, confirming the trends shown by the bulk density in the fresh state. Again, the best results were shown by mortars of the RFA* + h and RFA* + P families. The mixtures that incorporated commercial plasticizer exhibited the highest dry bulk densities for each percentage of substitution, in agreement with the observations of Pereira et al. [8] and Cartuxo et al. [11]. This is because these mixtures have the lowest values of W/C ratio and air content.

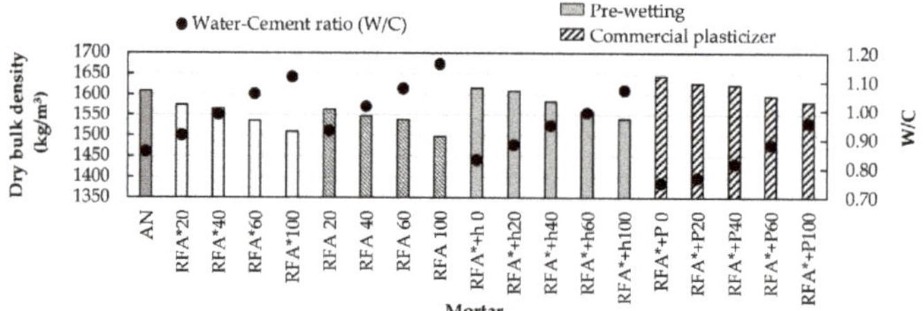

Figure 4. Dry bulk density and W/C ratio.

3.3.2. Compressive Strength

Regarding the characteristics of mortar in the hardened state, compressive strength is the most important because it largely determines durability. Figure 5 shows the compressive strength at 28 days and the W/C ratio for all mortar families.

Continuing with the trend of previously analyzed properties, compressive strength decreases as the percentage of sand substitution by RA increases, as can be seen in Figure 5. This is in agreement with observations by Cuenca-Moyano et al. [40] and Zhao et al. [42]. The decrease in the compressive strength of all mixtures as the percentage of substitution increases is due to the increase in the W/C ratio required to maintain workability. Examining the compressive strengths of the RFA* and RFA mixture families, it can be observed that the former reach slightly higher values. These results confirm the importance of obtaining RA through a deconstruction process. The mixtures made with pre-wetting of the aggregates (RFA* + h) exhibited a greater resistance for each substitution level than the RFA* and RFA families, which is indicative of the importance of pre-wetting the aggregate before making the mortar mixtures. The highest resistances were obtained for mixtures added with commercial plasticizer. This is because this family of mixtures required less water during its manufacture, so the W/C ratio was lower. In Figure 5, it can be seen that the control mixture meets the minimum resistance indicated for M5 mortar (5 MPa). The RFA* and RFA mixtures did not meet this requirement. On the other hand, of the pre-moistened mixtures, only those with substitution percentages ranging from 60% to 100% were below the 5 MPa resistance. On the other hand, it can be seen that in mixtures with commercial plasticizer added, only those with a 100% substitution do not meet the minimum resistance value for the type of mortar used.

According to these results, it is possible to replace up to 40% of NS with RFA*, with 80% pre-wetting without affecting the compressive strength of the mortar. If commercial plasticizer is used, up to 60% of NS can be replaced with good results.

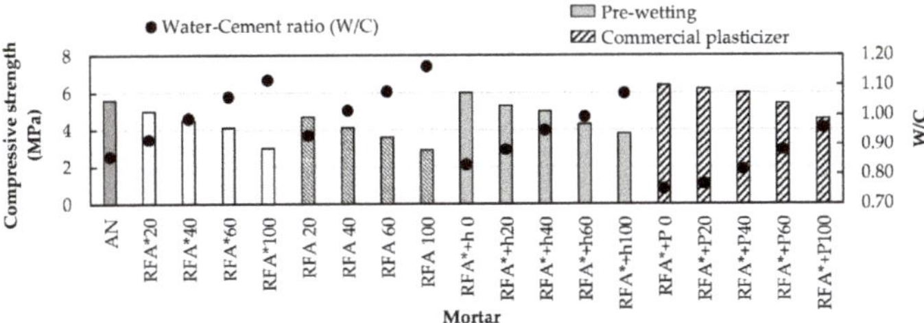

Figure 5. Compressive strength of the mortar at 28 days and W/C ratio.

3.3.3. Adhesive Strength

Adhesive strength is a fundamental property of masonry mortars. The characterization of this property was performed using the European standard UNE-EN 1015-12 [58]. Figure 6 shows the results obtained for each of the families of mixtures. In general, the same trend is observed as for the other properties studied so far: the adhesive strength is affected as the replacement ratio and the W/C ratio increase. These observations are consistent with those made by Amorim et al. [62] and Silva et al. [18]. Comparing with mixtures to which no plasticizer was added, it is noted that similar behaviors exist for the same replacement rates. The highest adhesive strength values were obtained for mixtures added with commercial plasticizer.

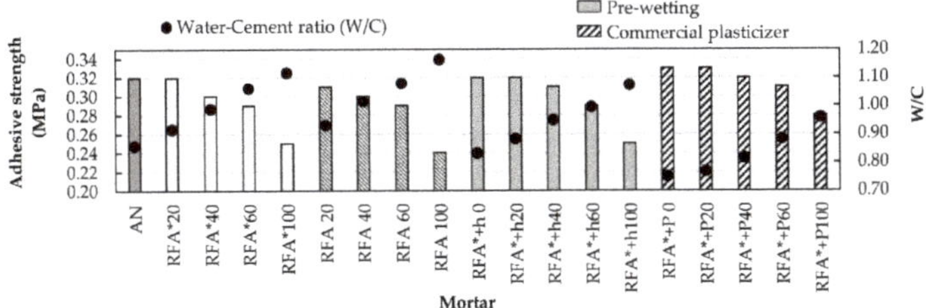

Figure 6. Mortar adhesion and W/C ratio.

One aspect to highlight is the fact that all mixtures with a 20% substitution percentage exhibit values similar to those of the reference mortar (NS). This seems to indicate that for this substitution percentage, there is no negative effect due to the incorporation of RA. On the other hand, Ledesma et al. [1], Neno et al. [13], Braga et al. [12], and Jiménez et al. [19] pointed out that using recycled fine concrete as a substitute for sand in a proportion between 10% and 20% increases adhesiveness with respect to mortar made with natural sand.

3.3.4. Water Absorption Due to Capillary Action

Water absorption due to capillarity is an important indicator of mortar durability. It is known that high absorption values are related to less durability. If the mortar has high water absorption, it will allow the appearance of humidity, as well as water carried particles and substances detrimental to mortar durability. Water absorption due to capillary action depends on mortar structure. The more compact it is, the smaller the pore network is. Consequently, denser mortars will have less water absorption [63].

Figure 7 shows water absorption values due to capillary action and water-cement ratio (W/C) for all the mixtures under observation. It is observed that mortar water absorption values increase as the W/C ratio and the NS replacement proportion by recycled aggregate increase [32,63]. As previously mentioned, if the amount of recycled aggregate is greater, the volume of water required will also be greater to maintain the consistency of the project (175 ± 5 mm), which increases the W/C ratio and consequently, mortars are created with a higher pore network that allows better water absorption.

RFA* + h and RFA* + P mixture families are shown to have the lowest absorption values due to their low water consumption. These mixtures exhibit a water consumption similar to the control mixture for substitution percentages of 20% and 60%, respectively. The mix made with the commercial additive and 20% RFA (RFA* + P20) exhibits less water absorption than the conventional mortar mix.

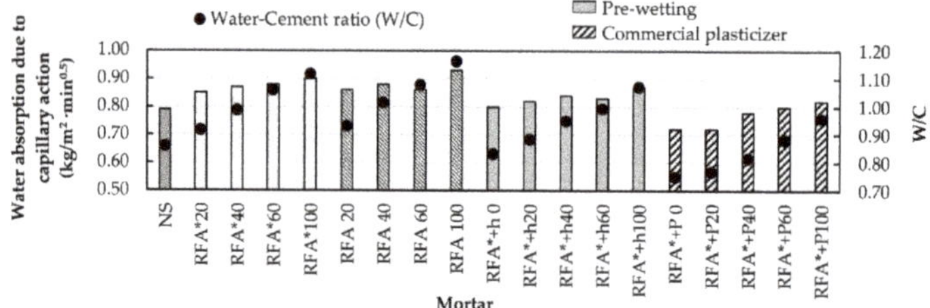

Figure 7. Water absorption values and W/C ratio.

4. Conclusions

In this research, the use of RA obtained from mortar elements as a substitute for natural sand in the preparation of masonry mortars was evaluated, as a proposal of a new field of application for these RAs. Two types of recycled aggregates were used in this research: RFA* and RFA. However, mixtures developed using aggregates obtained through a process specifically planned to recover debris for its use as a substitute for sand (RFA*) developed higher bulk density, adhesive strength, and compressive strength for each substitution percentage than those made with aggregates obtained through conventional demolition (RFA). The deconstruction technique used in this study is simple and competitive in terms of cost, since the RA was obtained directly on-site and transport costs were reduced.

The experimental results show that the bulk densities in the fresh and hardened state, as well as the compressive and adhesive strengths of the mortars, all exhibit the same tendency: they decrease in an almost linear way with the rate of replacement of NS by RA. Regarding the air content and water absorption due to capillary action in the mixtures, it is clear that they increase as the content of RA and the W/C ratio increase. The main agent that causes these behaviors is the high absorption potential of the RAs. As the number of RA increases, it is necessary to add more water to the mixture to achieve project consistency. This is detrimental to the quality of the mortar. It is then concluded that the properties of the recycled mortar are closely linked to the substitution ratio of NS by RA and the W/C ratio.

The pre-wetting of RA at 80% of its absorption capacity before mixing prevents excessive water absorption by the cement paste. Therefore, mixtures that were made with this procedure (RFA* + h) showed an improvement of the analyzed properties, relative to mortar made with RA at natural moisture. The mixtures that were added to by a commercial plasticizer (RFA* + P), which required a smaller amount of water to achieve project consistency than the rest of the mixtures. This brought a considerable improvement of mortar properties, both in fresh and hardened states. Consequently, mortars with the greatest density, greatest strength, best adhesive strength and lowest absorption capacity were those added with plasticizer, followed by those made with pre-wetting of the aggregates. It was observed that, for the first of these two mixtures, a 60% substitution value produces very similar results to those obtained with the reference mortar, so this percentage was established as optimum. Concerning mortars with pre-wetting of the aggregate, the optimum replacement percentage was 40%. As it was demonstrated, these two techniques contribute to reducing the W/C ratio and improving the compacity and the mechanic properties of mortar, also avoiding the appearance of cracks produced by high quantities of free water.

Due to the fact that mortar mixtures made with both techniques have characteristics similar to those of conventional mortars (with their respective optimum percentage of substitution), it is probable that both kinds of mortars have a similar maintenance cost. However, more detailed studies about this topic are needed.

The experimental results show that it is possible to reuse RAs coming from prefabricated mortar elements if adequate debris recovery techniques are established and are used in combination with procedures that reduce the amount of water required in the mixtures.

The use of this type of materials in conventional applications of masonry mortar (indoors and outdoors) will be very important in the near future, since this practice is closely related to sustainability of construction technology.

Author Contributions: Conceptualization, R.S.M.-O., E.M.-B., and S.A.D.; Investigation, Á.B.-J. and E.D.A.-M.; Methodology, R.S.M.-O., E.M.-B., E.D.A.-M., and S.A.D.; Writing—original draft preparation, R.S.M.-O.; Writing—review and editing, R.S.M.-O. and E.M.-B.; Supervision, Á.B.-J., F.M.-H., and S.A.D.; project administration, R.S.M.-O.; funding acquisition, E.M.-B., F.M.-H., and E.D.A.-M. All authors have read and agreed to the published version of the manuscript.

Funding: This research received no external funding.

Acknowledgments: The authors would like to thank Universidad Juárez Autónoma de Tabasco (Mexico) for its important support.

Conflicts of Interest: The authors declare no conflict of interest.

References

1. Ledesma, E.F.; Jiménez, J.R.; Fernández, J.; Galvín, A.P.; Agrela, F.; Barbudo, M.A. Properties of masonry mortars manufactured with fine recycled concrete aggregates. *Constr. Build. Mater.* **2014**, *71*, 289–298. [CrossRef]

2. Jiménez, C.; Barra, M.; Valls, S.; Aponte, D.; Vázquez, E. Durability of recycled aggregate concrete designed with the Equivalent Mortar Volume (EMV) method: Validation under the Spanish context and its adaptation to Bolomey methodology. *Mater. Constr.* **2013**, *64*, 64. [CrossRef]

3. Sas, W.; Dzięcioł, J.; Głuchowski, A. Estimation of Recycled Concrete Aggregate's Water Permeability Coefficient as Earth Construction Material with the Application of an Analytical Method. *Materials* **2019**, *12*, 2920. [CrossRef] [PubMed]

4. Manzi, S.; Mazzotti, C.; Bignozzi, M. Short and long-term behavior of structural concrete with recycled concrete aggregate. *Cem. Concr. Compos.* **2013**, *37*, 312–318. [CrossRef]

5.   Martín-Morales, M.; Sánchez-Roldán, Z.; Valverde-Palacios, I.; Valverde-Espinosa, I.; Zamorano, M. Study of potential advantages of pre-soaking on the properties of pre-cast concrete made with recycled coarse aggregate. *Mater. Constr.* **2016**, *66*, e076. [CrossRef]

6.   De Brito, J.; Agrela, F.; Silva, R.V. Legal regulations of recycled aggregate concrete in buildings and roads. In *New Trends in Eco-efficient and Recycled Concrete*; Elsevier BV: Amsterdam, The Netherlands, 2019; pp. 509–526.

7.   Nili, M.; Sasanipour, H.; Aslani, F. The Effect of Fine and Coarse Recycled Aggregates on Fresh and Mechanical Properties of Self-Compacting Concrete. *Materials* **2019**, *12*, 1120. [CrossRef]

8.   Pereira, P.; Evangelista, L.; De Brito, J.M.C.L. The effect of superplasticisers on the workability and compressive strength of concrete made with fine recycled concrete aggregates. *Constr. Build. Mater.* **2012**, *28*, 722–729. [CrossRef]

9.   Mefteh, H.; Kebaili, O.; Oucief, H.; Berredjem, L.; Arabi, N. Influence of moisture conditioning of recycled aggregates on the properties of fresh and hardened concrete. *J. Clean. Prod.* **2013**, *54*, 282–288. [CrossRef]

10.   Evangelista, L.; De Brito, J.M.C.L. Concrete with fine recycled aggregates: A review. *Eur. J. Environ. Civ. Eng.* **2013**, *18*, 129–172. [CrossRef]

11.   Cartuxo, F.; De Brito, J.M.C.L.; Evangelista, L.; Jiménez, J.R.; Ledesma, E.F. Increased Durability of Concrete Made with Fine Recycled Concrete Aggregates Using Superplasticizers. *Materials* **2016**, *9*, 98. [CrossRef]

12.   Braga, M.; De Brito, J.M.C.L.; Veiga, M.D.R.D.S. Incorporation of fine concrete aggregates in mortars. *Constr. Build. Mater.* **2012**, *36*, 960–968. [CrossRef]

13.   Neno, C.; De Brito, J.M.C.L.; Veiga, M.D.R.D.S. Using fine recycled concrete aggregate for mortar production. *Mater. Res.* **2013**, *17*, 168–177. [CrossRef]

14.   Saiz-Martínez, P.; Cortina, M.G.; Martinez, F.F. Characterization and influence of fine recycled aggregates on masonry mortars properties. *Mater. Constr.* **2015**, *65*, 058. [CrossRef]

15.   Ng, S.; Engelsen, C.J. Construction and demolition wastes. In *Waste and Supplementary Cementitious Materials in Concrete*; Elsevier BV: Amsterdam, The Netherlands, 2018; pp. 229–255.

16.   Chinchillas-Chinchillas, M.; Pellegrini-Cervantes, M.; Castro-Beltrán, A.; Rodríguez-Rodríguez, M.; Orozco-Carmona, V.; Peinado-Guevara, H. Properties of Mortar with Recycled Aggregates, and Polyacrylonitrile Microfibers Synthesized by Electrospinning. *Materials* **2019**, *12*, 3849. [CrossRef] [PubMed]

17.   Feng, P.; Chang, H.; Xu, G.; Liu, Q.; Zuquan, J.; Liu, J. Feasibility of Utilizing Recycled Aggregate Concrete for Revetment Construction of the Lower Yellow River. *Materials* **2019**, *12*, 4237. [CrossRef]

18.   Silva, J.; De Brito, J.M.C.L.; Veiga, M.D.R.D.S. Incorporation of fine ceramics in mortars. *Constr. Build. Mater.* **2009**, *23*, 556–564. [CrossRef]

19.   Jiménez, J.R.; Ayuso, J.; López, M.; Fernández, J.; De Brito, J.M.C.L. Use of fine recycled aggregates from ceramic waste in masonry mortar manufacturing. *Constr. Build. Mater.* **2013**, *40*, 679–690. [CrossRef]

20.   Silva, J.; De Brito, J.M.C.L.; Veiga, M.D.R.D.S. Recycled Red-Clay Ceramic Construction and Demolition Waste for Mortars Production. *J. Mater. Civ. Eng.* **2010**, *22*, 236–244. [CrossRef]

21.   Pacheco-Torgal, F.; Ding, Y.; Miraldo, S.; Abdollahnejad, Z.; Labrincha, J. The suitability of concrete using recycled aggregates (RAs) for high-performance concrete (HPC). In *Handbook of Recycled Concrete and Demolition Waste*; Elsevier BV: Amsterdam, The Netherlands, 2013; pp. 424–438.

22.   Rahal, K. Mechanical properties of concrete with recycled coarse aggregate. *Build. Environ.* **2007**, *42*, 407–415. [CrossRef]

23.   Cachim, P. Mechanical properties of brick aggregate concrete. *Constr. Build. Mater.* **2009**, *23*, 1292–1297. [CrossRef]

24.   Debieb, F.; Courard, L.; Kenai, S.; Degeimbre, R. Roller compacted concrete with contaminated recycled aggregates. *Constr. Build. Mater.* **2009**, *23*, 3382–3387. [CrossRef]

25.   Nezhad, A.A.; Ong, K.; Chandra, L. Economic and environmental assessment of deconstruction strategies using building information modeling. *Autom. Constr.* **2014**, *37*, 131–144. [CrossRef]

26.   Rao, M.C.; Bhattacharyya, S.K.; Barai, S.V. Demolition Techniques and Production of Recycled Aggregate. In *Springer Transactions in Civil and Environmental Engineering*; Springer Science and Business Media LLC: Berlin, Germany, 2018; pp. 39–63.

27.   Kumbhar, S.A.; Gupta, A.; Desai, D.B. Recycling and Reuse of Construction and Demolition Waste for Sustainable Development. *Int. J. Sustain. Dev.* **2013**, *6*, 83–92. Available online: https://ssrn.com/abstract=2383436. (accessed on 11 April 2020).

28. Coelho, A.; De Brito, J.M.C.L. Conventional demolition versus deconstruction techniques in managing construction and demolition waste (CDW). In *Handbook of Recycled Concrete and Demolition Waste*; Elsevier BV: Amsterdam, The Netherlands, 2013; pp. 141–185.

29. Mas, B.; Cladera, A.; Del Olmo, T.; Pitarch, F. Influence of the amount of mixed recycled aggregates on the properties of concrete for non-structural use. *Constr. Build. Mater.* **2012**, *27*, 612–622. [CrossRef]

30. Ferreira, L.; De Brito, J.M.C.L.; Barra, M. Influence of the pre-saturation of recycled coarse concrete aggregates on concrete properties. *Mag. Concr. Res.* **2011**, *63*, 617–627. [CrossRef]

31. Medina, C.; Juan-Valdés, A.; Frías, M.; De Rojas, M.I.S.; Moran, J.M.; Guerra, M.I. Caracterización de los hormigones realizados con áridos reciclados procedentes de la industria de cerámica sanitaria. *Mater. Constr.* **2011**, *61*, 533–546. [CrossRef]

32. Martinez, I.; Etxeberria, M.; Pavón, E.; Diaz, N. A comparative analysis of the properties of recycled and natural aggregate in masonry mortars. *Constr. Build. Mater.* **2013**, *49*, 384–392. [CrossRef]

33. Evangelista, L.; De Brito, J.M.C.L. Durability performance of concrete made with fine recycled concrete aggregates. *Cem. Concr. Compos.* **2010**, *32*, 9–14. [CrossRef]

34. Martínez-García, R.; Guerra-Romero, M.I.; Pozo, J.M.M.D.; De Brito, J.; Juan-Valdés, A. Recycling Aggregates for Self-Compacting Concrete Production: A Feasible Option. *Materials* **2020**, *13*, 868. [CrossRef]

35. González, J.G.; Rodríguez-Robles, D.; Valdés, A.J.; Pozo, J.M.M.D.; Guerra-Romero, M.I. Influence of Moisture States of Recycled Coarse Aggregates on the Slump Test. *Adv. Mater. Res.* **2013**, *742*, 379–383. [CrossRef]

36. Kim, J.H.; Robertson, R.E. Prevention of air void formation in polymer-modified cement mortar by pre-wetting. *Cem. Concr. Res.* **1997**, *27*, 171–176. [CrossRef]

37. Abdollahnejad, Z.; Mastali, M.; Falah, M.; Luukkonen, T.; Mazari, M.; Illikainen, M. Construction and Demolition Waste as Recycled Aggregates in Alkali-Activated Concretes. *Materials* **2019**, *12*, 4016. [CrossRef] [PubMed]

38. Barbudo, M.A.; De Brito, J.M.C.L.; Evangelista, L.; Bravo, M.; Agrela, F. Influence of water-reducing admixtures on the mechanical performance of recycled concrete. *J. Clean. Prod.* **2013**, *59*, 93–98. [CrossRef]

39. Zega, C.J.; Di Maio, Á.A. Use of recycled fine aggregate in concretes with durable requirements. *Waste Manag.* **2011**, *31*, 2336–2340. [CrossRef]

40. Cuenca-Moyano, G.M.; Martín-Morales, M.; Valverde-Palacios, I.; Valverde-Espinosa, I.; Zamorano, M. Influence of pre-soaked recycled fine aggregate on the properties of masonry mortar. *Constr. Build. Mater.* **2014**, *70*, 71–79. [CrossRef]

41. Cabral, A.E.B.; Schalch, V.; Molin, D.C.C.D.; Ribeiro, J.L.D. Mechanical properties modeling of recycled aggregate concrete. *Constr. Build. Mater.* **2010**, *24*, 421–430. [CrossRef]

42. Zhao, Z.; Rémond, S.; Damidot, D.; Xu, W. Influence of fine recycled concrete aggregates on the properties of mortars. *Constr. Build. Mater.* **2015**, *81*, 179–186. [CrossRef]

43. UNE-EN 13139/AC:2004. *Áridos Para Mortero*; Asociación Española de Normalización (Aenor): Madrid, España, 2004.

44. ASTM C150/C150M-19a. *Standard Specification for Portland Cement*; ASTM International: West Conshohocken, PA, USA, 2019.

45. ASTM C-595/C-595M-19. *Standard Specification for Blended Hydraulic Cements*; ASTM International: West Conshohocken, PA, USA, 2019.

46. UNE-EN 933-1:2012. *Ensayos Para Determinar Las Propiedades Geométricas De Los áridos. Parte 1: Determinación De La Granulometría De Las Partículas. Método Del Tamizado*; Asociación Española de Normalización (Aenor): Madrid, España, 2012.

47. UNE-EN 933-8:2012. *Ensayos Para Determinar Las Propiedades Geométricas De Los áridos. Parte 8: Evaluación De Los Finos. Ensayo Del Equivalente De Arena*; Asociación Española de Normalización (Aenor): Madrid, España, 2012.

48. UNE-EN 1097-6:2014. *Ensayos Para Determinar Las Propiedades Mecánicas y Físicas De Los áridos. Parte 6: Determinación De La Densidad De Partículas y La Absorción De Agua*; Asociación Española de Normalización (Aenor): Madrid, España, 2014.

49. UNE-EN 1744:2013. *Ensayos Para Determinar Las Propiedades Químicas De Los áridos. Parte 1: Análisis Químico*; Asociación Española de Normalización (Aenor): Madrid, España, 2013.

50. Dapena, E.; Alaejos, P.; Lobet, A.; Pérez, D. Effect of Recycled Sand Content on Characteristics of Mortars and Concretes. *J. Mater. Civ. Eng.* **2011**, *23*, 414–422. [CrossRef]

51. ASTM C494/C494M-17. *Standard Specification for Chemical Admixtures for Concrete*; ASTM International: West Conshohocken, PA, USA, 2017.

52. ASTM C1017/C1017M-13. *Standard Specification for Chemical Admixtures for Use in Producing Flowing Concrete*; ASTM International: West Conshohocken, PA, USA, 2013.

53. Fonseca, N.; De Brito, J.; Evangelista, L. The influence of curing conditions on the mechanical performance of concrete made with recycled concrete waste. *Cem. Concr. Compos.* **2011**, *33*, 637–643. [CrossRef]

54. UNE-EN 1015-6:1999/A1:2007. *Métodos De Ensayo De Los Morteros Para Albañilería. Parte 6: Determinación De La Densidad Aparente Del Mortero Fresco*; Asociación Española de Normalización (Aenor): Madrid, España, 2007.

55. UNE-EN 1015-7:1999. *Métodos De Ensayo De Los Morteros Para Albañilería. Parte 7: Determinación Del Contenido En Aire En El Mortero Fresco*; Asociación Española de Normalización (Aenor): Madrid, España, 1999.

56. UNE-EN 1015-10:2000. *Métodos De Ensayo De Los Morteros Para Albañilería. Parte 10: Determinación De La Densidad Aparente En Seco Del Mortero Endurecido*; Asociación Española de Normalización (Aenor): Madrid, España, 2000.

57. UNE-EN 1015-11:2000. *Métodos De Ensayo De Los Morteros Para Albañilería. Parte 11: Determinación De La Resistencia a Flexión y a Compresión Del Mortero Endurecido*; Asociación Española de Normalización (Aenor): Madrid, España, 2000.

58. UNE-EN 1015-12:2016. *Métodos De Ensayo De Los Morteros Para Albañilería. Parte 12: Determinación De La Resistencia a La Adhesión De Los Morteros De Revoco y Enlucido Endurecidos Aplicados Sobre Soportes*; Asociación Española de Normalización (Aenor): Madrid, España, 2016.

59. UNE-EN 1015-18:2003. *Métodos De Ensayo De Los Morteros Para Albañilería. Parte 18: Determinación Del Coeficiente De Absorción De Agua Por Capilaridad Del Mortero Endurecido*; Asociación Española de Normalización (Aenor): Madrid, España, 2003.

60. Vegas, I.; Ibanez, J.; Lisbona, A.; De Cortazar, A.S.; Frías, M. Pre-normative research on the use of mixed recycled aggregates in unbound road sections. *Constr. Build. Mater.* **2011**, *25*, 2674–2682. [CrossRef]

61. Leite, M.; Filho, J.G.L.F.D.; Lima, P.R.L. Workability study of concretes made with recycled mortar aggregate. *Mater. Struct.* **2013**, *46*, 1765–1778. [CrossRef]

62. Amorim, L.V.; Lira, H.D.L.; Ferreira, H.C. Use of Residential Construction Waste and Residues from Red Ceramic Industry in Alternative Mortars. *J. Environ. Eng.* **2003**, *129*, 916–920. [CrossRef]

63. Cuenca-Moyano, G.M.; Martín-Pascual, J.; Martín-Morales, M.; Valverde-Palacios, I.; Zamorano, M. Effects of water to cement ratio, recycled fine aggregate and air entraining/plasticizer admixture on masonry mortar properties. *Constr. Build. Mater.* **2020**, *230*, 116929. [CrossRef]

 *materials*

Article

Use of Mining Waste to Produce Ultra-High-Performance Fibre-Reinforced Concrete

Jesús Suárez González [1,*], Iñigo Lopez Boadella [1], Fernando López Gayarre [1], Carlos López-Colina Pérez [1], Miguel Serrano López [1] and Flavio Stochino [2]

[1] Poliytechnic School of Engineering, Campus de Viesques, University of Oviedo, 33203 Gijón, Spain; inigo2208@hotmail.com (I.L.B.); gayarre@uniovi.es (F.L.G.); lopezpcarlos@uniovi.es (C.L.-C.P.); serrano@uniovi.es (M.S.L.)

[2] Department of Civil, Environmental and Architectural Engineering, University of Cagliari, 09100 Cagliari, Italy; fstochino@unica.it

[*] Correspondence: suarezg@uniovi.es

Received: 1 May 2020; Accepted: 26 May 2020; Published: 28 May 2020

Abstract: This research work analyses the influence of the use of by-products from a fluorite mine to replace the fine fraction of natural aggregates, on the properties of Ultra-High-Performance Fibre-Reinforced Concrete (UHPFRC). Replacing natural aggregates for different kinds of wastes is becoming common in concrete manufacturing and there are a number of studies into the use of waste from the construction sector in UHPFRC. However, there is very little work concerning the use of waste from the mining industry. Furthermore, most of the existing studies focus on granite wastes. So, using mining sand waste is an innovative alternative to replace natural aggregates in the manufacture of UHPFRC. The substitutions in this study are of 50%, 70% and 100% by volume of 0–0.5 mm natural silica sand. The results obtained show that the variations in the properties of consistency, compressive strength, modulus of elasticity and tensile strength, among others, are acceptable for substitutions of up to 70%. Therefore, fluorite mining sand waste is proved to be a viable alternative in the manufacturing of UHPFRC.

Keywords: mining waste; ultra-high performance fibres reinforced concrete; compressive strength; flexural strength

---

1. Introduction

In 1973, the overexploitation of natural resources and environmental degradation due to the waste generated in industry, mining, etc. motivated the European Union to launch its Environment Action Program. This program is currently in its seventh edition. The latest resolutions, passed in 2013, prioritize increasingly efficient use of resources to make the European Union an intelligent, sustainable and inclusive economy. Construction is one of the most affected sectors, so the use of waste as a resource is currently a priority in the research lines of many institutions in this sector, particularly in relation to the manufacturing of concrete.

Ultra High Performance Concrete (UHPC) emerged in the early 1990s, when P. Richard [1] developed a new type of concrete that offered a compressive strength of over 200 MPa, a bending strength above 40 MPa, and some ductility. In 2002 the first recommendations for the structural use of these concretes ware published in France [2]. In 2005, noteworthy research work was carried out in Germany, building the groundwork for the basic knowledge needed to develop a reliable, economically feasible UHPC [3]. UHPC has been developed further in recent decades and is valued for its high mechanical strength and very low porosity.

The incorporation of short steel fibres in these concretes improves the performance of the structures against flexotraction, controls and reduces cracking. These concretes are called Ultra-High-Performance

Fibre-Reinforced Concrete (UHPFRC). Abbas et al. [4] reported that the incorporation of short steel fibres, increases the load against the first crack and the maximum load, as well as making it possible to double or triple the tensile strength of UHPC. Although the Association Française de Génie Civil considers UHPC as that having a compressive strength higher than 150 MPa [2], it is very common to classify concrete with a compressive strength of over 100MPa as UHPC. In this regard, F. Canovas qualifies concrete as having very-high strength when it has a compressive strength higher than 90 MPa and as having ultra-high strength when its compressive strength is higher than 125 MPa at 28 days [5].

Ecologically, however, UHPC has problems in terms of sustainability, since it uses very fine fractions of natural aggregates, especially silica sands, which require significant energy for grinding. Therefore, there is a need to find and supply alternative materials to avoid or reduce their use. The use of recycled materials or construction by-products as a total or partial substitution of natural aggregates in the search for a more sustainable UHPC is currently being researched. Ambily el al. [6] study the use of copper slags as an alternative to the finest aggregate in UHPC. With 100% substitution of the fine aggregate, variations in flexural strength are negligible, and despite the slight reported loss of compressive strength, the final values are above 150 MPa. Zhu et al. [7] and Zhao et al. [8] use iron tailings as an alternative to natural aggregate. They observed that up to 40% substitution, the mechanical properties of the UHPC were coMParable to those of the control concrete. In addition, Zhao et al. [8] showed that if the maximum size of the iron tailings was under 1.20 mm, a substitution of 60% is viable.

Glass waste is another type of material that can be used to substitute aggregates. Yang et al. [9] and Soliman and Tagnit-Hamou [10] have shown that using recycled glass to replace the natural sand gives promising mechanical properties in UHPC. For a ratio of 50% substitution, the slump and the compressive strength are coMParable to the control UHPC mix [10]. Following this same line of research, Zegardlo et al. [11] have shown that ceramic waste is another alternative to natural aggregate in the manufacture of UHPC. The results show that the compressive strength improves by 24% and the tensile strength rises to 34%.

The abovementioned studies focus on the use of waste from the construction sector. However, there are only a few studies that focus on incorporating waste from the extractive industry in the manufacture of concrete, specifically HPC and UHPC. The studies in which granite cutting waste is used as a substitution for fine aggregate in UHPC is relevant in this area. Sarbjeet et al. [12] showed that a substitution of 40% of the aggregate had a positive iMPact on the flexural strength while for a ratio of 70% the results were coMParable to those of the control concrete. López Boadella et al. [13] observed an increase in flexural strength and tensile strength by substituting 35% of micronized quartz with fine granite cutting waste. Similar results were provided by Kala [14] who found that the incorporation of granite fines as a partial replacement for sand has a positive effect on the properties of HPC in relation to compressive strength, flexural strength and tensile strength.

Mining, which is one of the greatest waste-generating activities [15,16] is the focus of this research. This a study of the feasibility of the utilization of by-products from a fluorite (CaF_2) mine as a replacement for natural aggregates in the manufacture of UHPFRC. This silica sand comes from the mine's washing installations. Once treated and washed, it could be used in the manufacture of UHPC. However, it cannot be used in conventional concrete due to the high percentage of fines. To the best of the authors' knowledge, there are no previous studies on the use of this type residue. The studies than come close are those already mentioned, which refer to granite fines. So, the main objective of this research is the innovative use of this by-product as a substitute for natural aggregates in the manufacture of UHPFRC.

2. Experimental Study

2.1. Description of the Materials Used

Type I 42.5 R/SR Ordinary Portland Cement (OPC, Lafarge-Holcim, Madrid, Spain) was used. As additions for the mix, two materials were used: densified silica fume (Elkem Microsilica®940, Elkem, Barcelona, Spain) with a 0.15 µm average particle size and micronized quartz with a maximum particle size of 40 µm. The steel fibres (Arcelor Mittal, Asturias, Spain) used in the research had a diameter of 0.2 mm and a length of 13 mm. As natural aggregates, two types of siliceous sands (Sílices La Cuesta, Asturias, Spain) with granulometric fractions 0–0.5 mm and 0.5–1.6 mm were used. In order to achieve optimum workability, a polyxycarboxylate superplasticizer was used. (Sika ViscoCrete-225 Powder, Sika, Madrid, Spain). Finally, as a substitute for the finest fraction of the sand, waste mining sand (WMS) (Minersa, Ribadesella, Spain), which is generated in the process of obtaining fluoride or fluorite spar in fluorite mines, was used (Figure 1a,b).

(a) Fluorite mine washing installations   (b) Waste draining system

Figure 1. Obtaining residual mine sand.

Table 1 shows a summary of the main properties of all the materials used. The greatest amount of fine particles present in the WMS is in the sand equivalent test.

Table 1. Properties of the materials.

Properties	Aggregate 0–0.5 mm	Aggregate 0.5–1.6 mm	Micronized Quartz	Silica Fume	Waste Mining Sand
Bulk density (kg/m^3)	2616	2616	2609	2300	2717
Sand equivalent	97	97	-	-	28
Absorption 24 h	0.28%	0.53%	-	-	-
Humidity (%)	0%	0%	< 0.2%	< 3.00%	0%

The chemical composition of the WMS is determined by X-ray fluorescence. The results obtained are shown in Table 2, where L.O.I stands for Loss on Ignition. The second most significant component is CaO, at 10%. The percentages of Al_2O_3 and Fe_2O_3 are approximately 1.5%, higher than those of the 0–0.5 mm silica sand.

Table 2. Chemical analysis (%) of the waste mining sand and the natural sand it replaces.

Material	SiO_2	CaO	Fe_2O_3	Al_2O_3	MgO	K_2O	Na_2O	TiO_2	P_2O_5	MnO	L.O.I
Waste mining sand	71.65	10.8	1.66	1.5	1.21	0.57	0.13	0.11	0.03	0.01	8.82
Aggregate 0–0.5 mm	99.34	< 0.1	< 0.13	0.30	< 0.1	< 0.1	< 0.03	< 0.1	< 0.1	-	0.23

Figure 2 shows the particles size distribution of the materials used. A difference between the granulometry of the 0–0.3 mm WMS, and the fraction of 0–0.5 mm silica sand which it substitutes, can be seen. The granulometric distribution of silica fume is a consequence of its densification. Its maximum particle size is 0.15 µm.

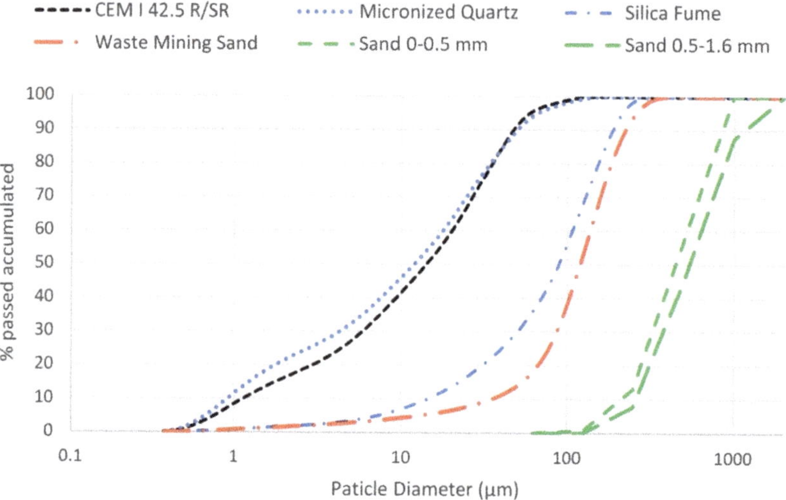

Figure 2. Grading curves of the materials.

2.2. Mix Design

To develop this research, the starting point was a self-coMPacting reference concrete with a compressive strength of more than 117 MPa. From this control mix, the experimental program was developed with replacements of 50%, 70% and 100% of 0–0.5 mm of silica sand for the same volume of WMS. Substitutions below 50% were not done as the results obtained for these percentages were satisfactory.

Table 3 shows the proportions of each of the UHPFRC mixes manufactured.

Table 3. Dosages of Ultra High Performance Fibre Reinforced Concrete (UHPFRC) (kg/m^3).

Material	Control	50% WMS	70% WMS	100% WMS
Cement	800	800	800	800
Sand 0–0.5 mm	302	151	91	-
Waste mining sand	-	161	225	321
Sand 0.5–1.6 mm	565	565	565	565
Micronized quartz	225	225	225	225
Silica fume	175	175	175	175
Water	175	175	175	175
Superplasticizer	10	10	10	10
Steel fibres	160	160	160	160

Figure 3 shows the grading curves of the total aggregate of the different UHPFRC mixes. As the percentage of substitution of the finest fraction of the sand for WMS increases, the percentage of particles with a size between 50 µm and 300 µm increases. This increase generates grading curves with a more continuous progression, given that the depression that occurs in the range between 50 µm and 1 mm for the control concrete is corrected. Because of this, the curves resulting from the use of mine sand are more similar to the ideal packing curve proposed by authors such as Andreasen or Funk and Dinger [17].

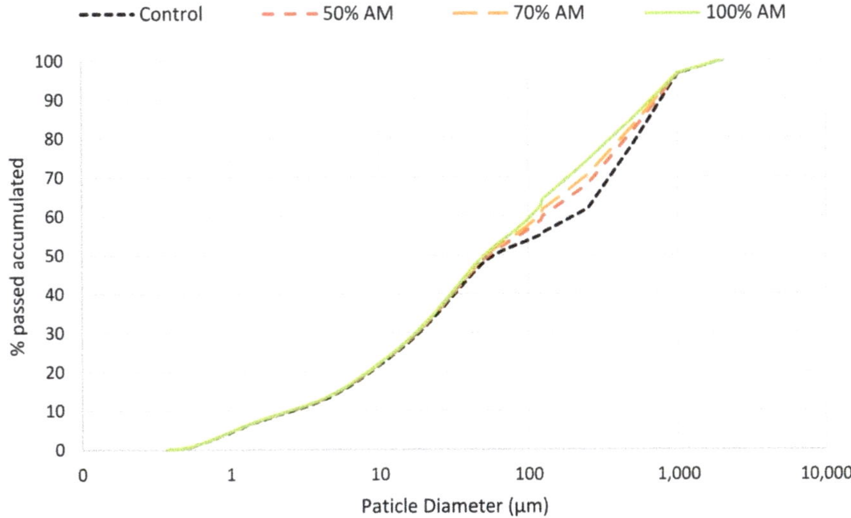

Figure 3. Grading curves of UHPFRC.

2.3. Development of the Experimental Program

The mix procedure is as follows: first the two parts of silica sand and/or the WMS are introduced in the mixer, followed by the micronized quartz, the silica fume and finally the cement. They are mixed for 30 s and then the whole amount of water is poured in. Mixing continues for 2 min more, after which the superplasticizer additive and the steel fibres are added. The whole mixing process ends after 25 min.

Nine specimens were manufactured for each mix, according to the specifications of standard UNE-EN 12390-1 [18]: three Ø15 × 30 cm cylindrical specimens, three 10 × 10 × 10 cm cubic specimens and three 10 × 10 × 40 cm prismatic specimens. The specimens were cured in a humid chamber at a temperature of 20 ± 2 °C and a relative humidity of 95% for 28 days following the specifications of the UNE-EN 12390-2 standard [19].

The following properties were measured: the consistency of fresh UHPFRC according to the NF P 18-470 standard [20], the density of hardened UHPFRC following the UNE-EN 12390-7 standard [21], the modulus of elasticity (UNE- EN 12390-13) [22], compressive strength (UNE-EN 12390-3) [23], flexural strength (NF P 18-470) [20] and tensile strength (NF P standard 18-470) [20].

Table 4 shows a summary of the results obtained in this study. All values, except slump, correspond to the average value of the three results obtained in each of the tests performed.

Table 4. Results of the experimental program.

Properties	Control	Difference (%)		
		50% WMS	70% WMS	100% WMS
Slump (cm)	25.0	16.0%	4.0%	20.0%
Density (kg/m^3)	2410.0	−1.2%	−1.7%	−2.1%
Compressive strength (MPa)	117.2	9.9%	11.8%	11.2%
Modulus of elasticity (GPa)	45.2	−3.5%	−5.8%	−6.4%
Flexural strength (MPa)	23.0	−7.0%	−3.5%	−17.4%
Tensile strength (MPa)	8.7	9.2%	−1.1%	−20.7%

3. Analysis of Results

3.1. Consistency of Fresh UHPFRC

Figure 4 shows the results obtained. In the three mixes there is an increase in the consistency of fresh UHPFRC, of between 4% and 25%. This improvement may be related with the variation in the granulometry of WMS (0–0.3 mm), in relation to the fraction of substituted silica sand (0–0.5 mm). It may also be related to the variation experienced by the grading curve of the mixtures in the range 0.1–0.5 mm when introducing the mine sand, since a more continuous distribution of the particle size is observed. This will improve packaging and reduce water requirements [24].

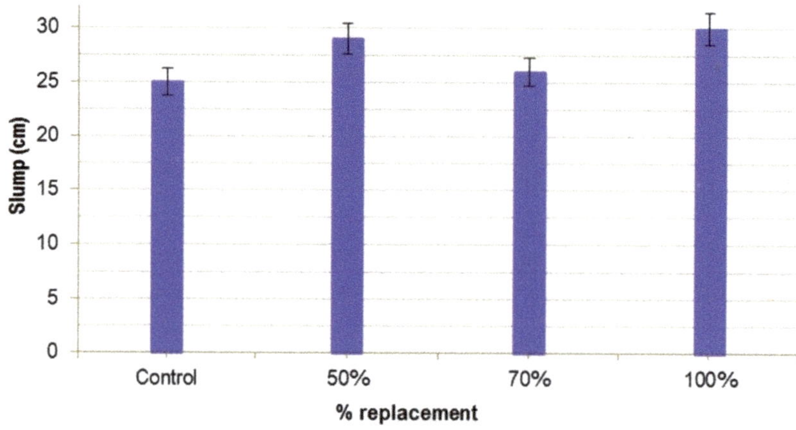

Figure 4. Slump of UHPFRC.

Differences between these results and those obtained by other authors can be explained by the characteristics of the type of waste used. Soliman and Tagnit-Hamou [10] observed an improvement in slump flow by incorporating glass powder, with slightly larger average particle diameter, as partial or total substitution of quartz sand. Similar results were obtained by Pyo et al. [25] when substituting the finest silica sand with waste generated in a tungsten mine. For substitutions of 50% and 100% there is an improvement in the flow of fresh concrete greater than 30%. However, Kou [26] and Zhao et al. [8] observe a decrease in flowability in their tests when using other types of waste. Kou [26] shows a loss in the flowability of UHPFRC when discarded fly ash is incorporated as a substitution for silica sand. Kou attributes this loss of flowability to a greater amount of fines present in fly ash. Zhao et al. [8] observed that substituting 100% of the natural aggregate with iron ore tailings produced a significant decrease in the flowability of the UHPC due to the more angular and irregular shape of the tailings.

3.2. Density of the Hardened UHPFRC

Figure 5 shows the values obtained for the different substitution percentages. As the percentage of WMS increases, there is a minimal density reduction, around 2% for a 100% substitution. These variations in density are so slight as to be within the range of variability of the results obtained, as shown in the error bars in Figure 5. Therefore, the use of WMS has a negligible influence on the density of hardened UHPFRC.

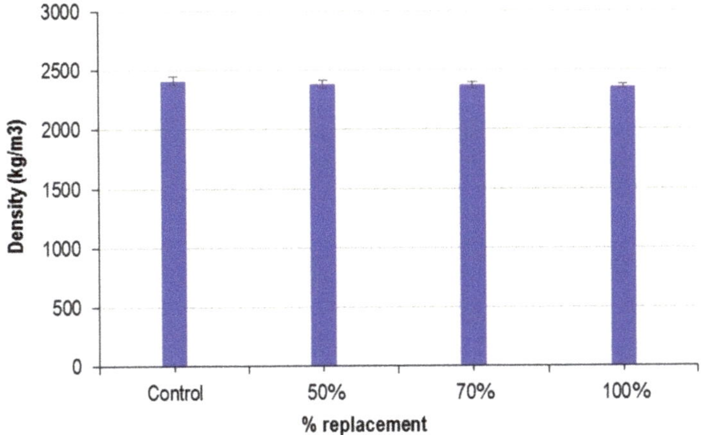

Figure 5. Density of UHPFRC.

3.3. Compressive Strength

There is an increase in strength for all degrees of substitution, as shown in Figure 6. This may be due to the difference between the maximum aggregate size of silica sand (0.5 mm) and that of sand from the fluorite mine (0.3 mm) since, especially in cement-rich concrete. The smaller the maximum particle size, the greater the compressive strength of the concrete [27–29].

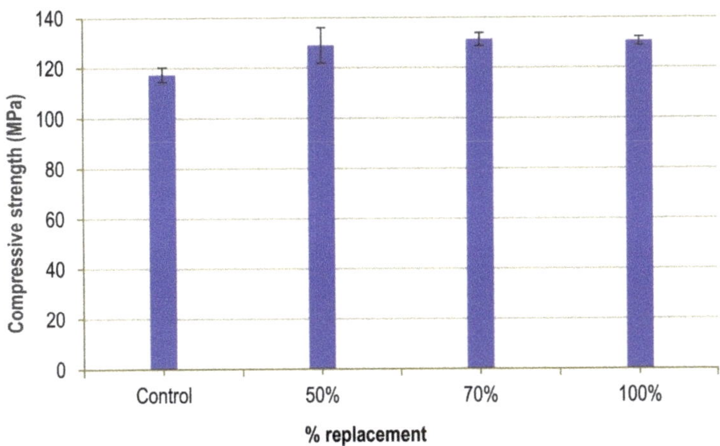

Figure 6. Compressive strength of UHPFRC.

These increases in the compressive strength of concrete are in line with the results obtained by Zegardlo, [11] who obtained an increase in strength of 24.7% when using ceramic waste as an alternative to natural aggregate. He attributes this improvement to the formation of mechanical hooks in the interfacial transition zone (ITZ), which generate better adhesion with the cement paste and greater strength. Similar results were obtained by Zhu [7] who observed an increase in compressive strength by incorporating 0–1.20 mm iron ore tailings to substitute 0–4.75 mm siliceous sands, independently of the percentage of substitution. However, other authors obtain less favourable results, depending on the waste used: Soliman and Tagnit-Hamou report a loss of compressive strength of 15% for a 100% substitution in their study using glass waste as an alternative to quartz sand [10].

4. Elasticity Modulus

Figure 7 displays the effect of WMS on the elasticity modulus of the UHPFRC. A slight and progressive decrease in the modulus of elasticity values can be seen as the substitution percentage increases. However, this difference is small, not exceeding 6%.

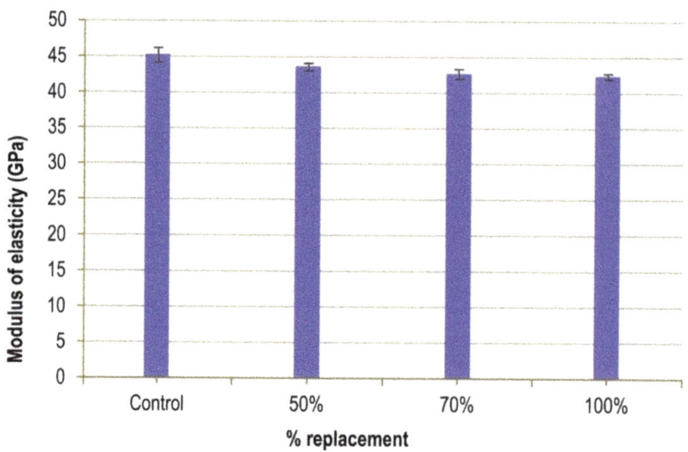

Figure 7. Modulus of Elasticity of UHPFRC.

This low influence on the modulus of elasticity is also reflected in the results obtained by Alsalman [30], when substituting sand with fly ash. According to the author, this may be due to an excess of fly ash, which makes it impossible for 100% of these particles to be hydrated. This counteracts the positive effect of hydrated ash.

Gonzalez-Corominas and Etxeberria [31] also observed a decrease in the modulus of elasticity (around 5%), although their study refers to low substitution percentages (less than 30%) and uses ceramic waste to manufacture high performance concrete (HPC). They relate this reduction to the lower density of the concrete derived from using ceramic waste.

4.1. Flexural Strength

The results regarding flexural strength, shown in Figure 8, are stable up to 70% substitution, when they rise slightly. In general, the trend is declining and drops to 17% for a substitution of 100%. For lower percentages the variation is smaller, less than 7%. The variability of the results may be due to the randomness of the distribution of steel fibres.

The explanation may be in the inferior quality of mine sand, since it has 10% CaO, as opposed to the natural sand it substitutes, which is almost 100% SiO_2. This could counteract the advantageous results produced by the improvement in the grading curve for high percentage substitutions, because of the lower mechanical resistance of the mine sand. In general, the result is similar to that obtained by Zhao et al. [8] and Zhu et al. [7] who use iron ore tailings as an alternative to natural aggregate. The concrete loses 18% of its flexural strength when 100% of the natural aggregate is substituted, while for the rest of the percentages the variations are very small. These results are also similar to those obtained by Taha [32], who uses glass waste as an alternative to natural aggregate in UHPC without fibres, although the results of flexural strength are not coMParable due to the lack of fibres. However, the results obtained by this author show a decrease in flexural strength of 21%, when 100% of the fine aggregate is substituted. Taha considers that this loss of strength may be a consequence of small cracks present in the particles of the glass waste, or the presence of organic materials, which can degrade over time, creating hollows in the microstructure of the concrete.

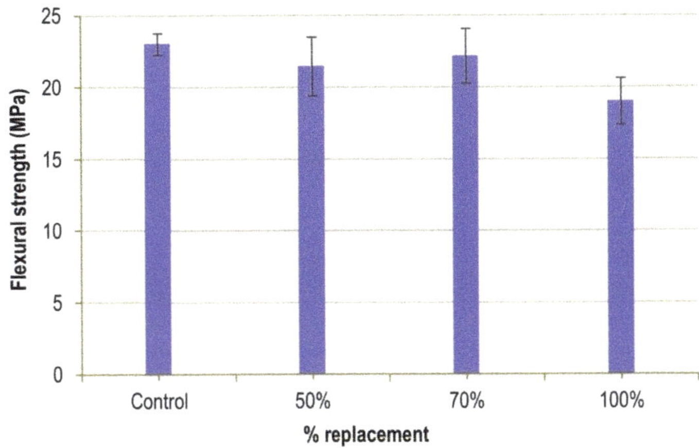

Figure 8. Flexural strength of UHPFRC.

4.2. Tensile Strength

In order to determinate the tensile strength of UHPFRC the stress-strain curves from the flexural tests have been taken as the starting point as seen in Figure 9.

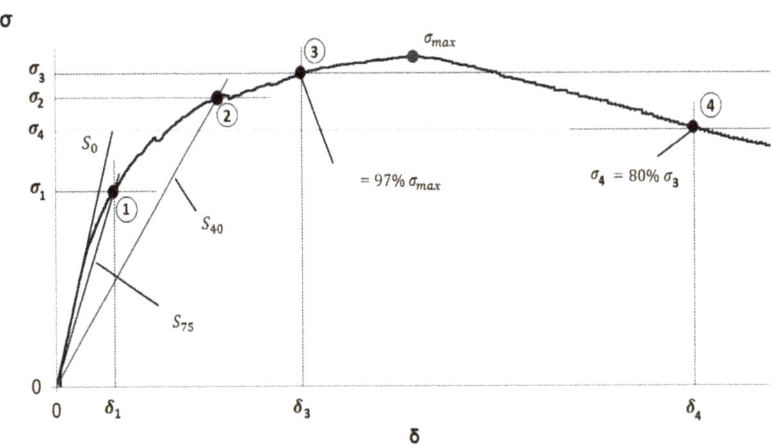

Figure 9. Stress–strain curves and key points.

From the curve obtained, the key points [33] necessary to determine the tensile behaviour were determined, using the following points:

Point 1 is the intersection of the curve test with a line of a slope equal to 0.75 of the slope of the elastic zone of the curve.

Point 2 is the intersection of the curve test with a line of a slope equal to 0.40 of the slope of the elastic zone of the curve.

Point 3 is the point of the ascending zone of the curve, with 97% of the highest stress.

Based on these points, the tensile strength of the UHPFRC can be obtained by the following equations [33]:

$$E = 2.40 \, h \, m \tag{1}$$

$$f_t = \frac{\sigma_{75}}{1,63}\left(\frac{\sigma_{75}}{\sigma_{40}}\right)^{0.19} \tag{2}$$

$$\varepsilon_{t,u} = \frac{f_t}{E}\left(7.65 \, \frac{\delta_{loc}}{\delta_{75}} - 10.53\right) \tag{3}$$

$$\varepsilon_{t,\,el} = f_t / E \tag{4}$$

$$\alpha = \varepsilon_{t,u} / \varepsilon_{t,el} \tag{5}$$

$$f_{t,u} = \alpha^{-0.18}\left(2.46 \, \frac{\sigma_{loc}}{\sigma_{75}} - 1.76\right)f_t \tag{6}$$

Where the pair of values (σ_{75}, δ_{75}), (σ_{40}, δ_{40}), and (σ_{loc}, δ_{loc}) correspond respectively with points 1, 2 and 3 in the curve of Figure 9. Equation (1) represents the value of elasticity modulus (E) in function of the slope (m) of the elastic region of the stress-strain curve and the thickness (h) of the specimen in mm. Equation (2) refers to the crack resistance (f_t) of the matrix reinforced with fibres, Equation (3) calculates the peak deformation ($\varepsilon_{t,u}$), Equations (4) and (5) refers to the standardised parameters $\varepsilon_{t,el}$ and α, and Equation (6) represents the value of the ultimate tensile strength ($f_{t,u}$) in function of the parameters previously defined.

The resulting tensile strengths can be seen in Table 4 and Figure 10. An initial increase in tensile strength values is followed by a decrease for high substitution percentages. For a 50% substitution, an increase in resistance of 9.5 MPa is achieved, which represents an increase of 9%, whereas for a 100% substitution, there is a loss of 20%. For a substitution of 70%, there is hardly any variation with respect to the control UHPFRC. As the error bars show, a high variability of the experimental results obtained for each substitution percentage is clear. This variation may be due to the random distribution of the fibres in the UHPFRC.

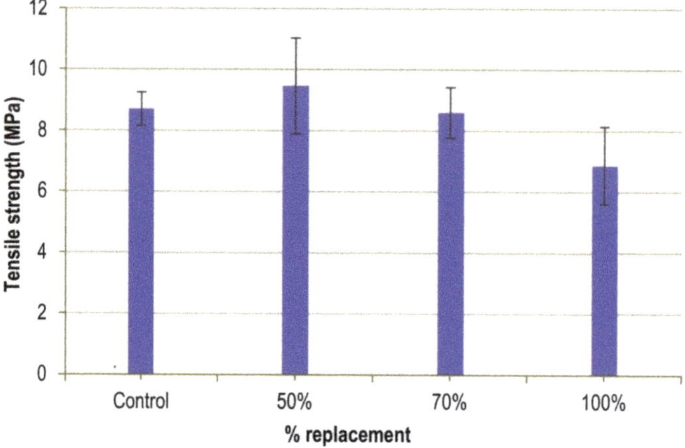

Figure 10. Tensile strength of UHPFRC.

The tensile strength results are similar to those of flexural strength, with an improvement for low substitution percentages followed by a drop that reaches 20% for a 100% substitution. The possible explanation, as in flexural strength, is the inferior quality of the mine sand used. The result obtained is similar to that of Taha [32] when analysing the influence of substituting glass waste for sand in the UHPC without fibres, although pointing out that the absence of fibres is an important factor. This author observes very little variation for 50% substitution, while for 100%, he reports a loss of tensile strength of 5%. As mentioned in the previous section, Taha considers that this loss may be due to small cracks present in the particles of the glass waste.

5. Conclusions

In this research, two main findings have been identified, which could explain the results obtained in the laboratory tests. The first is that the grading curve of the mixtures shows a more continuous distribution of the particle size, which improves packaging and reduce water requirements. The second is that the quality of mining sand is slightly inferior given that it contains 10% CaO as opposed to the natural sand it substitutes, which is almost 100% SiO_2. The overall consequence is that the improvement in packaging leads to an improvement in mechanical properties up approximately 70% of replacement, at which point the lower quality of waste mining sand counteracts the advantages of the better packaging.

In view of the results obtained in this work, it can be established that waste sand from fluorite mines (WMS) is a viable alternative as a partial substitution of silica sand in the manufacturing of Ultra High Performance Fibre Reinforced Concrete (UHPFRC). The most relevant conclusions are the following:

The incorporation of WMS improves the consistency of fresh UHPFRC due, probably, to its smaller particle size generating a lubrication effect of the mixture.

The density of UHPFRC with WMS is very similar to the density of the control concrete. The variations are very small, less than 2.5%, for any percentage of substitution.

The improved consistency of fresh UHPFRC with WMS and its smaller particle size favours the coMPactness of the concrete and causes an increase in compressive strength around 11% for all substitution percentages.

There is a slight decrease in the modulus of elasticity, which does not reach 6% for a 100% substitution with WMS, so the influence of using this type of waste on the modulus of elasticity of UHPFRC is negligible.

The results of flexural strength and tensile strength obtained show high variability for each substitution percentage. This may be a consequence of an inefficient distribution of the short steel fibres present in these concretes. However, the variations are small up to 70% substitution. Only for substitutions of 100% is there a significant loss in flexural strength and tensile strength, which reach values of 17% and 20%, respectively.

It can be held that the use of up to 70% of waste mine sand as a substitute for natural sand is a totally feasible option for manufacturing UHPFRC, given that it has a minimal effect on any of its mechanical properties. However, for a ratio of 100% of substitution there is a drop of over 15% in tensile and flexural strengths, making the use of this substitution percentage inadvisable.

Author Contributions: Conceptualization, M.S.L and F.L.G.; methodology, F.S. and I.L.B.; validation, F.S. and F.L.G. investigation, J.S.G. and C.L.-C.P.; resources, J.S.G. and I.L.B. writing—original draft preparation, I.L.B and J.S.G.; writing—review and editing, M.S.L. and J.S.G.; visualization, C.L.-C.P. and M.S.L.; supervision, C.L.-C.P. and F.L.G.; project administration, F.L.G.; funding acquisition, F.L.G. and J.S.G. All authors have read and agreed to the published version of the manuscript.

Funding: This research was funded by the Spanish Ministry of Economy and Competitiveness through the research project grant number BIA2016-78460-C3-2-R.

Acknowledgments: The authors also want to thank the support to carry out this study to ArcelorMittal, Elkem, Basf, Sika AG, Grupo Minersa, and the Ministry of Economy and Competitiveness of the Government of Spain. The support of CERIS research centre and Instituto Superior Técnico is also acknowledged.

Conflicts of Interest: The authors declare no conflict of interest.

References

1. Richard, P.; Cheyrezy, M. Composition of reactive powder concretes. *Cem. Concr. Res.* **1995**, *25*, 1501–1511. [CrossRef]

2. Bétons Fibrés à Ultra-Hautes Performances, Recommandations Provisoires, *Ultra High Performance Fibre-Reinforced Concretes*, Interim Recommendations, SETRA-AFGC, Groupe de travail BFUP: Paris, France, January 2002. (in French and English). Available online: http://dtrf.setra.fr/pdf/pj/Dtrf/0002/Dtrf-0002989/DT2989.pdf?openerPage=resultats&qid=sdx_q0 (accessed on 27 May 2020).

3. Schmidt, C.; Glotzbach, S.; Fröhlich, S.; Piotrowski, S. Sustainable building with UHPC—Coordinated research program in Germany. In Proceedings of the Hipermat 2012 3rd International Symposium on UHPC and Nanotechnology for High Performance Construction Materials, Kassel, Germany, 7–9 March 2012.

4. Abbas, S.; Soliman, A.M.; Nehdi, M.L. Exploring mechanical and durability properties of ultra-high performance concrete incorporating various steel fiber lengths and dosages. *Constr. Build. Mater.* **2015**, *75*, 429–441. [CrossRef]

5. Fernandez Cánovas, M. *Hormigón*; Colegio de Ingenieros de Caminos, Canales y Puertos: Madrid, Spain, 2013.

6. Ambily, P.; Umarani, C.; Ravisankar, K.; Prem, P.R.; Bharatkumar, B.; Iyer, N.R. Studies on ultra high performance concrete incorporating copper slag as fine aggregate. *Constr. Build. Mater.* **2015**, *77*, 233–240. [CrossRef]

7. Zhu, Z.; Li, B.; Zhou, M. The Influences of Iron Ore Tailings as Fine Aggregate on the Strength of Ultra-High Performance Concrete. *Adv. Mater. Sci. Eng.* **2015**, *2015*, 412878. [CrossRef]

8. Zhao, S.; Fan, J.; Sun, W. Utilization of iron ore tailings as fine aggregate in ultra-high performance concrete. *Constr. Build. Mater.* **2014**, *50*, 540–548. [CrossRef]

9. Yang, S.; Millard, S.; Soutsos, M.; Barnett, S.; Le, T. Influence of aggregate and curing regime on the mechanical properties of ultra-high performance fibre reinforced concrete (UHPFRC). *Constr. Build. Mater.* **2009**, *23*, 2291–2298. [CrossRef]

10. Soliman, N.; Tagnit-Hamou, A. Using glass sand as an alternative for quartz sand in UHPC. *Constr. Build. Mater.* **2017**, *145*, 243–252. [CrossRef]

11. Zegardlo, B.; Szeląg, M.; Ogrodnik, P. Ultra-high strength concrete made with recycled aggregate from sanitary ceramic wastes—The method of production and the interfacial transition zone. *Constr. Build. Mater.* **2016**, *122*, 736–742. [CrossRef]

12. Singh, S.; Nagar, R.; Agrawal, V.; Rana, A.; Tiwari, A. Sustainable utilization of granite cutting waste in high strength concrete. *J. Clean. Prod.* **2016**, *116*, 223–235. [CrossRef]

13. Boadella, I.L.; Gayarre, F.L.; González, J.M.S.; Gómez-Soberón, J.M.; Pérez, C.L.-C.; Serrano-Lopez, M.A.; De Brito, J. The Influence of Granite Cutting Waste on The Properties of Ultra-High Performance Concrete. *Materials* **2019**, *12*, 634. [CrossRef] [PubMed]

14. Kala, D. Effect of granite Powder on Strength Properties of Concret. *Int. J. Eng. Sci.* **2013**, *2*, 36–50.

15. Estadísticas Sobre Residuos—Statistics Explained. 2018. Available online: http://ec.europa.eu/eurostat/statistics-explained/index.php?title=Waste_statistics/es (accessed on 27 May 2020).

16. Estadísticas Sobre la Recogida y Tratamiento de Residuos. Encuesta Sobre Generación de Residuos en la Industria. 2014. Available online: http://www.ine.es/prensa/np1004.pdf (accessed on 27 May 2020).

17. Yu, R.; Spiesz, P.; Brouwers, H. Mix design and properties assessment of Ultra-High Performance Fibre Reinforced Concrete (UHPFRC). *Cem. Concr. Res.* **2014**, *56*, 29–39. [CrossRef]

18. EN 12390-1. *Testing Hardened Concrete, Part 1: Shape, Dimensions and other Requirements for Specimens and Moulds*; AENOR: Madrid, Spain, 2013. Available online: https://www.une.org/ (accessed on 27 May 2020).

19. UNE-EN 12390-2.*Testing Hardened Concrete, Part 2: Making and Curing Specimens for Strength Test*; AENOR: Madrid, Spain, 2009. Available online: https://www.une.org/ (accessed on 27 May 2020).

20. NF P 18-470. Bétons—Bétons Fibrés à Ultra Hautes Performances—Spécification, Performance, Production et Conformité, AFNOR: France. 2016. Available online: https://m.boutique.afnor.org/ (accessed on 27 May 2020).

21. EN 12390-7. *Testing Hardened Concrete, Part 7: Density of Hardened Concrete*; AENOR: Madrid, Spain, 2009; Available online: https://www.une.org/ (accessed on 27 May 2020).

22. EN 12390-13. *Testing Hardened Concrete, Part 13: Determination of Secant Modulus of Elasticity in Compression*; AENOR: Madrid, Spain, 2014; Available online: https://www.une.org/ (accessed on 27 May 2020).

23. EN 12390-3. *Testing Hardened Concrete, Part 3: Compressive Strength of Test Specimens*; AENOR: Madrid, Spain, 2009; Available online: https://www.une.org/ (accessed on 27 May 2020).

24. Vogt, C. Ultrafine Particles in Concrete: Influence of Ultrafine Particles on Concrete Properties and Application to Concrete Mix Design. Ph.D. Thesis, School of Architecture and the Built Environment Division of Concrete Structures, Stockholm, Sweden, 2010.

25. Pyo, S.; Tafesse, M.; Kim, B.-J.; Kim, H.-K. Effects of quartz-based mine tailings on characteristics and leaching behavior of ultra-high performance concrete. *Constr. Build. Mater.* **2018**, *166*, 110–117. [CrossRef]

26. Kou, S.-C.; Xing, F. The Effect of Recycled Glass Powder and Reject Fly Ash on the Mechanical Properties of Fibre-Reinforced Ultrahigh Performance Concrete. *Adv. Mater. Sci. Eng.* **2012**, *2012*, 263243. [CrossRef]

27. Albarwary, I.H.M.; AlDoski, Z.N.S.; Askar, L.K. Effect of Aggregate Maximum Size upon Compressive Strength of Concrete. *J. Univ. Duhok* **2017**, *20*, 790–797. [CrossRef]

28. Raheem, A.H.A.; Mahdy, M.; Mashaly, A.A. Mechanical and fracture mechanics properties of ultra-high-performance concrete. *Constr. Build. Mater.* **2019**, *213*, 561–566. [CrossRef]

29. Ibrahim, M.A.; Farhat, M.; Issa, M.A.; Hasse, J.A. Effect of Material Constituents on Mechanical and Fracture Mechanics Properties of Ultra-High-Performance Concrete. *ACI Mater. J.* **2017**, *114*, 453–465. [CrossRef]

30. Alsalman, A.; Dang, C.; Prinz, G.S.; Hale, W.M. Evaluation of modulus of elasticity of ultra-high performance concrete. *Constr. Build. Mater.* **2017**, *153*, 918–928. [CrossRef]

31. Gonzalez-Corominas, A.; Etxeberria, M. Properties of high performance concrete made with recycled fine ceramic and coarse mixed aggregates. *Constr. Build. Mater.* **2014**, *68*, 618–626. [CrossRef]

32. Taha, B.; Nounu, G. Properties of concrete contains mixed colour waste recycled glass as sand and cement replacement. *Constr. Build. Mater.* **2008**, *22*, 713–720. [CrossRef]

33. Martínez, J.A.L. Characterisation of the Tensile Behaviour of Uhpfrc by Means of Four-Point Bending Tests. Ph.D. Thesis, Universitat Politecnica de Valencia, Valencia, Spain, 2017.

Article

Complete Real-Scale Application of Recycled Aggregates in a Port Loading Platform in Huelva, Spain

Francisco Agrela [1,*], Francisco González-Gallardo [2], Julia Rosales [2], Javier Tavira [2], Jesús Ayuso [2] and Manuel Cabrera [2]

[1] Leonardo Da Vinci, Campus Rabanales, University of Córdoba, 14014 Córdoba, Spain
[2] Area of Construction Engineering, University of Cordoba, 14014 Cordoba, Spain; ir1gogaf@uco.es (F.G.-G.); jrosales@uco.es (J.R.); jtavira@ciccp.es (J.T.); ir1ayuje@uco.es (J.A.); manuel.cabrera@uco.es (M.C.)
* Correspondence: fagrela@uco.es

Received: 21 May 2020; Accepted: 8 June 2020; Published: 10 June 2020

Abstract: The application of recycled aggregates (RA) from construction and demolition waste and crushed concrete blocks is a very important challenge for the coming years from the environmental point of view, in order to reduce the exploitation of natural resources. In Spain, the use of these recycled materials in the construction of road bases and sub-bases is growing significantly. However, presently, there are few studies focused on the properties and behavior of RA in civil works such as road sections or seaport platforms. In this work, two types of RA were studied and used in a complete real-scale application. Firstly, recycled concrete aggregates (RCA) were applied in the granular base layer under bituminous superficial layers, and secondly mixed recycled aggregates (MRA) which contain a mix of ceramic, asphalt, and concrete particles were applied in the granular subbase layer, under the base layer made with RCA. Both RA were applied in a port loading platform in Huelva, applying a 100% recycling rate. This civil engineering work complied with the technical requirements of the current Spanish legislation required for the use of conventional aggregates. The environmental benefits of this work have been very relevant, and it should encourage the application of MRA and RCA in civil engineering works such as port platforms in a much more extended way. This is the first and documented real-scale application of RA to completely build the base and sub-base of a platform in the Huelva Port, Spain, replacing 100% of natural aggregates with recycled ones.

Keywords: seaport loading platform; recycled aggregates; civil infrastructures; structural granular layers; construction and demolition waste

---

1. Introduction

Several studies have demonstrated the feasibility of using recycled aggregates (RA) from construction and demolition waste (CDW) in structural road layers over last two decades [1,2]. Therefore, in recent years, the use of recycled aggregates in road and concrete applications has advanced considerably [3–8]. Although research focused on RA started in the 1990s, there are not enough specific technical regulations to promote the use of these types of aggregates [9]. This is why in Spain the Structural Concrete Instructions EHE-08 [10] are used for concrete construction, and the General Technical Specifications for Construction of Roads and Bridges [11] are used for road construction. Based on the proportions of the components of RA, De Brito, Agrela, and Silva (2019) [12], proposed a new classification of these that can be divided into:

- Recycled concrete aggregates (RCA): recycled aggregates which contain concrete particles larger than 85% or 90% of total dry mass. It would be possible to divide this category in two types, I and

II (Table 1), because there are some samples of RCA with more water absorption capacity and other different values in their properties.

- Mixed recycled aggregates (MRA): in this category there are three types, and only types I and II can be used in road layers (Table 1). These MRA-I and MRA-II are products obtained in the treatment of CDW, containing ceramic particles (Rb) between 15% and 40%.

Table 1. Classification of RA proposed for international application in road sections [12].

Types of RA Proposed	Composition *				Minimum Density (SSD)	Water Absorption Capacity (%)	Los Angeles Value (%)	Water-Soluble Sulphate (%)	Proposed Uses in Road Layers
	Rc+Ru (%)	Rb (%)	Ra (%)	Others (%)					
RCA-I	>90	<10	<5	<1	<0.7	<6	<35	<0.7	Concrete pavement, cement treated, or unbound granular subbases
RCA-II	>85	<15	<10	<3	<0.8	<8	<37	<0.8	Cement-treated or unbound granular subbases
MRA-I	>70	<30	<5	<5	<0.8	<8	<40	<0.8	Unbound granular subbases or capping of esplanades
MRA-II	>60	<40	<5	<8	<1.0	<12	<45	<1.0	Capping of esplanades or subgrades

* Rc: concrete and products thereof; Ru: unbound NA; Rb: ceramic bricks and tiles, calcium silicate masonry units; Ra: bituminous materials.

Other types are included in this proposal, MRA-III and RAA, but Table 1 only includes the most feasible RA to be used in road pavements, bases, sub-bases, and capping of esplanades, and additionally, showing the different applications of RA on roads.

An important value is that obtained with the Los Angeles Abrasion Test, which consists of producing an abrasive action by using standard steel balls that, when mixed with aggregates and rotate in a drum during a specific number of revolutions, also cause impact on aggregates. The percentage of wear of the aggregates due to friction with steel balls is determined and is known as the Los Angeles abrasion value, it is the difference in size before and after abrasion. The lower the percentage, the better the abrasion performance of the material.

RCA are the most widely used in granular unbound road layers. This material meets the requirements establish by the PG-3, usually showing the same characteristics as a natural aggregate [13].

Vegas et al. (2011) [14] conducted a pre-normative study to determine the physical, chemical, and mineralogical properties of MRA for their use in structural and unbound road layers. This study concluded that MRA with variable contents of concrete and ceramic particles could produce pozzolanic reactions, increasing the long-term mechanical properties of the structural road layer.

In recent years, RA have been applied in different studies for use on low-traffic roads and paths. MRA are generally used in road layers with reduced structural requirements, such as rural or pedestrian paths [13] and RCA are used in more demanding applications, such as on roads with a higher traffic volume [2]. Commonly, the different types of RA are partially applied. However, it is necessary to use these recycled materials in complete applications, in which all the conventional aggregates applied in structural layers can be replaced by RA to produce a significant reduction in environmental impact, maintaining the appropriate mechanical behavior and bearing capacity with respect to the conventional layers.

In this work, RCA and MRA have been applied in structural layers of seaport platforms. The RA have been applied over the entire surface of two layers the structural section. The design of port pavements is carried out according to their use and future activity, establishing the different calculation loads, their intensity of use and the type of existing subgrade. Specifically, in Spain, the recommendations in ROM 4.1-94 [15] and Standard 6.1-IC [16] are used for the design of port pavements.

In this work, two layers of 0.25 m of natural aggregates form of quarries were replaced by recycled aggregates for the port sub-base platform construction. Mechanical behavior, leaching properties, equivalent module, and deflections were studied to demonstrate the feasibility of using these materials in this type of application, which reduces the consumption of natural resources or raw materials, decreasing the ecological footprint.

2. Research Design

The current research aims to study and evaluate the physical, chemical, and mechanical behavior of recycled concrete aggregates (RA) for their use in a complete real-scale application in structural layers on a loading platform at the port "Ciudad de Palos" in Huelva. This platform is located in the area known as the outer harbor, on the left bank of the Ria Huelva (Huelva, Spain) and has an unpaved surface area of 28,500 m^2. An area of 8200 m^2 was chosen to develop this research (Figure 1).

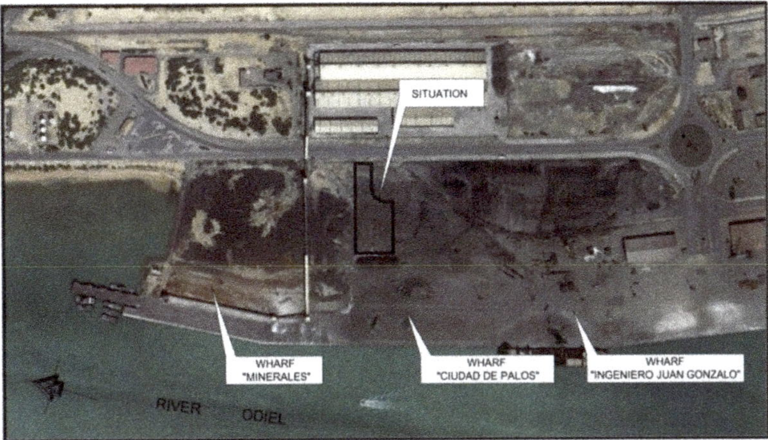

Figure 1. General view of the test area.

This study was divided into several stages, which are described below (Figure 2) in the chronological order that was followed during the research.

- Task 1. Study of MRA and RCA properties. In this initial phase, recycled aggregates (RCA and MRA) produced at the treatment plant were evaluated, studying the process of treatment applied at the recycling plant, the fines particles content, physic-chemical properties, or the potential contamination by leaching of the recycled aggregates. At the end of this phase, a solution was proposed to be applied in the real-scale application.
- Task 2. Mechanical behaviour of MRA and RCA was studied in order to establish the possibilities of using these recycled aggregates in the real-scale application. In this part of the research, compaction and bearing capacity were determined.
- Task 3. Execution of the loading platform at the Port of Huelva, specified in Figure 1, applying MRA and RCA in the base and sub-base structural layers of the platform.
- Task 4. Quality control of the execution processes, verifying that the specifications previously established and corroborated by laboratory tests were met. In this phase, density and water absorption in both granular structural layers of the loading platform were determined to confirm the correct application of MRA and RCA.
- Task 5. Medium- and long-term auscultation tests, using the application of an impact deflectometer. This study was carried out in the entire area of the loading platform, which was divided into 14 streets with a width of 5 m, numbered and delimited by measuring equipment. This allowed the study of the area of a longitudinal direction, obtaining an equivalent stretch of 1750 m in length.
- Task 6. Final evaluation of the application. After the execution of the application, and with a defined frequency, the basic parameters were evaluated.

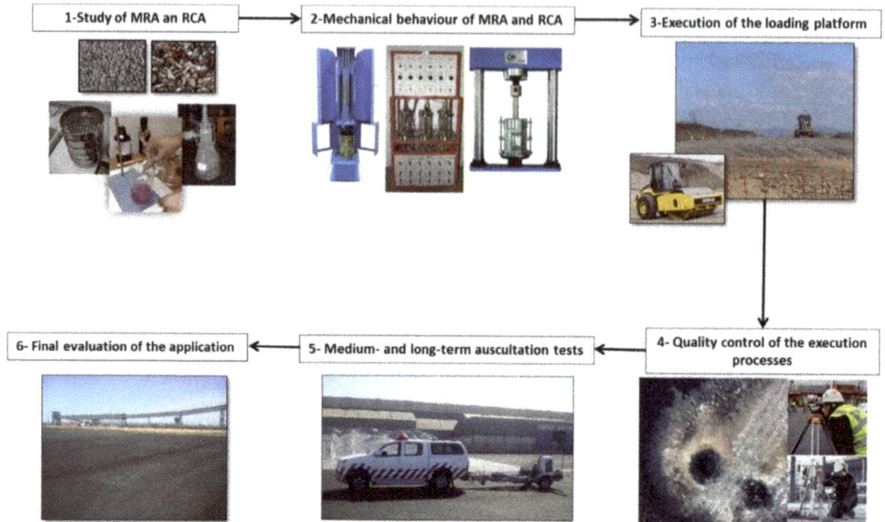

Figure 2. Research design.

This study has been developed in accordance with the technical specifications regulated by current legislation, PG3 and ROM-4.1/94, for both quality control of the materials, design and appropriate execution of the loading platform.

3. Recycled Materials

In this work, two types of recycled aggregates (RCA and MRA) were studied to be used as sub-base and base layers in the port loading platform.

Due to the appropriate mechanical properties of the recycled aggregates for their use in road layers compared to artificial gravel (artificial gravel (AG) is an aggregate of granular particles formed by stone material, it originates from the fragmentation of artificially crushed rocks), the authority of the Port of Huelva decided to modify a part of the structural section of the loading platform called "Ciudad de Palos". Two layers of AG were replaced by MRA and RCA. To carry out this replacement, characterization studies were initially performed on both RA, and the mechanical behavior was analyzed, comparing their results with AG (conventional materials in road construction).

On the one hand, mixed recycled aggregates (MRA) obtained from the treatment of construction and demolition waste (CDW) with variable contents of ceramic particles (Rb), in this case less than 30% were studied. The MRA used was composed of other different elements, including asphalt, particles with adhered mortar, natural aggregates, glass, gypsum, and other impurities. This material was used to make the sub-base in the loading platform section, and was classified as MRA-II, because the content of gypsum and other elements were reduced to less than 5%. It is very important to make an adequate selection at source of the CDW to achieve very low contents of different particles, concrete (Rc+Ru), ceramic (Rb), or even asphalt (Ra).

On the other hand, recycled concrete aggregates (RCA) were used and were composed of more than 85% particles of concrete and natural aggregates. This material was used to build the base of the loading platform. In this case it was classified as RCA-II, due to the content of natural aggregates and particles with adhered mortar, which are greater than 85% but less than 90% required to achieve the RCA-I type. It was possible to observe that gypsum particles and others such as plastic, glass, etc., presented less than 1%, resulting a suitable material with few impurities.

Both materials were obtained from different treatment processes at a recycling plant located in Huelva, Spain according to EN: 13242: 2003 (Aggregates for unbound and hydraulically bound materials for use in civil engineering).

Table 2 shows the results of the composition of recycled aggregates according to EN 933-11. In addition, they were classified in accordance with Table 1, proposed by de De Brito, Agrela, and Silva (2019) [12].

Table 2. Constituents of recycled aggregates according to EN 933-11

	Rc (%)	Ru (%)	Rb (%)	Ra (%)	Gypsum	Others (%)	Classification
MRA	25.2 72.1	46.9	25.6	4.6	0.8	0.7	MRA-I
RCA	21.02 88.58	67.56	8.3	2.6	0.2	0.3	RCA-II

The application of RCA and MRA on structural layers of the roads or port loading platforms could mean an improvement in its mechanical properties in the medium and long term [17]. Other authors demonstrated the self-cementation capacity of RCA because the non-hydrated cement particles existing in the old concrete particles, when entering in contact with water in their second cycle of life, could activate different chemical reactions of hydration, contributing to an improvement in the bearing capacity of the esplanade [18,19].

Table 3 shows the information concerning the properties tested in the laboratory according to the standards indicated.

Table 3. Physical and geometric properties of recycled aggregates

Properties	RCA	MRA	Required Limits—PG3	Test Method
Water-soluble sulphate content (SO$_3$%)	0.22	0.53	0.7	EN 1744-1
Acid-soluble sulphate content (SO$_3$%)	0.3	0.72	0.8	EN 1744-1
Total sulphate content (SO$_3$%)	0.55	0.88	2.5	EN 1744-1
Density-SSD (kg/dm^3)				EN 1097-6
0–4 mm	2.65	2.37	>2200	
4–31.5 mm	2.42	2.30	>2200	
Water absorption (%)				EN 1097-6
0–4 mm	3.31	9.09		
4–31.5 mm	5.59	10.79		
Plasticity	Non-plastic	Non-plastic	Non-plastic	EN ISO 17892-12
Particle size distribution (mm)	Percent passing (%)			EN 933-2
40 mm	100	100		
31.5 mm	91	93		
20 mm	76	80		
8 mm	59	60		
4 mm	49	48		
2 mm	40	40		
0.5 mm	23	12		
0.25 mm	14	3		
0.063 mm	6	1		
Flakiness index	15.3	12.4	<35	EN 933-3
Los Angeles coefficient	35.2	38	<40	EN 1097-2
Crushed and broken surfaces (%)	98.6	97.7	>40	EN 933-5
Sand equivalent	42	45	–	EN 933-8

PG-3 requires a maximum content of water-soluble sulfates for recycled aggregates from concrete demolitions (limit value 0.7%), with the results that both MRA and RCA do not exceed this value. As for the results of the tests of sand equivalent and slab index of both samples (RCA and MRA), the requirements established by the Spanish standard PG-3 [11] for fill of graded crushed aggregate are fulfilled.

The values of water absorption are slightly higher in the MRA samples with respect to RCA, which has been taken into account in the process of compaction of the layers.

On the other hand, the results obtained from the Los Angeles coefficient in both samples are slightly higher than the limit value of 35 that establishes the regulation for natural arid, being greater than the limit for recycled aggregates (<40).

The particle size distribution curves of both materials, RCA and MRA, are represented in Figure 3 together with the corresponding particle size distribution envelope of the artificial axes ZA-0/20 (granular mixtures) as indicated in PG-3 [11]. It can be seen that these curves present a particle size distribution within limits.

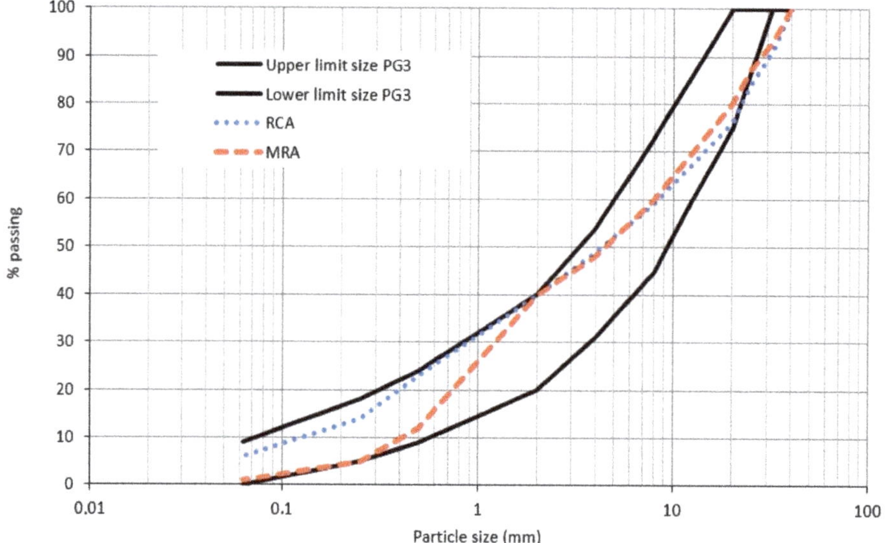

Figure 3. Particles' size distribution curves compared with the particle size distribution limits.

Based on the results obtained, RCA present better chemical and physical properties than MRA as it is usual. In general, both RA have properties similar to AG named in the Spanish standard as ZA-0/20.

4. Laboratory Tests and Results

4.1. Proctor Test/Moisture–Density Relationship

Two different types of tests were performed to identify the relationship between density and humidity of the different levels. For the subgrade we considered the standard Proctor test (PN), while for the subbase and base we used the modified Proctor test (PM).

The subgrade was constructed with selected soil, being necessary to obtain in situ 100% or more of dry density standard Proctor determined in laboratory according to UNE 103500:1994. The data for optimum moisture and maximum dry density were 5.5% and 1.79 ton/m^3 respectively.

The modified Proctor (MP) was performed according to UNE 103501:1994. This test is similar to the standard Proctor but in this case five layers of the soil were compacted in the modified Proctor mold, applying a higher compaction energy.

The data for maximum dry density and optimum moisture are shown in Figure 4.

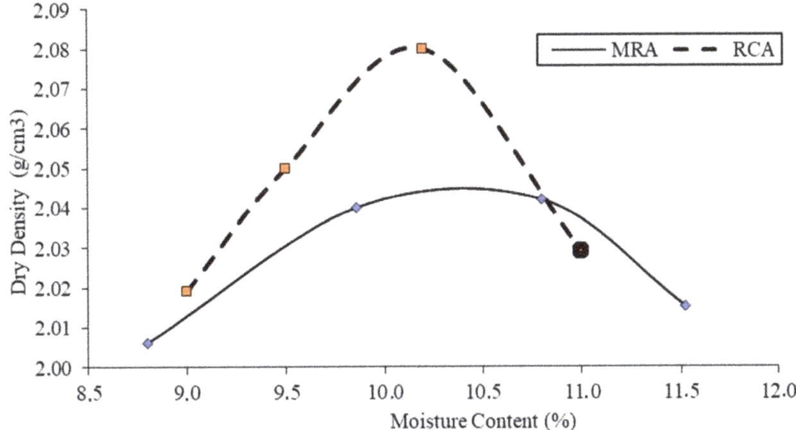

Figure 4. Moisture—dry density ratio.

The maximum density and optimum moisture content obtained for the layer built with RCA were 2.08 ton/m^3 and 10.2% respectively. The MRA series presented a lower density than the RCA (2.04 ton/m^3 and optimum humidity of 10.5%) samples mainly because the MRA showed a lower value of saturated dry surface density with respect to the RCA as shown in Table 2, according to Poon and Chan (2006) [1].

The high porosity found in MRA, especially in the mortar and ceramic materials, is the reason why the optimum moisture content of compaction is slightly higher than in the RCA series.

4.2. California Bearing Ratio (CBR)

The CBR method measures the shear strength of a soil or aggregate compacted in a laboratory under controlled conditions of density and moisture and is used to evaluate the bearing capacity of the soil as subgrade, subbase, and base of a port and a road pavements as well as land classification. The CBR value was determined according to EN 13286-47: 2012, and this test depends on moisture, density, and overload conditions.

Table 4 shows the results obtained in CBR tests for the three materials studied.

Table 4. Results of experimental tests performed in the laboratory

Properties	Subgrade	MRA	RCA	Test Method
Modified Proctor				UNE-103501
Maximun dry density (kg/dm^3)	–	2.04	2.08	
Optimum moisture content (%)	–	10.80	9.77	
Normal Proctor				UNE 103500
Maximun dry density (kg/dm^3)	1.79	–	–	
Optimum moisture content (%)	5.5	–	–	
C.B.R. index	28	50	74	EN 13286-47

From the obtained values, the subgrade is classified as an E3 type (very good) with selected soils with CBR > 20.

The values obtained from the CBR index for both materials (RCA and MRA) are considered suitable for granular layers of pavements.

4.3. Triaxial Test

From this test method, resistant parameters of fine-grained soils and aggregates are obtained as well as the stress–strain ratio through the determination of confined compressive strength and shear stress.

The sample is first hydrostatically loaded until the confinement pressure is reached, then an axial load is applied which increases progressively until the sample breaks. The containment pressure is kept constant.

In this case, the triaxial tests have been carried out according to EN ISO 17892-9:2018, with four cycles of loading and unloading and a confining pressure of 55 KPa for all the cases, which have allowed for determination of the resistance and the deformation parameters of RCA and MRA.

For the MRA sample, the loading–unloading steps are carried out with maximum and minimum values of load deviation of 290 and 25 KPa respectively, corresponding to 75% and 6% of the maximum deviation of breakage (383 KPa) previously determined with other test samples.

In the same way, we proceed with the RCA samples, being the maximum and minimum values of load deviation of 483 and 30 KPa, corresponding to 67% and 4% of the maximum deviation of breakage (721 KPa). These results obtained for the MRA and RCA are shown in Figure 5.

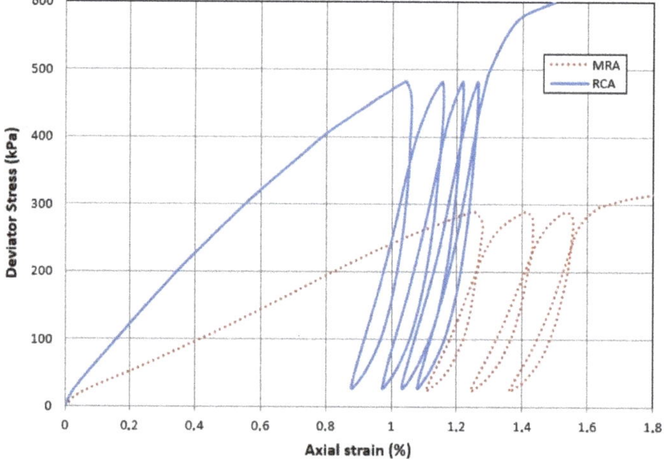

Figure 5. Strain–stress curves.

Table 5 shows the secant deformation modules obtained for the MRA and RCA samples. It can be seen that the deformation modules remain practically constant from the second load cycle, there being a ratio of 5 and 4 respectively with the first load stage.

Table 5. Results of experimental triaxial tests

Load Stage	MRA	RCA
	Secant Modulus of Deformation (MPa) (100–200 KPa)	Secant Modulus of Deformation (MPa) (150–300 KPa)
1st	24.4	48.1
2st	122.6	193.9
3st	117.8	208.3
4st	126.6	214.3

On the other hand, greater resistance and rigidity was observed in the RCA samples with respect to the MRA.

4.4. Leaching Test EN 12457-4:2004

An assessment procedure was carried out for heavy metals and inorganic anions present in RCA and MRA in accordance with the EU Landfill Directive. The test procedure consisted of a compliance test by leaching 90 g of dry sample in deionized water with a liquid/solid ratio of 10 (L/S = 10). The two materials (RCA and MRA) were classified according to the limits set by the Landfill Directive 2003/33/EC.

All samples were classified as inert, as can be seen in Table 6 compared to the legal limits.

Table 6. Leachate concentrations (mg/kg) for RCA and RMA by EN 12457-4

L/S = 10	RCA	MRA	Limit Values Directive 2003/33/EC
	(mg/kg)	(mg/kg)	Inert (mg/kg)
Cr	0.134	0.248	0.50
Ni	0.003	0.007	0.40
Cu	0.015	0.044	2.00
Zn	0.001	0.028	4.00
As	0.000	0.002	0.50
Se	0.000	0.008	0.10
Mo	0.056	0.112	0.50
Cd	0.000	0.000	0.04
Sb	0.000	0.021	0.06
Ba	0.076	0.185	20.00
Hg	0.000	0.000	0.01
Pb	0.000	0.000	0.50
Sulfates	398.0	863.0	1000

RCA and MRA showed very low values of heavy metals and sulfates in the leaching test. The low sulfate content, which is not usual in recycled aggregates, is due to a selection at the source in the treatment plant that reduces very substantially the gypsum content in the recycled aggregate.

5. Real-Scale Application, Test Program, and Results

5.1. Description of the Sections

A surface semiflexible was designed, formed by granular layers and bituminous materials.

Granular layers are composed of a layer of 25 cm of mixed recycled aggregate (MRA) and 25 cm of recycled concrete aggregate (RCA).

The bituminous layer have a total thickness of 20 cm composed of a bituminous base 8 cm (AC basis G 22), an intermediate layer of 8 cm (AC bin 22 S) and a surface layer of 4 cm (AC 16 surf S) (Figure 6).

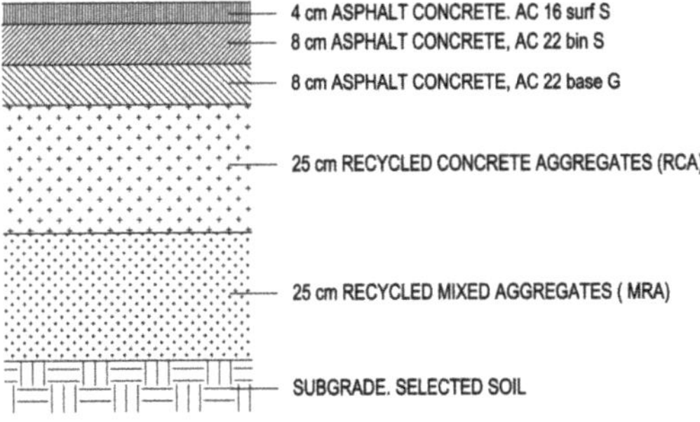

Figure 6. Cross sections of the port pavement.

5.2. Compaction of Layers

Nuclear density equipment was used to determine the density and moisture content of the soil in situ, according to ASTM D-6938. This test method is a rapid non-destructive technique that is used as acceptance tests of compacted soil layers as long as the material under test is homogeneous.

Multiple measurements of density and moisture content were taken in surface of subgrade and granular layers (10 in subgrade, 40 in MRA, and 21 in RCA) to perform a statistical analysis of the results.

The mean and standard deviation (SD) values of each level are included in Table 7.

Table 7. In situ assessments of density and moisture

Properties	Subgrade	MRA	RCA
Density (kg/dm^3)			
Mean	1.79	2.08	2.12
SD	20	50	60
Compaction (%)			
Mean	100.0	101.7	102.1
SD	1.11	2.59	2.95
Moisture content (%)			
Mean	5.81	10.75	10.30
SD	1.75	0.95	0.80

5.3. Deflection Measurements

Impact deflectometer testing is a method used to evaluate the support capacity of the subgrade, granular, and bituminous layers. The test involves the application of a dynamic load on a damper system that transmits to circular plate resting on the surface, allowing measurements of the deflections produced on its surface through several sensors aligned with the plate and adequate computer equipment according to ASTM D4694 [19].

The Dynatest HWD 8081 impact deflectometer was used for the development of the work according to General Technical Specifications for High-Performance Dynamic Monitoring Test [20]. This equipment records the vertical deflection of the surface under the point of application of the load and in six other points located at 30, 45, 60, 90, 120, and 150 cm respectively.

For the development of the work the study area has been divided into streets of 5 m wide, obtaining an equivalent length of 1750 m. Deflection measures have been taken every 10 m in all the layers that have been defined in Section 5.1 as long as it has been possible to measure them, taking into account the coordination of the equipment (deflectometers). At all test points, a previous impact of settlement has been made before the test impacts.

In Figure 7 a full aerial image of the experimental platform is presented, and Universal Transverse Mercator (UTM) coordinates at end and start of lanes are shown.

Deflection correction coefficients have been applied for the humidity of the subgrade and pavement temperature, according to the corrections established in Regulation 6.3-IC of the Ministry of Development and Standard NLT-338/07.

The tests were carried out on the finished surfaces (compacted and refined) of the subgrade, granular, and bituminous layers. Six months after completion of the work, in November 2014, a new measurement of the deflections on the tread layer was made to study its evolution.

Once the measurement work was completed, the results of the deflections in all its layers were achieved over a length of 710 m. Figures 8 and 9 represent these measurements, and in Table 7 the mean values and standard deviation are included for each loading platform layer.

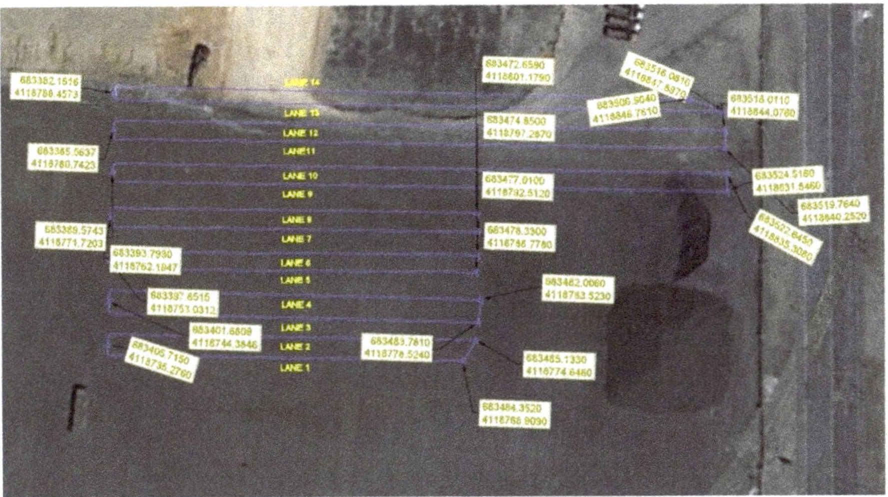

Figure 7. Lanes of the experimental platform.

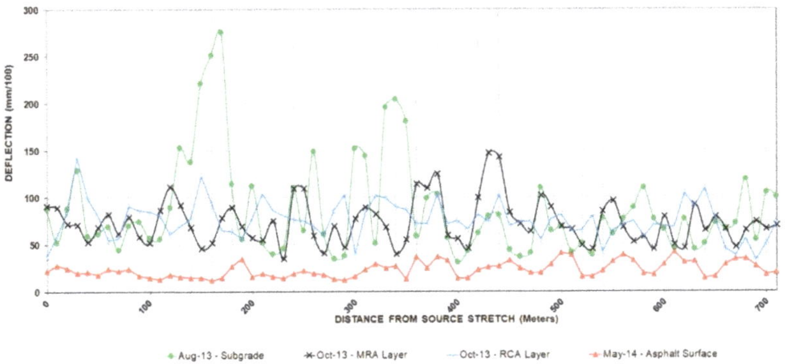

Figure 8. Deflection measurement.

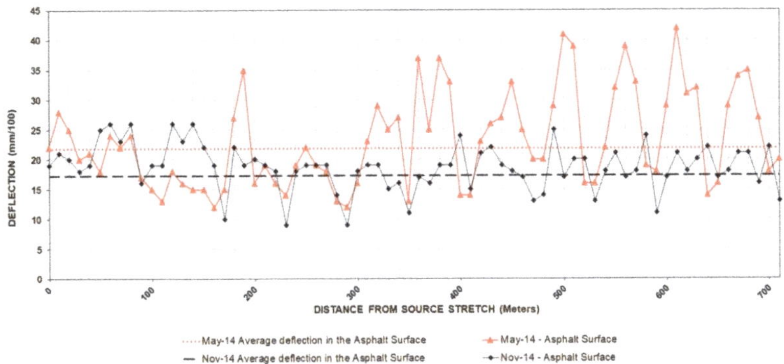

Figure 9. Deflection measurement in the asphalt concrete layers.

It can be seen that there is an improvement in the evolution of asphalt surface behavior: the mean deflection obtained in November is 23.5% less than compared to May, as well as an increase in uniformity as can be inferred from the standard deviation values.

According to the standard 6.3 IC: Rehabilitation of pavements, the completely finished pavement meets the requirements of homogeneous section and uniform behavior, obtaining an average value of deflections of 0.169 mm and a sample standard deviation of 0.045. On the other hand, the characteristic deflection obtained is 0.258 mm, which is equivalent to a 97.5% probability that the deflection will not be exceeded in the study section.

5.4. Equivalent and Inverse Modulus

Two different methodologies have been used to evaluating the structural capacities of the pavement from the deflections measured on the surface of the different layers: the equivalent modulus and modulus obtained by inverse calculation method.

On the one hand, the equivalent module Ev, which is obtained from the measurement of deflection that has been made of each layer and it represents an equivalent module value of all existing layers below the analyzed layer.

The bearing capacities of the various layers have been evaluated by the parameters of deflections and surface modules obtained with the first geophone according to the formulation proposed by Brown (1996).

$$E_v = \frac{2pa\left(1 - v^2\right)}{d} \tag{1}$$

where:

E_v: Equivalent modulus of the entire pavement system beneath the load plate
a: Radius of the FWD plate, 225 mm on granular layers and 150 mm on asphalt concrete.
pa: Pressure of FWD impact load under the load plate, 246.KPa on granular layers and 693.21 KPa on asphalt concrete.
d: Deflection at 0 mm of center of the FWD plate.
v: Poisson's ratio.

The results of equivalent modulus in different layers were included in Table 8. The relative standard deviation (RSD) has been considered to compare the variability of the results between the different layers and dates (Figures 10–12).

On the other hand, inverse calculation method is a backcalculation procedure commonly used to estimate the pavement layer moduli based on the non-destructive with impact deflectometer tests.

An elastic multilayer software called BISAR [21] was used. Theories of Burmister (1945) [22], and Acum and Fox (1951) [23], and a solution to determine stress and strain of Schiffman (1962) [24] are used in the algorithm of this software. Theoretical deflection was calculated for each layer and section using theoretical elastic modulus shown on Table 9.

Table 8. Statistical parameters of modulus equivalent, *Ev*

Properties	Subgrade (Aug-13)	MRA (Oct-13)	RCA (Oct-13)	Asphalt Surface (May-14)	Asphalt Surface (Nov-14)
Equivalent modulus					
Mean (MPa)	140.3	145.6	152.2	927.1	1178.5
RSD (%)	45.5	33.6	41.7	30.0	28.5

Table 9. Theoretical mechanical property and statistical parameter results

Layer	Date	Mean (MPa)	RSD (%)	Eck (MPa)	Thickness (m)	Theoretical Elastic Modulus (MPa)	Theoretical Deflection m/10^{-6}
Asphalt Layer	May-14	6480	16.5	5409	0.20	5600	410
	Nov-14	6364	16.0	5346			
Granular Base (RCA)	May-14	556	23.6	424	0.25	350	680
	Nov-14	652	20.7	516			
Subbase (MRA)	May-14	356	18.5	290	0.25	165	1110
	Nov-14	380	18.2	311			
Subgrade	May-14	305	19.3	246	2	55	
	Nov-14	240	19.6	192			

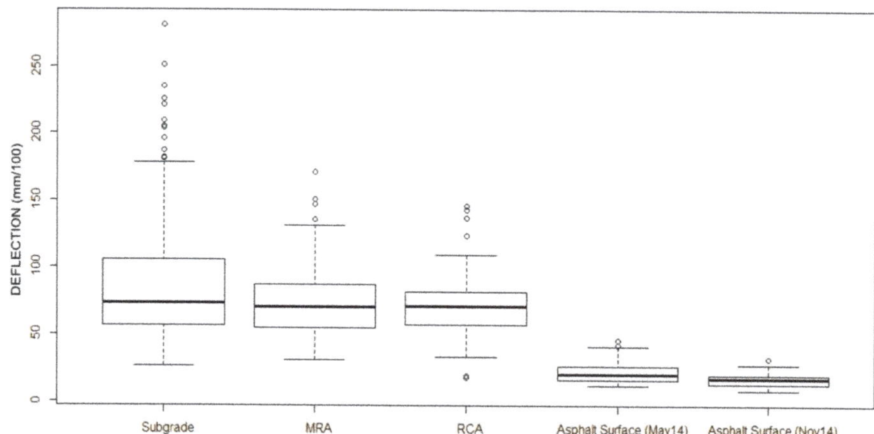

Figure 10. Boxplots showing deflection measurement in different layers.

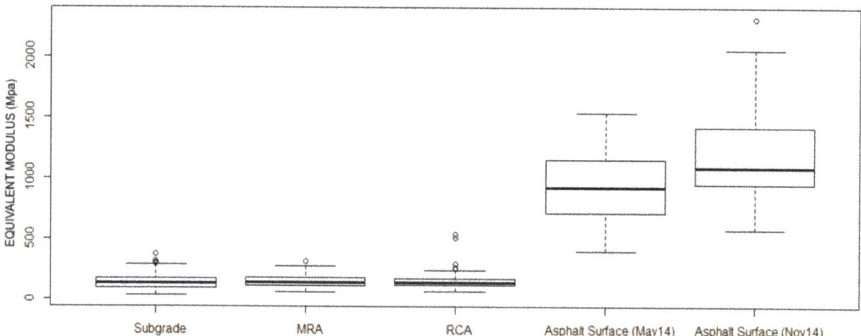

Figure 11. Boxplots showing evolution of the equivalent modulus in different layers.

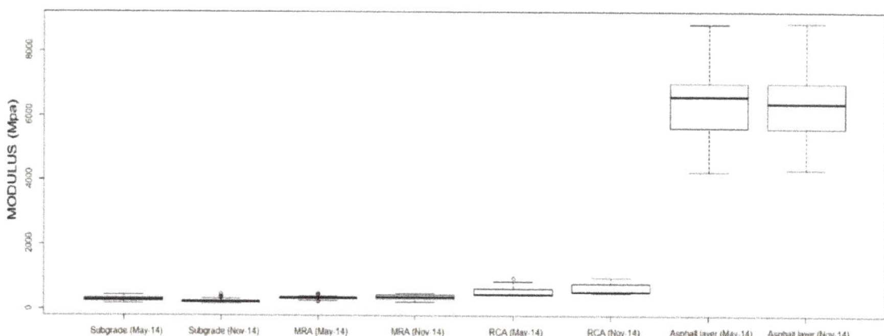

Figure 12. Boxplot of moduli back calculation FWD tests, 2014.

There is a significant evolution of the equivalent modulus on the completely finished asphalt surface: the mean of equivalent modulus obtained in November is 27.1% higher than those determined in May, as well as an increase in uniformity as can be inferred from the relative standard deviation.

Pavement instruction of Andalusia [11] shows a correlation between CBR and modulus, so subgrade moduli was determined from the values of the CBR tests of the existing soil. According to Huang (1993) [25] asphalt concrete have a 0.33 and granular layers and subgrade a 0.35 Poisson ratio.

Elastic moduli of the four layers shown were obtained using Evercalc [26]. The software recalculates the modules by an iterative process comparing the average data with the theoretical data. The process will run until it finds convergence with a limited error.

Table 9 shows the values of the modules obtained from the indirect calculation method and the results of some statistical parameters.

According to the results offered in Table 9, a decrease in the values obtained from the elastic modulus in the subgrade can be seen. A higher amount of rain registered in the area before the November test could be the reason for the decrease in the bearing capacity of this level. On the other hand, an increase of the elastic moduli of the MRA and RCA layers was observed (6.7% and 17.3% respectively), probably due to the cementation of some cement compounds existing in the materials of these layers. The elastic modulus of the asphalt layer has not changed significantly, although there is a slight increase in uniformity as can be seen in RSD values.

For all of the above, it can be concluded that the improvement of the equivalent modulus observed in the month of November was due to the increase in the bearing capacity of the MRA layer and especially the RCA layer.

Finally, it is necessary to emphasize the good bearing capacity that was obtained with the RCA and MRA layers. Effectively, RCA mean moduli (652 MPa) are greater than crushed stone moduli obtained from natural aggregates (500 MPa) and MRA mean moduli (380 MPa) are greater than natural selected soil (250 MPa) (Spanish general technical specifications for road construction, 2004).

6. Life Cycle Assessment

A study of the environmental impact caused by the production of each of the aggregates used in this study (MRA and RCA) was carried out by means of a life cycle assessment (LCA).

The aim was to check the environmental impact caused by the use of recycled aggregates instead of the methodology used in a traditional construction system. A comparative analysis of the production system used to produce MRA, RCA, and AG was carried out.

The environmental impact caused by the production process of each aggregate was quantified according to ISO 14040 (2006) and ISO 14044 (2006).

The LCA analysis was performed using the SimaPro® 8.0.2 software application and processed using the CML-IA method [27]. The evaluation methodology EN 15804 + A1 (CEN 2013) was used for the study, as it is the most suitable methodology in relation to sustainability for building products and services.

The boundaries considered in each of the production systems of the three types of aggregates analyzed are shown in Figure 13. The infrastructures and production systems of the quarry and the RCA and MRA recycling plants have been considered. The equipment and machinery necessary for production are known.

The life cycle impact assessment methodology combines 11 impact categories. These impact categories are: abiotic depletion of elements, abiotic depletion of fossil fuels, global warming, ozone layer depletion, human toxicity, fresh water aquatic ecotox, marine aquatic ecotoxicity, terrestrial ecotoxicity, photochemical oxidation, acidification, and eutrophication.

According to the different aggregates production processes shown in Figure 13, an inventory phase was carried out in which inputs (energy and raw material) and outputs (products, co-products, wastes, and emissions) of the particular production system of each aggregate were considered. The Ecoinvent database v.3.01 (allocation) [28] was used as secondary data for generic materials, energy, and transport.

Table 10 shows the characteristics of the equipment used in aggregate production.

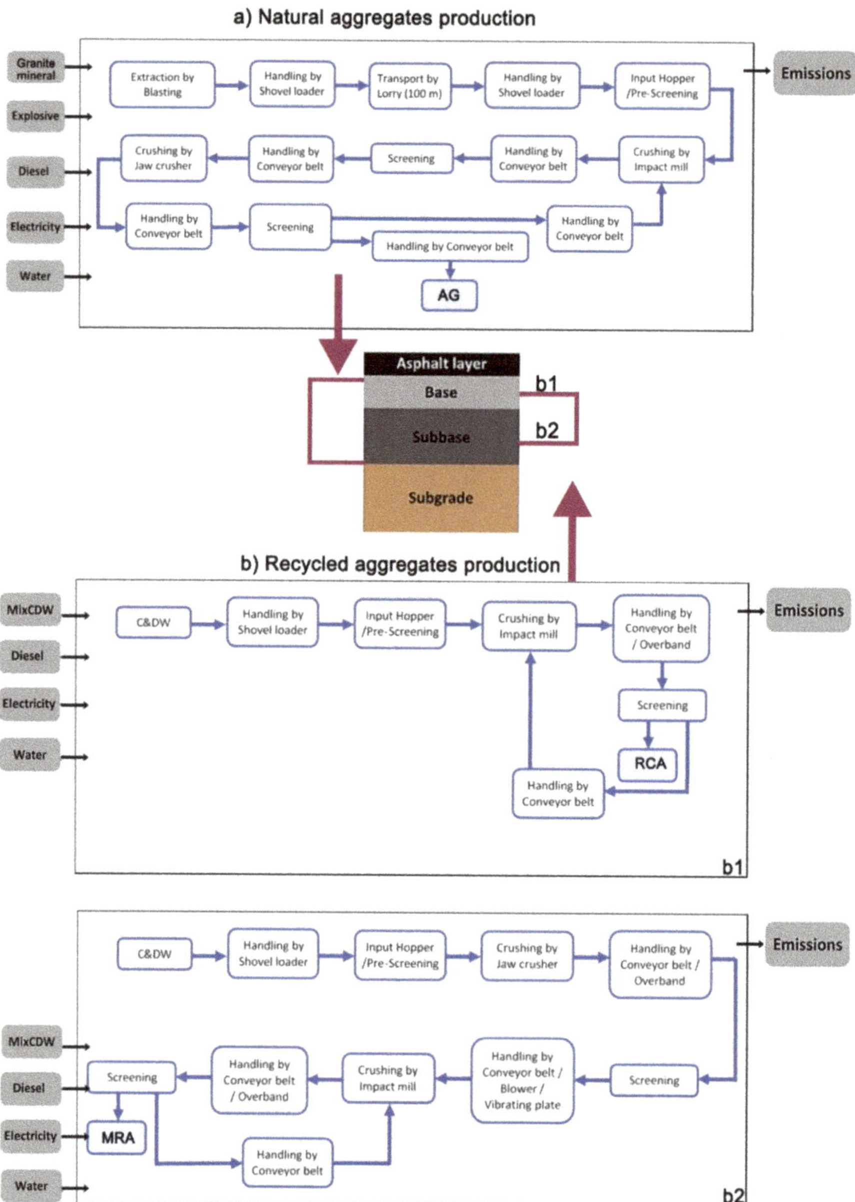

Figure 13. System boundaries for production of artificial gravel (AG), production of concrete and mix CDW recycled aggregate (RCA) and (MRA).

Table 10. Specifications of AG, RCA, and MRA production equipment

Process		Equipment	Amount	Power (kW)	Production (t/h)	Distance (km)
AG	Handling	Shovel loader	2	–	32.318	–
	Transport	Lorry 28 t	1	–	–	0.3
	Handling	Conveyor belt, 10 m	3	8	108.53	–
		Conveyor belt, 25 m	2	20	166.91	–
	Screening	Vibrating screen	3	18.5	225	–
	Crushing	Impact mill	1	122.06	400	–
		Jaw crusher	1	203.12	400	–
RCA	Handling	Shovel loader	1	–	32.318	–
	Crushing	Impact mill	1	75	250	–
	Handling	Conveyor belt, 5 m	2	4	112.1	–
		Overband	1	3.68	114.4	–
	Screening	Vibrating screen	1	22.08	250	–
MRA	Handling	Shovel loader	1	–	32.318	–
	Screening	Vibrating screen	3	22.08	250	–
		Vibrating plate	1	3	80	–
	Crushing	Jaw crusher	1	160	325	–
		Impact mill	1	75	250	–
	Handling	Conveyor belt, 5 m	3	4	112.2	–
		Conveyor belt, 10 m	1	7.36	112.2	–
		Overband	2	3.68	114.4	–
		Blower	1	14	144.74	–

Through the LCA, a comparison of the environmental impact caused in the production of the different aggregates was carried out. The results of the characterization for the production of 1 t of aggregates are shown in Table 11. In addition, the results of variation compared to the highest value in each of the impact indicators are shown (Figure 14).

Table 11. Characterization results of AG, RCA, and MRA (per 1 t)

Impact Category	Units	AG	MRA	(Δ%)	RCA	(Δ%)
Abiotic depletion	kg Sb eq.	2.32×10^{-6}	8.63×10^{-7}	−62.9	6.85×10^{-7}	−70.5
Abiotic depletion (fossil fuels)	MJ	23.9	13.9	−41.6	11.1	−53.3
Global warming (GWP100a)	kg CO_2 eq.	2.02	1.19	−41.1	0.938	−53.6
Ozone layer depletion (ODP)	kg CFC-11 eq.	1.99×10^{-7}	1.08×10^{-7}	−45.9	8.79×10^{-8}	−55.8
Human toxicity	kg 1,4-DB eq.	0.761	0.451	−40.7	0.345	−54.7
Fresh water aquatic ecotoxicity	kg 1,4-DB eq.	0.677	0.404	−40.2	0.302	−55.3
Marine aquatic ecotoxicity	kg 1,4-DB eq.	2.58×10^3	1.54×10^3	−40.3	1.14×10^3	−55.8
Terrestrial ecotoxicity	kg 1,4-DB eq.	5.13×10^{-3}	3.09×10^{-3}	−39.6	2.32×10^{-3}	−54.7
Photochemical oxidation	kg C_2H_4 eq.	4.43×10^{-4}	2.53×10^{-4}	−42.9	1.96×10^{-4}	−55.7
Acidification	kg SO_2 eq.	1.24×10^{-2}	6.95×10^{-3}	−43.8	5.49×10^{-3}	−55.6
Eutrophication	kg PO_4 eq.	3.82×10^{-3}	2.24×10^{-3}	−41.4	1.76×10^{-3}	−53.9

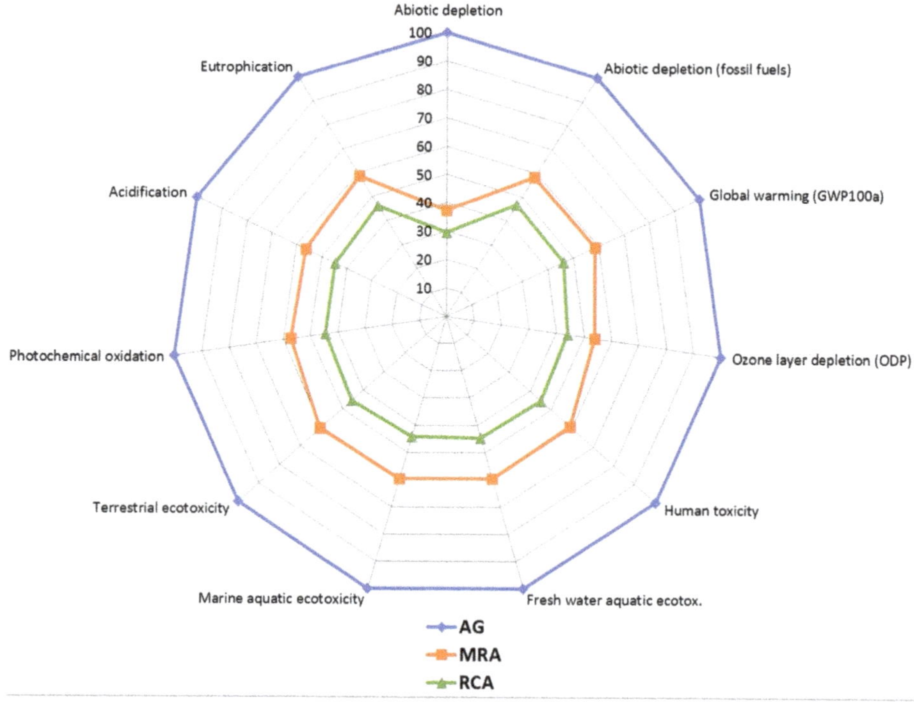

Figure 14. Comparative graph of aggregates production impact assessment.

AG production resulted in a greater impact in all categories evaluated. MRA production reduced impacts such as global warming or ozone layer depletion by 41.1% and 45.9% respectively with respect to AG production. This reduction was even greater in the production of RCA, where values for these same indicators were obtained of 53.6% and 55.8%.

The increase in the environmental impact caused by the production of AG with respect to recycled aggregates could be due mainly to the greater number of processes involved in their production, to the processes of extraction and consumption of natural resources.

This analysis shows the improved environmental impact caused by the production of recycled aggregates [29,30].

To determine the environmental loads derived from the real construction of the seaport loading platform by applying MRA and RCA in the base and sub-base section (Figure 6), the CO_2 emissions generated in the construction of the 8200 m^2 were analyzed (Table 12). The study was carried out by applying 25 cm of MRA and 25 cm of RCA compared to a traditional construction that would correspond to 50 cm of AG applied to the entire surface of the port's platform.

Table 12. CO_2 emissions generated by construction of 8200 m^2 port loading platform with RCA and MRA compared to traditional construction using AG

Impact Category	Units	Traditional Construction 50 cm (AG)	25 cm of Mixed Recycled Aggregate (MRA) and 25 cm of Recycled Concrete Aggregate (RCA)	(Δ%)
Global warming (GWP100a)	kg CO_2 eq.	1.92×10^4	9.11×10^3	−52.6%

The highest value for CO_2 emissions was for AG application in the construction of the port's loading platform. A traditional construction would generate 19.19 tons of CO_2. However, the application of RMA and RCA in the project section reduced CO_2 emissions by 52.6%. This demonstrates the environmental benefit generated by using recycled aggregates in construction, which cause less impact, both in production and in its later implementation.

7. Conclusions

This paper provides the results of research on two recycled materials (MRA and RCA) with different composition to be used as subbases and bases of port pavements, obtaining the following conclusions:

- The recycled materials analyzed, although they have somewhat lower densities than natural materials, especially the higher content of ceramic material (MRA), present physical, chemical, and mechanical properties very suitable for use as a subbase and bases for port platforms.
- The process of compaction of the layers with recycled materials requires pre-wetting and a more demanding control than with the use of natural materials.
- An important factor for using recycled aggregates is the content of acid-soluble sulfates. A percentage lower than 1% in SO_3 is recommended for layers without cement treatment.
- Regarding environmental issues, the leaching test classifies the RCA and MRA as inert materials.
- The layers of subbase and base with recycled materials investigated do not significantly improve the equivalent modulus of the subgrade when it has good properties (subgrade type E3), instead they contribute to providing homogeneity in their behavior against the deflectometer impact test.
- The recycled materials that have been used as a subbase layer (MRA) and base (RCA) have not only provided proper mechanical and deformational capabilities for use in the port pavement, but also an improvement in their properties over time.
- The production of MRA and RCA applied at the base and sub-base of the loading dock generates lower impacts than the production of AG as a result of the lower number of processes required for their treatment and production.
- The application of recycled materials (MRA and RCA) as base and sub-base layers, in addition to providing appropriate technical characteristics, leads to a large reduction in CO2 emissions in relation to the application of AG in these layers.

Therefore, it can be concluded that it is possible to use recycled aggregates in the construction of subbases and base layers of port pavement, providing an important environmental benefit, since it reduces the consumption of natural resources or raw materials, decreasing the ecological footprint.

Author Contributions: Conceptualization, M.C. and F.A.; methodology, F.G.-G.; software, J.R. and J.T.; validation, J.A., F.A; formal analysis, M.C. and F.G.-G.; investigation, F.A. and M.C.; data curation, F.G.-G. and M.C.; writing—original draft preparation, M.C. and F.G.-G.; writing—review and editing, M.C. and J.R.; visualization, J.A.; supervision, F.A. and M.C.; project administration, F.A.; funding acquisition, F.A. All authors have read and agreed to the published version of the manuscript."

Funding: This research received external and partial funding from Sando Company.

Acknowledgments: The authors want to thank Sando Company and the authors would also like to express gratitude for the help provided at all times in this work by the Port of Huelva, especially to Alfonso López Pazos, who is responsible for the infrastructure in the Port of Huelva.

Conflicts of Interest: The authors declare no conflict of interest.

References

1. Poon, C.S.; Chan, D. Feasible use of recycled concrete aggregates and crushed clay brick as unbound road sub-base. *Constr. Build. Mater.* **2006**, *20*, 578–585. [CrossRef]
2. Pérez, P.; Agrela, F.; Herrador, R.; Ordoñez, J. Application of cement-treated recycled materials in the construction of a section of road in Malaga, Spain. *Constr. Build. Mater.* **2013**, *44*, 593–599. [CrossRef]
3. Evangelista, L.; de Brito, J. Mechanical behaviour of concrete made with fine recycled concrete aggregates. *Cem. Concr. Compos.* **2007**, *29*, 397–401. [CrossRef]
4. Agrela, F.; Barbudo, A.; Ramírez, A.; Ayuso, J.; Carvajal, M.D.; Jiménez, J.R. Construction of road sections using mixed recycled aggregates treated with cement in Malaga, Spain. *Resour. Conserv. Recycl.* **2012**, *58*, 98–106. [CrossRef]
5. Silva, R.V.; Jiménez, J.; Agrela, F.; de Brito, J. Real-scale applications of recycled aggregate concrete. In *New Trends in Eco-efficient and Recycled Concrete*; Elsevier: Amsterdam, The Netherlands, 2019; pp. 573–589.
6. Silva, R.; de Brito, J.; Dhir, R. Use of recycled aggregates arising from construction and demolition waste in new construction applications. *J. Clean. Prod.* **2019**, *236*, 117629. [CrossRef]
7. Contrafatto, L.; Cosenza, R.; Barbagallo, R.; Ognibene, S. Use of recycled aggregates in road sub-base construction and concrete manufacturing. *Ann. Geophys.* **2018**, *61*, 223. [CrossRef]
8. Etxeberria, M.; Vázquez, E.; Marí, A.; Barra, M. Influence of amount of recycled coarse aggregates and production process on properties of recycled aggregate concrete. *Cem. Concr. Res.* **2007**, *37*, 735–742. [CrossRef]
9. Silva, R.V.; de Brito, J.; Dhir, R. Properties and composition of recycled aggregates from construction and demolition waste suitable for concrete production. *Constr. Build. Mater.* **2014**, *65*, 201–217. [CrossRef]
10. Ministerio de Fomento. *I. del Hormigón Estructural, I. EHE-08*; Ministerio de Fomento: Madrid, Spain, 2008.
11. Ministry of Development. *Order/FOM/2523/2014 Amending Certain Articles of the General Technical Specifications for Road and Bridge Works (PG-3), Relating to Basic Materials, Pavements, and Signs, Markings and Vehicle Restraint Systems*; (Orden/FOM/2523/2014 Por la Que se Actualizan Determinados Artículos del Pliego de Prescripciones Técnicas Generales para Obras de Carreteras y Puentes, Relativos a Materiales Básicos, a Firmes y Pavimentos, y a Señalización, Balizamiento y Sistemas de Contención de Vehículos); BOE No. 3; Ministry of Development: Madrid, Spain, 2015; pp. 584–1096. (In Spanish)
12. de Brito, J.; Agrela, F.; Silva, R.V. Legal regulations of recycled aggregate concrete in buildings and roads. In *New Trends in Eco-Efficient and Recycled Concrete*; Elsevier: Amsterdam, The Netherlands, 2019; pp. 509–526.
13. Jimenez, J.R.; Ayuso, J.; Agrela, F.; López, M.; Galvín, A.P. Utilisation of unbound recycled aggregates from selected CDW in unpaved rural roads. *Resour. Conserv. Recycl.* **2012**, *58*, 88–97. [CrossRef]
14. Vegas, I.; Ibañez, J.A.; Lisbona, A.; de Cortazar, A.S.; Frías, M. Pre-normative research on the use of mixed recycled aggregates in unbound road sections. *Constr. Build. Mater.* **2011**, *25*, 2674–2682. [CrossRef]
15. del Estado, M.P. ROM 4.1-94. In *Guidelines for the Design and Construction of Port Pavements*; Puertos del Estado (Government of Spain): Madrid, Spain, 1994.
16. Spanish Ministry of Public Works. *Order/FOM/3460 by Adopting the Standard 6.1 IC of Sections of the Road Instruction*; BOE: Madrid, Spain, 2003; Volume 297, pp. 44274–44292.
17. Tavira, J.; Jiménez, J.R.; Ayuso, J.; Sierra, M.J.; Ledesma, E.F. Functional and structural parameters of a paved road section constructed with mixed recycled aggregates from non-selected construction and demolition waste with excavation soil. *Constr. Build. Mater.* **2018**, *164*, 57–69. [CrossRef]
18. Poon, C.-S.; Qiao, X.; Chan, D. The cause and influence of self-cementing properties of fine recycled concrete aggregates on the properties of unbound sub-base. *Waste Manag.* **2006**, *26*, 1166–1172. [CrossRef] [PubMed]
19. Garach, L.; de Oña, J.; López, G.; Baena, L. Development of safety performance functions for Spanish two-lane rural highways on flat terrain. *Accid. Anal. Prev.* **2016**, *95*, 250–265. [CrossRef] [PubMed]
20. PPTG ADAR. General technical specifications for high-performance dynamic monitoring tests. In *Management of Andalusian Infraestructures*; Andalusian Goverment: Seville, Spain, 2004.
21. De Jong, D.L.; Peutz, M.G.F.; Korswagen, A.R. Computer program BISAR, layered systems under normal and tangential surface loads. *External Rep.* **1979**, No. AMSR 6.
22. Burmister, D.M. The general theory of stresses and displacements in layered soil systems. III. *J. Appl. Phys.* **1945**, *16*, 296–302. [CrossRef]
23. Acum, W.A.; Fox, L. Computation of load stresses in a three-layer elastic system. *Geotechnique* **1951**, *2*, 293–300. [CrossRef]

24. Schiffman, R.L. General analysis of stresses and displacements in layered elastic systems. In Proceedings of the International Conference on the Structural Design of Asphalt Pavements University of Michigan, Ann Arbor, MI, USA, 20–24 August 1962.

25. Huang, Y.H. *Pavement Analysis and Design*; Prentice Hall, Inc.: Sadr River, NJ, USA, 1993.

26. E.U.s. *Guide, Pavement Analysis Computer Software and Case Studies*; Washington State Department of Transportation: Washington, DC, USA, 2005.

27. Guinée, J.B.; Lindeijer, E. *Handbook on Life Cycle Assessment: Operational Guide to the ISO Standards*; Springer Science & Business Media: Berlin/Heidelberg, Germany, 2002.

28. Ecoinvent v3.0.1. *Life Cycle Inventories of Production Systems*; Swiss Centre for Life Cycle Inventories: Zurich, Switzerland, 2014.

29. Serres, N.; Braymand, S.; Feugeas, F. Environmental evaluation of concrete made from recycled concrete aggregate implementing life cycle assessment. *J. Build. Eng.* **2016**, *5*, 24–33. [CrossRef]

30. Rosado, L.P.; Vitale, P.; Penteado, C.S.G.; Arena, U. Life cycle assessment of natural and mixed recycled aggregate production in Brazil. *J. Clean. Prod.* **2017**, *151*, 634–642. [CrossRef]

MDPI
St. Alban-Anlage 66
4052 Basel
Switzerland
Tel. +41 61 683 77 34
Fax +41 61 302 89 18
www.mdpi.com

Materials Editorial Office
E-mail: materials@mdpi.com
www.mdpi.com/journal/materials